NEW HORIZONS FROM MULTI-WAVELENGTH SKY SURVEYS

INTERNATIONAL ASTRONOMICAL UNION

UNION ASTRONOMIQUE INTERNATIONALE

NEW HORIZONS FROM MULTI-WAVELENGTH SKY SURVEYS

PROCEEDINGS OF THE 179TH SYMPOSIUM OF THE
INTERNATIONAL ASTRONOMICAL UNION,
HELD IN BALTIMORE, U.S.A.,
AUGUST 26–30, 1996

EDITED BY

BRIAN J. McLEAN
DANIEL A. GOLOMBEK
JEFFREY J. E. HAYES

and

HARRY E. PAYNE

Space Telescope Science Institute,
Baltimore, MD, U.S.A.

KLUWER ACADEMIC PUBLISHERS

DORDRECHT / BOSTON / LONDON

A C.I.P. Catalogue record for this book is available from the Library of Congress.

ISBN 0-7923-4802-8 (HB)

*Published on behalf of
the International Astronomical Union
by
Kluwer Academic Publishers, P.O. Box 17, 3300 AA Dordrecht, The Netherlands.*

*Sold and distributed in the U.S.A. and Canada
by Kluwer Academic Publishers,
101 Philip Drive, Norwell, MA 02061, U.S.A.*

*In all other countries, sold and distributed
by Kluwer Academic Publishers Group,
P.O. Box 322, 3300 AH Dordrecht, The Netherlands.*

Printed on acid-free paper

All Rights Reserved
© *1998 International Astronomical Union*

No part of the material protected by this copyright notice may be reproduced or utilized in any form or by any means, electronic or mechanical including photocopying, recording or by any information storage and retrieval system, without written permission from the publisher.

Printed in the Netherlands

These Proceedings are dedicated to
George O. Abell and *Albert G. Wilson*
for their invaluable contribution in completing
the *first* Palomar Observatory Sky Survey

Contents

Preface xix
List of Participants xx
Conference Photograph xxv

Part 1. General Sky Survey Reviews

THE PROSPECTS OF LARGE SURVEYS IN ASTRONOMY . . . 3
    M. Harwit
SELECTION EFFECTS AND BIASES 11
    M. Disney
RADIO SURVEYS 19
    J.J. Condon
INFRARED SURVEYS: A GOLDEN AGE OF EXPLORATION . . 27
    C.A. Beichman
PHOTOGRAPHIC SKY SURVEYS 41
    I.N. Reid
ELECTRONIC OPTICAL SKY SURVEYS 49
    T.A. McKay
ULTRAVIOLET SKY SURVEYS 57
    N. Brosch
GAMMA RAY SKY SURVEYS 69
    N. Gehrels
PROSPECTS FOR FUTURE ASTROMETRIC MISSIONS 79
    P.K. Seidelmann

Part 2. Survey Projects

WIDE-FIELD LOW-FREQUENCY IMAGING WITH THE VLA . 89
    N.E. Kassim, D.S. Briggs and R.S. Foster
THE MIYUN 232 MHZ GENERAL CATALOGUE: 91
    B. Peng, W. Yuan, R. Nan and Y. Yan
KILOMETER-SQUARE AREA RADIO SYNTHESIS TELESCOPE 93
    B. Peng and R. Nan

RADIO CONTINUUM SURVEYS ABOVE 2 GHZ 95
  J.L. Jonas

THE MPIFR RADIO CONTINUUM SURVEYS AND THEIR
    WWW DISTRIBUTION 97
  E. Fürst, W. Reich, B. Uyaniker and R. Wielebinski

THE UTR-2 VERY LOW FREQUENCY SKY SURVEY
    AND ITS MAIN RESULTS 100
  K.P. Sokolov

THE RATAN GALACTIC PLANE SURVEY 103
  S.A. Trushkin

DENIS: A DEEP NEAR INFRARED SOUTHERN SKY SURVEY . 106
  N. Epchtein

SIRTF SURVEYS AND LEGACY SCIENCE 109
  G.G. Fazio and D.P. Clemens

THE EUROPEAN LARGE AREA ISO SURVEY: ELAIS 112
  S. Oliver, M. Rowan-Robinson, C. Cesarsky, L. Danese,
  A. Franceschini, R. Genzel, A. Lawrence, D. Lemke, R. McMahon,
  G. Miley, J-L. Puget and B. Rocca-Volmerange

ASTRONOMY ON THE MIDCOURSE SPACE EXPERIMENT .. 115
  S.D. Price, E.F. Tedesco, M. Cohen, R.G. Walker, R.C. Henry,
  M. Moshir, L.J. Paxton and F.C. Witterborn

THE WIDE-FIELD INFRARED EXPLORER (WIRE) MISSION . 118
  D.L. Shupe, P.B. Hacking, T. Herter, T.N. Gautier, P. Graf,
  C.J. Lonsdale, G.J. Stacey, S.H. Moseley, B.T. Soifer, M.W. Werner
  and J.R. Houck

FIRST OBSERVATIONS WITH THE 1.5 M RC TELESCOPE AT
    MAIDANAK OBSERVATORY 121
  B.P. Artamonov

SOME RESULTS OF THE BATC CCD COLOR SURVEY 123
  J-S. Chen

THE AIST-STRUVE SPACE PROJECT SKY SURVEY 125
  M.S. Chubey, I.M. Kopylov, D.L. Gorshanov, I.I. Kanayev,
  V.N. Yershov, A.E. Il'In, T.R. Kirian and M.G. Vydrevich

ASPA—SKY PATROL FOR THE FUTURE 127
  C. La Dous and P. Kroll

WIDE FIELD OPTICAL IMAGING AT CTIO 129
    A.R. Walker

LAMOST PROJECT 131
    Y. Chu and Y H. Zhao

THE AAT'S 2-DEGREE FIELD PROJECT 135
    K. Taylor, R.D. Cannon and Q.A. Parker

SIMULTANEOUS SPECTRA OF A COMPLETE SAMPLE
    OF SOURCES FROM THE PMN SURVEY 139
    M.G. Mingaliev, A.M. Botashev and V.A. Stolyarov

AROPS : THE ASIAGO RED OBJECTIVE-PRISM SURVEY . . . 142
    U. Munari and R. Passuello

EUV SOURCES AND TRANSIENTS DETECTED BY THE ALEXIS
    SATELLITE 144
    D. Roussel-Dupré, T. Pfafman, J. Bloch and J. Theiler

THE UIT SURVEY OF THE ULTRAVIOLET
    SKY BACKGROUND. 147
    W.H. Waller and T.P. Stecher

Part 3. The Interstellar Medium

PHYSICAL PROCESSES IN THE LARGE SCALE ISM FROM
    DUST OBSERVATIONS 153
    F. Boulanger

RELATIONS BETWEEN STAR FORMATION AND
    THE INTERSTELLAR MEDIUM 165
    Y. Fukui and Y. Yonekura

A DENIS SURVEY OF STAR FORMING REGIONS 172
    E. Copet

TRACING THE INTERSTELLAR MEDIUM IN OPHIUCHUS
    ACROSS 14 ORDERS OF MAGNITUDE IN FREQUENCY 175
    S.W. Digel, S.D. Hunter and S.L. Snowden

A NEW CO SURVEY OF THE MONOCEROS OB1 REGION . . . 177
    R.J. Oliver, M.R.W. Masheder and P. Thaddeus

AN Hα SURVEY OF THE GALACTIC PLANE 179
    Q.A. Parker and S. Phillipps

AN IMAGING SURVEY OF THE GALACTIC H-ALPHA
    EMISSION WITH ARCMINUTE RESOLUTION 182
  B. Dennison, J.H. Simonetti, G.A. Topansa and C. Kellerher

INTERSTELLAR CIRRUS OBSERVED IN BALMER Hα 184
  P.R. McCullough

THE MARSEILLE OBSERVATORY Hα SURVEY:
    COMPARISONS WITH CO, 6 CM AND IRAS DATA ... 186
  D. Russeil, P. Amram, Y.P. Georgelin, Y.M. Georgelin, M. Marcelin,
  A. Viale, E. Le Coarer and A. Castets

A LARGE-SCALE CO IMAGING OF THE GALACTIC CENTER 189
  T. Oka, T. Hasegawa, F. Sato, H. Yamasaki, M. Tsuboi and
  A. Miyazaki

SURVEY OF CORRELATED FIR, HI, CO, AND
    RADIO-CONTINUUM EMISSION FEATURES IN THE
    MULTI-PHASE MILKY WAY 191
  W.F. Wall and W.H. Waller

SURVEY OF FINE-SCALE STRUCTURE IN THE
    FAR-INFRARED MILKY WAY 194
  W.H. Waller, F. Varosi, F. Boulanger and S.W. Digel

Part 4. Galactic Structure

FORMATION AND EVOLUTION OF THE MILKY WAY 199
  S.R. Majewski

KINEMATICS AND THE GALACTIC POTENTIAL 209
  O. Bienaymé

USING MULTI-WAVELENGTH ALL SKY INFORMATION TO UN-
    DERSTAND LARGE SCALE GALACTIC STRUCTURE. . 217
  R.L. Smart, R. Drimmel and M.G. Lattanzi

STRUCTURE AND KINEMATICAL PROPERTIES OF THE
    GALAXY AT INTERMEDIATE GALACTIC LATITUDES 221
  D.K. Ojha, O. Bienaymé and A.C. Robin

PROPER MOTIONS IN THE BULGE: LOOKING THROUGH
    PLAUT'S LOW EXTINCTION WINDOW 223
  R.A. Méndez, R.M. Rich, W.F. van Altena, T.M. Girard, S. van den
  Bergh and S.R. Majewski

GALACTIC STRUCTURE WITH THE APS CATALOG OF THE
  POSS I 225
  J.A. Larsen
GALACTIC STRUCTURE WITH GSC-II MATERIAL 228
  A. Spagna, M.G. Lattanzi, G. Massone, B. McLean and B.M. Lasker
INVESTIGATION OF GALACTIC STRUCTURE
  WITH DENIS STAR COUNTS 231
  S. Ruphy
STARCOUNTS IN THE HUBBLE DEEP FIELD:
  FEWER THAN EXPECTED, MORE THAN EXPECTED . 234
  R.A. Méndez, G. De Marchi, D. Minniti, A. Baker and W.J. Couch
MULTIWAVELENGTH MILKY WAY:
  AN EDUCATIONAL POSTER 237
  D. Leisawitz, S.W. Digel and S. Geitz

Part 5. Extra-Galactic Astronomy

WEAK GRAVITATIONAL LENSING—THE NEED FOR
  SURVEYS 241
  P. Schneider
THE EVOLUTION OF QUASARS AND THEIR CLUSTERING . 249
  P.S. Osmer
TOWARDS A BETTER UNDERSTANDING OF
  ACTIVE GALACTIC NUCLEI 257
  P. Padovani
MEASURES OF GALACTIC AND INTERGALACTIC MASS IN
  CLUSTERS 266
  D. Windridge, S. Phillipps and M. Birkinshaw
THE SCALING RELATIONS FOR CLUSTERS OF GALAXIES . . 268
  J. Annis
THE RADIO PROPERTIES OF X-RAY SELECTED
  EXTRAGALACTIC OBJECTS 270
  L. Ramírez-Castro, I. Pérez-Fournon and F. Cabrera-Guerra

UNCOVERING ULTRA-LUMINOUS GALAXIES
IN THE IRAS FSC THROUGH RADIO AND OPTICAL
CROSS-IDENTIFICATION 273
G. Aldering

EARLY RESULTS FROM AN HST IMAGING SURVEY OF THE
ULTRALUMINOUS IR GALAXIES 275
K.D. Borne, H. Bushouse, L. Colina and R.A. Lucas

THE K-BAND LUMINOSITY FUNCTION OF GALAXIES 278
J.P. Gardner, R.M. Sharples, C.S. Frenk and B.E. Carrasco

THE K-BAND WIDE FIELD SURVEY: UNDERSTANDING THE
LOCAL GALAXIES........................ 281
J. Huang

MULTI-COLOR SURFACE PHOTOMETRY OF
NEARBY GALAXIES 285
T. Ichikawa, N. Itoh and K. Yanagisawa

MUTI-COLOR IMAGING OF CLUSTERS OF GALAXIES WITH
MOSAIC CCD CAMERAS 287
*S. Okamura, M. Doi, N. Kashikawa, W. Kawasaki, Y. Komiyama,
M. Sekiguchi, K. Shimasaku, M. Yagi and N. Yasuda*

QSO COLOR SELECTION IN THE SDSS 291
H.J. Newberg and B. Yanny

THE CALAR ALTO DEEP IMAGING SURVEY 293
*H. Hippelein, S. Beckwith, R. Fockenbrock, J. Fried, U. Hopp,
C. Leinert, K. Meisenheimer, H.-J. Röser, E. Thommes and C. Wolf*

THE CALAR ALTO DEEP IMAGING SURVEY: FIRST RESULTS 296
*E. Thommes, K. Meisenheimer, R. Fockenbrock, H. Hippelein and
H-J. Röser*

HAMBURG/SAO SURVEY OF EMISSION-LINE GALAXIES .. 299
*V. Lipovetsky, D. Engels, A. Ugryumov, U. Hopp, G. Richter,
Y. Izotov, A. Kniazev and C. Popescu*

KISS: A NEW DIGITAL SURVEY FOR
EMISSION-LINE OBJECTS 302
*A. Kniazev, J. Salzer, V. Lipovetsky, T. Boroson, J. Moody,
T. Thuan, Yu. Izotov, J. Herrero and L. Frattare*

HOW TO FIND BL LAC OBJECTS IN THE RASS 305
P. Nass

THE WARPS X-RAY SURVEY OF GALAXIES,
 GROUPS, AND CLUSTERS 308
  D. Horner, C.A. Scharf, L.R. Jones, H. Ebeling, E. Perlman,
  M. Malkan and G. Wegner

THE WARPS BLAZAR SURVEY 310
  E.S. Perlman, P. Padovani, L. Jones, P. Giommi, A. Tzioumis,
  J. Reynolds and R. Sambruna

RESULTS FROM ASCA SKY SURVEYS 312
  Y. Ogasaka, Y. Ueda, Y. Ishisaki, T. Kii, T. Takahashi,
  K. Makishima, H. Inoue, K. Ohta, T. Yamada, T. Miyaji and
  G. Hasinger

Part 6. Large Scale Structure

LARGE SCALE STRUCTURE OF THE UNIVERSE 317
  N.A. Bahcall

LARGE SCALE STRUCTURE IN THE LYα FOREST 329
  G. Williger, A. Smette, C. Hazard, J. Baldwin and R. McMahon

A SEARCH FOR LARGE VOIDS FROM COMBINED SAMPLES OF
 GALAXY CLUSTERS AND GALAXIES 332
  K.Y. Stavrev

LOW LUMINOSITY GALAXY DISTRIBUTION IN
 LOW DENSITY REGIONS 335
  U. Hopp

CLUSTERING PROPERTIES OF FAINT BLUE GALAXIES . . . 337
  M.W. Kümmel and S.J. Wagner

THE K-BAND HUBBLE DIAGRAM FOR X-RAY SELECTED
 BRIGHTEST CLUSTER GALAXIES 339
  R.G. Mann and C.A. Collins

MULTIWAVELENGTH STUDY OF THE SHAPLEY
 CONCENTRATION 342
  S. Bardelli, E. Zucca, G. Zamorani, G. Vettolani and R. Scaramella

A RICH CLUSTER REDSHIFT SURVEY FOR LARGE-SCALE
 STRUCTURE STUDIES 344
  D. Batuski, K. Slinglend, J. M. Hill, S. Haase, C. Miller, K. Michaud

THE ESO SLICE PROJECT (ESP) REDSHIFT SURVEY 346
    G. Vettolani, E. Zucca, A. Cappi, R. Merighi, M. Mignoli, G. Stirpe, G. Zamorani, H. MacGillivray, C. Collins, C. Balkowski, V. Cayatte, S. Maurogordato, D. Proust, G. Chincarini, L. Guzzo, D. Maccagni, R. Scaramella, A. Blanchard and M. Ramella

INPUT CATALOGUE FOR THE 2DF QSO REDSHIFT SURVEY 348
    R.J. Smith, B.J. Boyle, T. Shanks, S.M. Croom, L. Miller and M. Read

REDSHIFT SURVEY OF 951 IRAS GALAXIES
    IN THE SOUTHERN MILKY WAY 351
    N. Visvanathan and T. Yamada

A DEEP 20 CM RADIO MOSAIC OF THE ESP GALAXY
    REDSHIFT SURVEY 353
    I. Prandoni, L. Gregorini, P. Parma, G. Vettolani, H.R. de Ruiter, M.H. Wieringa and R.D. Ekers

RADIO EMISSION FROM HIGH REDSHIFT GALAXIES:
    VLA OBSERVATIONS OF THE HUBBLE DEEP FIELD . 356
    E.A. Richards

EXPOSING NEW COMPONENTS OF THE X-RAY BACKGROUND WITH MULTI-WAVELENGTH SKY SURVEYS 358
    E.C. Moran

Part 7. Data Processing Techniques

STATISTICAL METHODOLOGY FOR
    LARGE ASTRONOMICAL SURVEYS 363
    E.D. Feigelson and G.J. Babu

DIGITAL COLOUR MAPPING OF THE SKY FROM
    SUPERCOSMOS 371
    H.T. MacGillivray

COLOUR EQUATIONS FOR TECHPAN FILMS 374
    D.H. Morgan and Q.A. Parker

AN OBJECTIVE APPROACH TO SPECTRAL
    CLASSIFICATION 376
    A.J. Connolly and A.S. Szalay

GSPC-II: A CATALOG OF PHOTOMETRIC CALIBRATORS FOR
    THE SECOND GENERATION GUIDE STAR CATALOG.. 379
    M. Postman, B. Bucciarelli, C. Sturch, T. Borgman, R. Casalegno, J. Doggett and E. Costa

TECHNIQUES FOR SCHMIDT PLATE REDUCTIONS WITH APPLICATION TO GSC1.2 381
    J.E. Morrison and S. Röser

WADING THROUGH THE QUAGMIRE OF SCHMIDT-PLATE COORDINATE SYSTEMATICS 384
    C.-L. Lu, I. Platais, T.M. Girard, V. Kozhurina-Platais, W.F. Van Altena, C.E. López and D.G. Monet

PROGRESS IN WIDE FIELD CCD ASTROMETRY 386
    N. Zacharias

LINKING THE RADIO AND OPTICAL FRAMES WITH MERLIN 389
    S.T. Garrington, R.J. Davis, L.V. Morrison and R.W. Argyle

Part 8. Catalogues

OVERVIEW OF THE TYCHO CATALOGUE 395
    E. Høg, C. Fabricius, V.V. Makarov, D. Egret, J.L. Halbwachs, G. Bässgen, V. Großmann, K. Wagner, A. Wicenec, U. Bastian and P. Schwekendiek

MAIN PROPERTIES OF THE HIPPARCOS CATALOGUE 399
    F. Mignard

THE ASTROGRAPHIC CATALOGUE 406
    D.H.P. Jones

COMPLETION OF THE STERNBERG ASTRONOMICAL INSTITUTE ASTROGRAPHIC CATALOGUE PROJECT . 409
    V. Nesterov, A. Gulyaev, K. Kuimov, A. Kuzmin, V. Sementsov, U. Bastian and S. Röser

SALVAGING AN ASTROMETRIC TREASURE 415
    M. Hiesgen, P. Brosche, A. Ortiz Gil, J. Colin, A. Fresneau, M. Geffert, H.T. MacGillivray, S. Hecht, E. Kallenbach, M. Odenkirchen, C. Schäffel and H.-J. Tucholke

A NEW HIGH DENSITY, HIGH PRECISION ASTROMETRIC CATALOG 418
    M.I. Zacharias, G.L. Wycoff, G.G. Douglass, T.E. Corbin and N. Zacharias

CONTENTS, TEST RESULTS, AND DATA AVAILABILITY FOR GSC 1.2 420
    S. Röser, J. Morrison, B. Bucciarelli, B. Lasker and B. McLean

THE 488,006,860 SOURCES IN THE USNO-A1.0 CATALOG . . . 422
    B. Canzian

CATALOGING OF THE DIGITIZED POSS-II:
    INITIAL SCIENTIFIC RESULTS 424
    S.G. Djorgovski, R.R. de Carvalho, R.R. Gal, M.A. Pahre,
    R. Scaramella and G. Longo

THE SECOND GUIDE STAR CATALOGUE 431
    B. McLean, G. Hawkins, A. Spagna, M. Lattanzi, B. Lasker,
    H. Jenkner and R. White

THE ROSAT ALL-SKY SURVEY BRIGHT SOURCE CATALOG . 433
    W. Voges, B. Aschenbach, Th. Boller, H. Bräuninger, U. Briel,
    W. Burkert, K. Dennerl, J. Englhauser, R. Gruber, F. Haberl,
    G. Hartner, G. Hasinger, M. Kürster, E. Pfeffermann, W. Pietsch,
    P. Predehl, C. Rosso, J.H.M.M. Schmitt, J. Trümper and
    U. Zimmermann

Part 9. Multi-Wavelength Cross Identification

CROSS WAVELENGTH COMPARISON OF IMAGES AND
    CATALOGS 437
    J.G. Bartlett and D. Egret

THE HAMBURG IDENTIFICATION PROGRAM OF ROSAT ALL-
    SKY SURVEY SOURCES 444
    N. Bade, L. Cordis, D. Engels, D. Reimers and W. Voges

CROSS-CORRELATION OF LARGE SCALE SURVEYS:
    RADIO-LOUD OBJECTS IN THE ROSAT ALL SKY SUR-
    VEY 447
    W. Brinkmann, W. Yuan and J. Siebert

IDENTIFICATION OF A COMPLETE SAMPLE OF NORTHERN
    ROSAT ALL-SKY SURVEY X-RAY SOURCES 449
    J. Krautter, I. Thiering, F.-J. Zickgraf, I. Appenzeller, R. Kneer,
    W. Voges, A. Serrano and R. Mujica

THE OPTID DATABASE: DEEP OPTICAL IDENTIFICATIONS
    TO THE IRAS FAINT SOURCE SURVEY 450
    C. Lonsdale, T. Conrow, T. Evans, L. Fullmer, M. Moshir,
    T. Chester, D. Yentis, R. Wolstencroft, H. MacGillivray and D. Egret

Part 10. Databases

EXPLORING TERABYTE ARCHIVES IN ASTRONOMY 455
 A.S. Szalay and R.J. Brunner

THE WIDE-FIELD PLATE DATABASE:
 A NEW TOOL IN OBSERVATIONAL ASTRONOMY . . . 462
 M.K. Tsvetkov, K.Y. Stavrev, K.P. Tsvetkova, E.H. Semkov,
 A.S. Mutafov and M.-E. Michailov

SKYVIEW: THE MULTI-WAVELENGTH SKY ON THE INTERNET 465
 T. McGlynn, K. Scollick and N. White

A WWW DATABASE OF APS POSS IMAGES 467
 C.S. Cornuelle, G. Aldering, A. Sourov, R.M. Humphreys, J. Larsen
 and J. Cabanela

THE ALADIN INTERACTIVE SKY ATLAS 469
 F. Bonnarel, H. Ziaeepour, J.G. Bartlett, O. Bienaymé, M. Crézé,
 D. Egret, J. Florsch, F. Genova, F. Ochsenbein, V. Raclot, M. Louys
 and P. Paillou

THE SDSS SCIENCE ARCHIVE 471
 R.J. Brunner

THE GSC-I AND GSC-II DATABASES:
 AN OBJECT-ORIENTED APPROACH 474
 G. Greene, B. McLean, B. Lasker, D. Wolfe, R. Morbidelli and
 A. Volpicelli

SURVEYS IN THE ADC ARCHIVE 478
 N.G. Roman

GLOBAL NETWORK ACCESS AND PUBLICATION
 OF SURVEY DATA 480
 N.E. White

Part 11. Conference Summary and Resolutions

THE NATURE AND SIGNIFICANCE OF SURVEYS 489
 V. Trimble

SUMMARY TALK : MULTI-WAVELENGTH SKY SURVEYS . . . 493
 O. Lahav

Conference Resolutions 500

Author index 503

Preface

This volume contains the papers presented at IAU Symposium 179 on "New Horizons from Multi-Wavelength Sky Surveys" which was held at The Johns Hopkins University from the 26th through the 30th of August 1996. This meeting was organized under the auspices of the "Wide-Field Imaging" working group of IAU Commission 9.

This was the latest conference in the rapidly advancing and continually evolving field of wide field astronomy. We have come a long way from the original conferences in Geneva (May 1989, "Proceedings of the First Conference on Digitised Sky Surveys," Bull. d'Information du CDS, No.37) and Edinburgh (June 1991, "Digitised Optical Sky Surveys," Astrophys.Space Science Library,vol 174), which dealt primarily with photographic sky surveys and the digitization of these optical data for quantitative analysis. The technological advances in instrumentation, detector development and computing have brought digital surveys and surveys at other wavebands to the forefront as shown by the conference in Potsdam (August 1993, "Astronomy from Wide Field Imaging," IAU Symposium 161). More recently, the working group is highlighting the importance of spectroscopic and multi-wavelength observations by sponsoring two meetings, one in Athens (May 1996, "Wide-Field Spectroscopy," Astrophys.Space Science Library,vol 212), and this one in Baltimore. This is a good indication of how broad and active a field this has become.

There were 192 registered participants at the meeting representing 16 countries from all parts of the globe and 150 papers or posters were presented. This volume presents the written versions for most of these contributions.

This conference was sponsored by the International Astronomical Union, the Space Telescope Science Institute, The Johns Hopkins University, The University of Maryland and the Goddard Space Flight Center. We are very grateful to the sponsoring institutions for their generous support.

A conference of this size cannot possibly succeed without the efforts and dedication of a large number of people including the Science and Local Organising Committees, the local staff who kept things running smoothly, and all the participants; we are indebted to them all.

Brian McLean
Space Telescope Science Institute

January 1997

List of Participants

Greg Aldering, University of Minnesota, USA
James Annis, Experimental Astrophysics Group, Fermilab, USA
Kentaro Aoki, National Astronomical Observatory of Japan, Japan
Brent A. Archinal, U.S. Naval Observatory, USA
Robert W. Argyle, Royal Greenwich Observatory, England
Boris Artamonov, Sternberg Astronomical Institute, Russia
Norbert Bade, Hamburger Sternwarte, Germany
Neta Bahcall, Princeton University, USA
Sandro Bardelli, Osservatorio Astronomico di Trieste, Italy
James Bartlett, Observatoire Astronomique de Strasbourg, France
David J. Batuski, University of Maine, USA
Robert Becker, Lawrence Livermore National Laboratory, USA
Charles Beichman, Caltech/JPL, USA
Olivier Bienayme, Observatoire Astronomique de Strasbourg, France
Francois Bonnarel, Observatoire Astronomique de Strasbourg, France
Kirk Borne, Hughes STX - Goddard Space Flight Center, USA
Francois Boulanger, Institut d'Astrophysique Spatiale, France
Wolfgang Brinkmann, MPI für Extraterrestrische Physik, Germany
Noah Brosch, Wise Observatory, Israel
Robert J. Brunner Johns Hopkins University, USA
Blaise Canzian, USRA/U.S. Naval Observatory, USA
Daniela Carollo, Osservatorio Astronomica di Torino, Italy
Jian-sheng Chen, Beijing Astronomical Observatory, China
Cynthia Cheung, NASA/GSFC, USA
Yaoquan Chu, University of Science and Technology, China
Angie Clarke, Space Telescope Science Institute, USA
James Condon, National Radio Astronomy Observatory, USA
Andrew Connolly, Johns Hopkins University, USA
Kem Cook, Lawrence Livermore National Laboratory, USA
Eric Copet, Observatoire de Paris, France
Thomas E. Corbin, US Naval Observatory, USA
Chris Cornuelle, University of Minnesota, USA
Reinaldo de Carvalho, California Institute of Technology, USA
Brian Dennison, Physics Dept., Virginia Tech, USA
Eric Deul, Sterrewacht Leiden, Netherlands
Mark Dickinson, Space Telescope Science Institute, USA
Seth Digel, Hughes STX, NASA/GSFC, USA
Michael Disney, College of Cardiff, University of Wales, Wales
S. George Djorgovski California Institute of Technology, USA
Jesse Doggett, Space Telescope Science Institute, USA

Daniel Egret, CDS, Observatoire de Strasbourg, France
Nicolas Epchtein, Observatoire de Paris, France
Giovanni G. Fazio, Harvard-Smithsonian Astrophysical Observatory, USA
Eric Feigelson, Penn State University, USA
Paul Feldman, Johns Hopkins University, USA
Holland Ford, Johns Hopkins University, USA
Wolfram Freudling, Space Telescope - ECF, Germany
Ernst F. G. Fürst, Max-Planck-Institut für Radioastronomie, Germany
Yasuo Fukui, Dept. of Physics and Astrophysics, Nagoya Univ., Japan
Roy Gal, California Institute of Technology, USA
Lorraine Garcia, Space Telescope Science Institute , USA
Jonathan P. Gardner, University of Durham, England
Stephen Gauss, U.S. Naval Observatory, USA
Neil Gehrels, NASA/GSFC, USA
Marvin E. Germain, U.S. Naval Observatory, USA
Daniel Golombek, Space Telescope Science Institute, USA
Gretchen Greene, Space Telescope Science Institute, USA
Loretta Gregorini, Istituto di Radioastronomia, CNR, Italy
Herbert Gursky, Naval Research Laboratory, USA
Martin Harwit, USA
Guenther Hasinger, Astrophysikalisches Institut Postsdam, Germany
Michael Hauser, Space Telescope Science Institute, USA
George Hawkins, Space Telescope Science Institute, USA
Jeffrey J.E. Hayes, Space Telescope Science Institute, USA
David Helfand, Columbia University, USA
George Helou, NED/IPAC, USA
Richard C. Henry, Johns Hopkins University, USA
Martin Hiesgen, Observatoire Astronomique de Strasbourg, France
Robert Hindsley, U.S. Naval Observatory, USA
Erik Hoeg, Copenhagen University Observatory, Denmark
Ulrich Hopp, Universitaetssternwarte Munich, Germany
Don Horner, Goddard Space Flight Center, USA
Jiasheng Huang, Univ of Hawaii - Institute for Astronomy, USA
Takashi Ichikawa, Kiso Observatory, University of Tokyo, Japan
Helmut Jenkner, Space Telescope Science Institute, USA
Justin Jonas, Rhodes University, South Africa
Derek Jones, Royal Greenwich Observatory, England
Namir E. Kassim, Naval Research Laboratory, USA
Anne Kinney, Space Telescope Science Institute, USA
Stephen Knapp, Johns Hopkins University, USA
Alexei Y. Kniazev, Special Astrophysical Observatory, Russia
Joachim Krautter, Landessternwarte Heidelberg, Germany

Martin W. Kümmel, Landessternwarte Heidelberg, Germany
Kip Kuntz, GSFC and University of Maryland, USA
Andrei Kuzmin, Sternberg Astronomical Institute, Russia
Constanze la Dous, Sonneberg Observatory, Germany
Ofer Lahav, Institute of Astronomy, Cambridge, England
Vicki Laidler, Space Telescope Science Institute, USA
Jeff Larsen, University of Minnesota, USA
Barry Lasker, Space Telescope Science Institute, USA
David Leisawitz, Goddard Space Flight Center, USA
Carol Lonsdale, California Institute of Technology, USA
Harvey T. MacGillivray Royal Observatory Edinburgh, Scotland
Steven Majewski, University of Virginia, USA
Robert Mann, Imperial College, England
Michael R.W. Masheder, University of Bristol, England
Peter R. McCullough, Univ of Illinois, Astronomy Dept, USA
Thomas A. McGlynn, NASA Goddard Space Flight Center/USRA, USA
Timothy A. McKay, University of Michigan, USA
Brian McLean, Space Telescope Science Institute, USA
Michael Meakes, Space Telescope Science Institute,, USA
Rene A. Mendez-Bussard, European Southern Observatory, Germany
Francois Mignard, OCA/CERGA, France
Richard S. Miller, Los Alamos National Laboratory, USA
Bruno Milliard, Laboratoire d'Astronomie Spatiale du CNRS, France
Marat Mingaliev, Special Astrophysical Observatory, Russia
Edward C. Moran, IGPP/LLNL, USA
David Morgan, Royal Observatory Edinburgh, Scotland
Jane Morrison, Astronomishes Rechen Institut, Germany
Ulisse Munari, Osservatorio Astronomico di Padova, Italy
Petra Nass, Max-Planck Institut für Extraterrestrische Physik, Germany
Heidi Newberg, Fermilab, USA
Yasushi Ogasaka, NASA/GSFC/JSPS, USA
Tomoharu Oka, The Institute of Physical and Chemical Research, Japan
Sadanori Okamura, Dept. of Astronomy, University of Tokyo, Japan
Seb Oliver, Imperial College of Sci. Tech. and Medicine, England
Patrick S. Osmer, The Ohio State University, USA
Paolo Padovani, Dipartimento di Fisica, Italy
Quentin A. Parker, Anglo-Australian Observatory, Australia
R. Bruce Partridge, Haverford College, USA
Bo Peng, Beijing Astronomical Observatory, China
Eric Perlman, Space Telescope Science Institute, USA
Imants Platais, Yale University, USA
Theodore Pohler, Johns Hopkins University, USA

Marc Postman, Space Telescope Science Institute, USA
Isabella Prandoni, IRA-CNR, Bologna, Italy
Steven Price, Phillips Lab., USA
Luis Ramirez-Castro, Instituto de Astrofisica de Canarias, Spain
Neill Reid, Palomar Observatory, USA
Eric Richards, University of Virginia, USA
Nancy G. Roman, Hughes STX, USA
Diane Roussel-Dupree, ALEXIS Satellite Project-LANL, USA
Stephanie Ruphy, DESPA Observatoire de Paris, France
Delphine Russeil, Observatoire de Marseille, France
Jane Russell, National Science Foundation, USA
Roberto Scaramella, Osservatorio Astronomico di Roma, Italy
Cheryl Schmidt, Space Telescope Science Institute, USA
Marion Schmitz, NED/IPAC/CALTECH, USA
Peter Schneider, Max-Planck-Institut für Astrophysik, Germany
Ethan Schreier, Space Telescope Science Institute, USA
Keith A. Scollick, NASA Goddard Space Flight Center, USA
P. Kenneth Seidelmann, U.S. Naval Observatory, USA
Robin Shelton, NASA/Goddard Space Flight Center, USA
David Shupe, IPAC/Jet Propulsion Laboratory, USA
Richard Smart, Osservatorio Astronomico di Torino, Italy
Robert J. Smith, Institute of Astronomy, Cambridge, England
Konstantin P. Sokolov, Institute of Radio Astronomy, Ukraine
Allesandro Spagna, Osservatorio Astronomico di Torino, Italy
Konstantin Stavrev, Bulgarian Academy of Sciences, Bulgaria
Raphael Steinitz, Ben Gurion University, Israel
Gordon Stewart, Leicester University, England
Conrad Sturch, Space Telescope Science Institute, USA
Alex Szalay, Johns Hopkins University, USA
Larry G. Taff, NASA Headquarters/JPL, USA
Eduard Thommes, Max-Planck-Institut für Astronomie, Germany
Gregory Topasna, Physics Dept., Virginia Tech, USA
Virginia Trimble, University of Maryland, USA
Joachim Truemper, MPI fuer Extraterrestrische Physik, Germany
Sergei Trushkin, Special Astrophysical Observatory, Russia
Milcho Tsvetkov, Bulgarian Academy of Sciences, Bulgaria
Sean E. Urban, US Naval Observatory, USA
Meg Urry, Space Telescope Science Institute, USA
Azita Valinia, NASA/Goddard Space Flight Center, USA
Piet van der Kruit, Rijksuniversiteit Groningen, Holland
Frank Varosi, Hughes STX Corp, USA
Giampaolo Vettolani, Instituto di Radiastronomia CNR, Italy

Natarajan Visvanathan, MSSSO, Australia
Michael Vogeley, Space Telescope Science Institute, USA
Stefan J. Wagner, Landessternwarte Heidelberg, Germany
Alistair Walker, Cerro Tololo Inter-American Observatory, Chile
Jasper V. Wall, Royal Greenwich Observatory, England
William F. Wall, Instituto Nacional de Astrofisica, Mexico
William H. Waller, Hughes STX and NASA/GSFC, USA
Masaru Watanabe, Univ of Tokyo, School of Science, Japan
Gart Westerhout, Baltimore, USA
Nicholas E. White, NASA/Goddard Space Flight Center, USA
Richard L. White, Space Telescope Science Institute, USA
Gerard Williger, NASA/Goddard Space Flight Center, USA
David Windridge, Univ of Bristol - Goldney Hall, England
David Wolfe, Space Telescope Science Institute, USA
Marion I. Zacharias, USRA/USNO, USA
Norbert Zacharias, USRA/USNO, USA
Yongheng Zhao, Beijing Astronomical Observatory, China
Elena Zucca, Osservatorio Astronomico di Bologna, Italy

Figure 1. Participants of IAU Symposium 179, Baltimore, USA, August 26-30, 1996

Part 1. General Sky Survey Reviews

THE PROSPECTS OF LARGE SURVEYS IN ASTRONOMY

M. HARWIT
511 H Street, SW, Washington, DC, 20024-2725

Abstract. Major astronomical discoveries of the past have been closely linked with the implementation of strikingly powerful new observational techniques in astronomy. Surveys, in general, were not as likely to lead to the discovery of new phenomena. In recent years, however, many far-reaching instrumental advances have been incorporated into observatories in space, whose mission, in part, was to conduct unbiased surveys. Here, we provide a review of some of the successes of earlier surveys, with the purpose of identifying the relative extent to which increased instrumental capabilities, as contrasted to increased sky coverage or numbers of sources observed, may be expected to lead to the discovery of new phenomena. A similar comparison can be made of the extent to which these two approaches—emphasis on instrumental capabilities, as contrasted to emphasis on sky coverage— also contribute to an increase in astrophysical understanding. Here, the distinction needs to be made between discovering a new phenomenon and understanding its underlying processes.

1. The Nature of Astronomical Discovery

The almost ubiquitous role of powerful new technologies in revealing new cosmic phenomena is by now well established (Harwit 1981). Access to entirely new wavelength domains in the electromagnetic spectrum, led to the discovery of a microwave background dating back to the first million years in the evolution of the cosmos; galaxies far more luminous at infrared wavelengths than in all other spectral ranges combined; massive X-ray haloes surrounding the central galaxies of large clusters; and bursts of gamma rays, possibly emanating from sources more luminous than any others known in the universe.

Access to extremely high spatial resolving powers in radio astronomy led first to the discovery of quasi-stellar radio sources (quasars) and, as radio-interferometric resolving powers increased even further, to the discovery of superluminal sources. Still in the radio regime, great strides in spectral resolving power gave us stellar and interstellar masers, while improvements in time resolution, by many orders of magnitude, showed up the existence of pulsars. Similar discoveries resulted from thousand-fold improvements in spatial, spectral and temporal resolving powers in other wavelength domains.

Such discoveries often were made virtually as soon as the novel capabilities essential to making the requisite observations were introduced into astronomy. Seldom were these initial exciting findings followed by the discovery of other comparably significant phenomena. The rule of thumb appeared to be, that new phenomena were discovered quickly following the introduction into astronomy of techniques that went beyond the state of the art by three or more orders of magnitude—either in access to new spectral wavelength regimes or, within a given regime, to increased spatial, spectral or temporal resolving power.

It is not as though a new technique becomes useless after its initial flash of success. Rather, it joins an increasingly comprehensive kit of astronomical tools that help us to probe and analyze the emission from astronomical sources to improve our understanding of them.

There is reason to believe that large surveys, capable of increasing the number of sources studied by many orders of magnitude, may similarly uncover new phenomena. In particular, rare phenomena or faint sources overshadowed by millions of others that are more prevalent or luminous might be discovered only in this fashion. If the number of sources studied can be increased a thousand or a million-fold, major new phenomena may be expected to show up. The comprehensive surveys now being planned often are designed to measure hundreds of millions of sources—many orders of magnitude more, in most spectral ranges, than available before. The chances of discovering rare phenomena thus come into range.

2. Monochromatic, Polychromatic, or Multispectral Surveys

Monochomatic surveys, such as the early optical surveys of the nineteenth century provided useful sky charts and catalogues that listed accurate positions and approximate magnitudes of sources. Such surveys limited the insights one could gain on differences characterizing the different sources. Point sources could be identified only by their different magnitudes, unless they were repeatedly observed, which was time consuming until photographic plates were introduced in the latter part of the century. In 1899,

S. I. Bailey of Harvard University made photographic observations of the globular clusters Messier 5 from a site in Arequipa, Peru. One might not think of this as a survey, but he was simultaneously observing somewhere between a hundred thousand and a million stars. In comparing different plates, he was surprised to see that the magnitude of quite a number of stars had changed. He had discovered the existence of short-period cluster variables.

In the 1960s, the two-micron infrared sky survey conducted by Neugebauer and Leighton (1969) was seminal in identifying whole new varieties of sources of radiation that emitted powerfully in the near infrared. But the survey's full implications could be understood only by identifying individual sources, largely with the aid of far deeper optical surveys, particularly with the comprehensive Palomar Sky Survey plates that covered the same portions of the sky.

Roughly forty years apart, the Palomar Sky Survey and the Hipparcos mission have provided important two-color capabilities in the optical range. The two-color discrimination was helpful for many purposes. But positional identification was perhaps the most useful feature of the Palomar plates, as far as radio, infrared, or X-ray astronomers were concerned. They helped to identify a source as a star, a planetary nebula, or a galaxy, and often could provide more detailed descriptions of the nature of the object within its class. In contrast, the Hipparcos mission's greatest benefit for most astronomers is the unprecedented positional accuracy it has provided. Its positional and parallax accuracy are in the range of 0.0013 and 0.0015 arc seconds. Its proper motion measures on roughly 100,000 stars are accurate to 0.0018 arc seconds per year.

Did the Palomar plates or the Hipparcos missions lead to the discovery of new phenomena? By themselves, probably not, though they are massive surveys. But their contributions to understanding are undeniable. The Hipparcos mission has already shown that an accurate measure of distance provides a far better defined Herzsprung-Russell diagram for stars local to the solar neighborhood. Stellar properties apparently vary significantly less among these stars than might have been previously thought. Once a star's B−V coloration is known, its absolute magnitude on the main sequence or on the red giant branch, is quite narrowly defined (ESA 1996).

The two-micron survey's successors, the DENIS survey just starting in the south, and the all-sky 2-MASS survey about to begin in the north, will each conduct observations at three near-infrared wavelengths, and will detect in excess of a hundred million stars and 500,000 galaxies. All sightings will be cross-correlated with other wavelength ranges. The spectrophotometric classifications to be obtained in this way, may provide the characterizing features that will permit us to discover new phenomena.

Polychromatic surveys have greater potential for differentiation and discovery of new kinds of sources than their monochromatic counterparts. The IRAS four-color survey provided us with clear differentiation between infrared sources bright predominantly at the shorter 12 μm or at the longer 100 μm end of the spectral range. Even here, however, additional insights and true discoveries came about only through comparison to observations obtained in entirely different spectral ranges. Thus, we became aware of the existence of ultraluminous, $L \geq 10^{13} L_\odot$ extragalactic sources, mostly identified with colliding galaxies undergoing a burst of star formation, and dust clouds around stars, possibly remnants of proto-planetary dust clouds. In both these instances, the pre-existence of much more comprehensive optical surveys was essential to the recognition that something new might be at play.

3. Specialized Capabilities for Revealing Unprobed Characteristics

It is becoming clear that, at the current state of development in astronomy, where virtually every part of the electromagnetic spectrum has been at least superficially probed, it is no longer possible to make discoveries in a single spectral band alone. Such early discoveries as quasi-stellar radio sources, far more compact than the largely extended sources previously known, or radio sources that pulsated at incredibly regular rates, or a ubiquitous X-ray background that was clearly different from point-like X-ray sources, or a bright microwave background—all of which were remarkable in their own rights without even identifying corresponding optical features—may soon become matters of the past. Within a given spectra range, the only remaining, broad-band investigations that might uncover new phenomena would be extremely high energy gamma ray observations or a search for extremely fast pulsations. Somewhat surprisingly, the optical band is one domain in which opportunities for fast timing have never existed. Atmospheric interference makes short term observations from the ground unreliable, and no spacecraft mission has incorporated that capability. A survey with such instrumental capabilities could well uncover previously unknown phenomena.

4. Uneven Coverage

The difficulty at the moment is the very uneven sensitivity and angular and spectral resolution capabilities in the different spectral ranges. The Compton Gamma Ray Observatory (CGRO) has permitted us to make great strides forward. But, in terms of the number of sources observed, it has only brought us to the point to which the Uhuru X-ray survey had brought

X-ray astronomy two decades ago. The IRAS survey gave us access to hundreds of thousands of sources, but its spatial resolution was only at the arc minute level, a far cry from optical or radio spatial resolving powers. In the far-infrared and X-ray domains advances in spectral resolving power are providing increasing capabilities in identifying grains and molecular, atomic and ionic species across a wide set of temperature and pressure ranges. Our recognition and understanding of the physical and chemical processes at work are thus vastly improved. By itself, high resolution spectroscopy can also be an engine of discovery. The early detection of interstellar and circumstellar masers through high-resolution radio spectroscopy is one case in point. But the analytic virtues of spectroscopy ultimately may prove far more significant than the ability to discover startling new phenomena.

Each generation builds more powerful instruments, but so far the mismatch between capabilities in the different spectral regions is enormous. For CGRO scientists the obvious frustration is the inability to locate the tantalizing gamma-ray bursts with the arc second or sub-arc second spatial precision that would be needed to identify potential sources on optical or radio maps.

5. Discovering Rare Phenomena

The discovery of phenomena that are exceedingly rare, occurring perhaps even less frequently than supernovae or gamma-ray bursts, may require new kinds of surveys. To date, we have not undertaken long-duration surveys, where evolutionary effects or seldom-occurring, faint, nearby phenomena could be discovered. Most of our surveys have stopped after only a few years. Sometimes their endurance is linked to the life-work of an individual; in other instances, novel techniques overtake observations already in hand, making them outdated.

A trade-off exists between long-duration surveys that await rare occurrences, and greater sensitivity investigations, which enable us to detect these same occurrences at greater distances and, thereby, at more frequent intervals. This trade-off may, however, have limited applicability. If gamma-ray bursts turn out to be cosmological, as now seems likely, and if they actually have a threshold energy below which they are not generated, then increasing sensitivity will not be of help in registering larger numbers of bursts. If there exist phenomena that occur thousands or millions of times less frequently than gamma-ray bursts but still play powerful roles in cosmic evolution, then we may need to mount surveys that persist many decades or centuries to discover them. We have not yet learned how to do that.

Gamma-ray astronomers have been fortunate in having a relatively empty sky against which to observe their bursts. Imagine how much more difficult it would be to mount a survey to search for rare pulsations in a sky filled with hundreds of millions of sources all variable, at some level, intrinsically or because of interstellar scintillation.

6. Data Rates

Consider the search for a faint set of sources that might radiate appreciably in only two broad spectral domains, perhaps the infrared and the optical domains. Let the source flare up for an hour, and only an hour once every century, and let it then be no more luminous than a star of one solar luminosity. And finally, let there be only a thousand of these sources in the Galaxy, only ten of which might be visible at optical wavelengths—the rest being obscured by dust.

The bit rate required to discover these sources is readily calculated. We might wish to comb the sky with pixel sizes one arcsecond on a side. That would mean searching through $\sim 5 \times 10^{11}$ pixels in an all-sky survey. One of these stars might, therefore, be seen to flare up, on average once in ten years, or $\sim 10^5$ hours. To be sure that we appropriately characterize the signal, we might wish to make ten measurements on the flare-up in each of the two wavelength bands, during the hour the flare-up lasts, for a total of twenty measurements per hour on the source. In turn, this means that each pixel would need to be observed twenty times per hour, for ten years, before a source of the new kind was discovered. Let twenty bits of information suffice to characterize the range of intensities and positional information for each source, and we see that the total required bit rate is 2×10^{14} per hour, or 50 gigabits per second. This is a quite impossible rate, right now.

Of course, it is clear that both the bit rate and the archiving problem can be drastically reduced by proper filtering and data compression. One might choose to make routine measurements that are far coarser, say covering 100 square arc seconds at a time. Only if the total brightness of that area differed appreciably from the expected signal for the patch, based on a complete star map kept in memory, would a more detailed look be taken at this area. This would already reduce the incoming bit rate to a more acceptable gigabit per second level. The rate of transmitting information to a central archive might be further cut by many orders of magnitude, since only differences from the standard maps would need to be transmitted and archived. Still, the example illustrates some of the steps that will be needed in large surveys if rare events are to be discovered in the enormous masses of data that could be expected in principle. Clever compression techniques can also cut the cost of data analysis.

What I have just described bears relation to the MACHO project. Here one looks for a sudden flaring up of relatively faint sources in arbitrary parts of the sky. One is seeing remote sources lensed by foreground objects which might or might not be as luminous as the lensed source. That project has already registered many terrabytes of information. It is still relatively simple. There may be other phenomena where a source would become apparent only at high spectral resolution, for example if there were short duration bursts of cosmic masers at gamma frequencies. These might not make themselves apparent against the ubiquitous gamma ray background unless searched for with high spectral resolution techniques. To add to the difficulties, if these sources were distant, their systematic red shift would require some form of high resolution multifrequency detection. All this is possible, but it is not likely to become practicable with a single instrument that gathers all these data simultaneously.

7. Optimization Schemes

We will need to work out strategies to optimize deep searches. We are not yet sure, even now, whether numerous small but dedicated space missions, can be more effective than observatory class complexes that can cross-correlate a variety of types of information—spatial, spectral, temporal and polarization. And if we do choose the more specialized missions, how should we decide on the most propitious characteristics. Should astrophysical theory determine their make-up, as for example in the SWAS mission to search for and map water vapor, molecular oxygen and a few other chemical species of high astrophysical interest? Or should we build in additional breadth—and at what additional cost—to assure that nothing unexpected has been overlooked? These are important considerations and it is somewhat surprising that little thought has been given to optimization schemes. To date, we have opted almost entirely for either the general observatory class missions, or else for highly dedicated missions, like Hipparcos, radio surveys for CO or atomic hydrogen, or a search for general relativistic effects such as the Gravity Probe-B.

We do not currently have anything like an optimum search strategy. We may need to consider the best ways of sparse sampling in the temporal, spatial, and spectral domains. These will have to be designed to permit a search for patterns—correlated signals—within and among observations in these domains.

Sparse sampling leaves open the possibility that some rare phenomenon will remain hidden. How can we minimize the risk of that without taking up a far more expensive, exhaustive search?

Methods of multiplexing, to permit a variety of data to be simultaneously recorded and processed, may provide advantages (c.f. Martin Harwit and Neil J. A. Sloane, 1979). What are the best ways of doing this without prejudging the types of patterns that may emerge?

Just as we in astronomy adopted hand-me-down hardware from the military, throughout the half century since World War II, we may now, in the wake of the Cold War, see what optimization schemes the military and the CIA have developed in their search through enormous data banks they must have accumulated. We have already adopted many of their techniques, but I expect that there is a great deal more that we could take over wholesale without too much expenditure of effort.

Perhaps no optimum strategy exists, and perhaps we are already doing the best we could be expected to. But it might be useful to mount an investigation where such questions could be discussed. It would be different from the ten-year national academy studies, in that there would be no winning mission to emerge. Rather, the investigation would aim to articulate a general philosophy, most probably independent of wavelength range, and therefore not hotly contested by astronomers with different wavelength interest and preferences. It might be a philosophy that would take into account our astrophysical understanding at any given epoch in the development of astronomy, and feed that back to arrive at a progressively more informed way of choosing the next set of optimized steps.

Spending some thought in developing a broad philosophy of this kind, with suggestions to science administrators for sensibly applying it, could turn out to be one of the most effective ways of getting our money's worth from expensive surveys.

Acknowledgements

The author's work is supported by two grants from NASA.

References

Harwit, M. "Cosmic Discovery", Basic Books, New York, 1981
Neugebauer, G. and Leighton, R.B. "Two Micron Sky Survey", NASA SP-3047, Washington, D.C., 1969.
ESA Report to COSPAR 1996, page 36
Harwit, M. and Sloane, N.J.A. "Hadamard Transform Optics", Academic Press, 1979

SELECTION EFFECTS AND BIASES

M. DISNEY
Department of Physics and Astronomy
University of Wales College of Cardiff

1. Introduction

A respected astronomer recently suggested, because the whole spectrum has been explored, that the heroic days of astronomy are now over. I will try, by referring to the narrow optical and 21-cm windows with which I am familiar, to argue that the he was quite wrong. We still haven't covered the whole spectrum, and in any case prejudices, biases and selection effects could still veil much of the truth from sight.

2. Human Bias

Human bias can grossly affect the universe we perceive. For instance: (a) Our life-spans are very limited. We are therefore prone to believe, because we hope it is true, that we are rapidly converging upon some true picture of the universe. The notion that millennia may have to pass, that observations lasting for centuries will have to be done, before the picture emerges, are not ones we readily entertain. (b) We are mostly educated as, and therefore tend to think like, physicists. But astronomy is an altogether different subject in which half our major discoveries are still serendipitous. An instance of this is "Mad Big-Telescope Disease," a mind-set borrowed from the Super-Collider school, which believes that only gargantuan machines are capable of new discoveries. In rebuttal simply think of: Expansion of the Universe (24″); Clusters (18″); QSOAL's (84″); Lenses (84″); Large-scale flows (60″); Extra Solar planets (74″); MACHO's (50″) ... (c) As a species we are far better equipped to exchange information with each other than fathom it out for ourselves. Virtually all our knowledge is thus second-hand. Any misconception in the network will be difficult to identify, viz. the Flat Earth, and flat Space. (d) As astronomers we have special problems with Occam's razor. It urges us to null hypotheses such as: "If you can't detect it then assume it's not there" or "If you don't see it vary, assume it's constant." Yes, but ...

3. Economic Bias

Economic bias can slant one's ideas dramatically. For instance in $F(\nu)$ units the quasar 3C 273 appears a strong radio source, but in energy terms (i.e., $\nu F(\nu)$ units) its radio output is trifling compared to the UV. The reason we think of quasars, and other phenomena, predominantly as radio sources may have to do with an economic bias—the cost per unit area of telescopes at different wavelengths, which are, in (1996) \$US m^{-2}

Radio	mm	Optical	NIR	All other wavelengths
10^{3-4}	10^5	10^6	10^6	10^9 (i.e., Space)

4. Wavelength Bias

Have we carried out a first reconnaissance of the whole spectrum, and, if not, how far have we still to go? I now argue that we are only half way there.

Figure 1 shows much of the Cosmos plotted in $\nu F(\nu)$ units. Some sources are broad-band—e.g., (d) Crab Nebula; (f) Cyg-A; (g) BL-Lac; (h) 3C-273—while the rest are narrow-band or thermal. (a) is a cool bright star, (b) a nearby Elliptical and (c) a nearby Spiral. A crude sampling of the whole spectrum would suffice to find the broad-band sources but the thermal sources could easily elude such a crude inspection. Until we are sure that no class of luminous thermal sources is still evading our census we are still open to surprises. The question then becomes "How many surveys are required to turn up the brighter thermal sources at any temperature?" I claim that the answer is about 19.

To see why, consider the Planck spectrum, which has the same identical shape in Figure 1 at whichever wavelength range you choose to locate it— see the 2.7° CMB spectrum at lower left. How many such Planck spectra will fit across the entire 15 dex frequency range from radio to γ-rays? The answer will depend on how deep you take each spectrum to be from crest to base. One could argue as follows. A preliminary survey of the sky at a new wavelength should detect a few thousand sources, that being the maximum number you could hope to follow up in other windows. But isotropic sources tend to follow the relation $N(>S) \sim S^{-3/2}$ where S is flux. For an N of thousands you therefore want $S_{min}/S_{max} \sim 1/300 = 10^{-2.5}$. This sets the depth of a Planck function in Figure 1 at 2.5 dex. One can read off the width of the function at this depth; it is about 1.5 dex. In other words

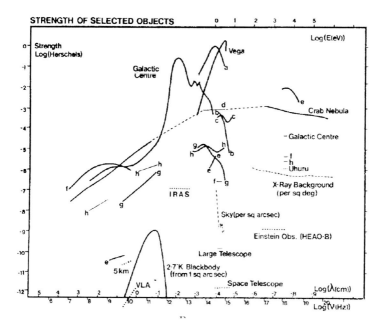

Figure 1. Prominent cosmic objects in $\nu F(\nu)$ units—in which only the relative ordinate scale is important—see text for labels (Disney and Sparks 1982).

neighbouring surveys for thermal sources, each containing several thousand sources, only overlap in the frequency domain when they are less than 1.5 dex in frequency apart. To be certain no new kinds of sources are missed at the margins we probably need 50 per cent overlap in the frequency domain. *Thus neighbouring surveys for a few thousand sources each, may miss new classes of thermal objects if the surveys are situated more than 0.75 dex in frequency apart.*

Dividing the total spectral range of 15 dex by 0.75 dex yields a requirement for 20 surveys in all, less one for the UV curtain. Thus 19 (The argument is not strictly applicable longward of the CMB—where thermal sources must be hidden below sky).

The nature of the 19 windows must be such that

$$10^{0.75} = 5.6 = \frac{\nu_{n+1}}{\nu_n} = \frac{\lambda_n}{\lambda_{n+1}} = \frac{E_{n+1}}{E_n} = \frac{T_{n+1}}{T_n}$$

There should be 5 surveys longward of the CMB, 5 more between the CMB and the UV curtain (0.1μ) and 9 high energy surveys between 0.1 keV and 1GeV. For instance, between the CMB and UV curtain we need surveys at roughly:

$\lambda(\mu)$	300μ	60μ	12μ	2.5μ	0.5μ	0.1μ
$\sim T°(K)$	12°	60°	300°	1500°	6500°	36,000°
Survey	None	IRAS	IRAS	Caltech	POSS	None

Counting the ones that have been done is not easy, particularly at higher energy, because the spectral purities are not always so clear and the positions not adequate for identification. My rough estimate is: Optical (1), IRAS (2), Radio (3–4); X-ray (1–3) and γ-Ray (1–3). We have between 11 and 6 more all-sky surveys to do before we can say we've had a good first look at the cosmos. There are obvious gaps at 300μ (12°K), 0.1μ (36,000°K), 5mm, and 100 keV.

5. Luminosity Bias

Knowing what classes of object inhabit the universe at a given wavelength is not the same thing as knowing which are significant. For that purpose you need, as a first step, to know the relative numbers of intrinsically luminous and less luminous sources—the so-called "Luminosity Function" (LF). And that is far harder to come by for it requires distances to large numbers of objects. The problem is that the Visibility $V(L)$ of a source of luminosity L, that is to say the maximum Volume in which it could lie and still be detectable in a flux-limited survey, rises as $L^{3/2}$. Hence luminous sources may be grossly over-represented in such surveys; viz naked-eye stars.

LF's are frequently represented as a power-law of the form $N(L).dL \sim L^{-\alpha}.dL$ where the index α may be a slowly varying function of L. At high L α must be > 2 for luminosity convergence; at low L $\alpha < 1$ for number-convergence. Somewhere between $\alpha \sim 1.5$, and one generally finds that sources in the LF close to the point where $\alpha \sim 1.5$ are much the most conspicuous in a flux-limited survey (galaxy astronomers call them "$L^\star$" galaxies); the more luminous ones are too rare and the less luminous too faint to figure prominently. LF surveys containing a few thousand sources invariably reveal a slope ~ 1.5, which tells one very little about the source population, everything about lack of dynamic-range in such surveys. If $O(L) \equiv$ number of sources observed per logarithmic interval in luminosity L, which is what you want in order to measure the LF, then it is easy to show that $dlogO(L)/dlogL = 5/2 - \alpha$. One can then ask how far down the LF you can measure if you observe N distances in total. If you want to go down to $L/L^\star = 10^{-y}$ and you want 10^x objects in the faintest bin—for statistical accuracy, then you must find distances to $N > 10^x 10^{y(5/2-\alpha)}$ sources. For 20% accuracy in the lowest logL bin you need to observe the following numbers of sources to go 10^y below $L^\star$:

	y=1	y=1.6	y=2	y=2.5	y=3	y=4
α=1.0	1,000	4,000	10,000	30,000	10^5	10^6
α=1.5	250	1,000	2,500	7,000	25,000	250,000
α=2.0	100	400	1,000	3,000	10,000	100,000

Such numbers warn us that making a census of the universe, even at a single wavelength, will be a hard slog, and that claims to know the LF, based on a thousand or so sources, are naive. Quite apart from statistics, intrinsically sub-luminous objects can only be found close by. But if our neighbourhood is not typical, we will be left with an incurably biased view.

6. Surface Brightness Bias

Looking out into the night from a lighted room one sees only other similarly illuminated areas. The darkened buildings and mountains will be hidden beneath the local glare. But we do live in such a lighted room, close to a bright star in a spiral arm of a giant galaxy. From here most structures in the universe may be invisible, or at best very difficult to see, at any wavelength.

Take optical galaxies. It has long been known (Freeman 1970; Disney 1976) that the vast majority of catalogued galaxies have surface brightnesses (SB's) almost identical to the terrestrial sky. Is this an extraordinarily fortuitous coincidence? Or does it mean that the majority of galaxies, even in our neighbourhood, still remains to be found? Which explanation is less incredible? The same SB constancy is true for ellipticals, for dwarfs as well as giants, for blue galaxies as well as red (Disney and Phillips 1985, even though the SB is measured in B!) and, almost incredibly (since SB $\propto (1+z)^4$), for galaxies in the Hubble Deep Field at a median redshift of one (Jones and Disney 1996). What is going on?

It is possible to calculate the Visibility of a galaxy as a function of its SB, and the result is dramatic. Galaxies will be included in a catalogue only if their apparent angular sizes Θ and apparent luminosities l, measured at some limiting isophote, exceed some minimum values Θ_{min} and l_{min}. For a galaxy of given luminosity these isophotal values will be sensitive functions of its SB Σ. If Σ is too high the galaxy will be both physically and apparently small; if too low most of its light will fall below the limiting isophote, implying a faint apparent l. If one calculates the "Visibility" $V(L,\Sigma)$ of a galaxy, that is to say the maximum volume within which it can lie, yet still exceed the catalogue limits, then $V(L,\Sigma) = L^{3/2}\Lambda(\Sigma)$ where $\Lambda(\Sigma)$ is a uniquely pointed function of Σ, centred at a value defined by $\Sigma_{CAT} \equiv l_{min}/\pi\Theta_{min}^2$. $\Lambda(\Sigma)$ (Disney and Phillipps 1983) is so peaked because it is the lower envelope of two plunging curves, one defined by $\Theta = \Theta_{min}$

and the other by $l = l_{min}$, which intersect at Σ_{CAT}. The fact that most catalogued galaxies have SB's very close to this preferred value could then be a selection effect. Credence is lent to this by the observations of Davies et al. (1994) of 918 spirals in the ESO catalogue with redshifts. They find an uncanny fit between the numbers in each SB bin, and the median volumes from which they are drawn. What else could they be observing but a dramatic selection effect which is hiding most of the galaxies, even in our neighbourhood, below the glare of the sky ?

When first suggested this conjectured world of "Iceberg Galaxies" was quickly ruled out by HI observers (e.g., Shostack 1977) who claimed they would have picked up their 21-cm signatures in "off-beams" while observing optical galaxies. But their claim was based on the unreasonable implied assumption that galaxies of low surface density would nevertheless have high enough HI columns for them to be observed. But it is easy to show (e.g., Disney and Banks 1996) that 21-cm observations also have a SB, or in this case column-density, limit:

$$N_{HI} \geq 10^{18} T_{sys} \sqrt{\Delta V (kmsec^{-1})/t_{obs}(sec)} \text{ cm}^{-2}$$

(where ΔV is the profile velocity-width) which is independent of telescope size—because small telescopes have bigger beams. And 21-cm observers have traditionally used integration times (Briggs 1990) far too short to reach down to the columns (10^{18} cm^{-2}) expected of Icebergs with normal M_{HI}/L_B's.

Imagine a Shadow World consisting of the same galaxies, at the same distances, as we see in the UGC, but where each Shadow Galaxy is 10 times larger in diameter. The shadow M31 would be almost 40 degrees across. The mean SB of these shadows would be 100 below sky, and their N_{HI}'s about 100 below current survey limits. The 10^4 shadows, each more than 15' across, would cover 1% of the sky. To be sure of finding about 25 of them in a clustered universe one would need to survey $\sim$1% of the sky (i.e., 12 Schmidt plates, 20,000 CCD frames, 10,000 big radio-telescope beams] to extremely deep limits of 28Bμ, $N_{HI} = 10^{18}$ cm^{-2}). We certainly haven't searched for such a possible world seriously so far, and it is unlikely we will find it serendipitously except as QSOALs (where we may have seen it already—Phillipps et al. 1991) and as the very occasional "Crouching Giant"—that is to say a giant Iceberg whose higher SB bulge has already been misclassified as a dwarf. Predicted in 1980 (Disney), Bothun et al. (1987) found the first Crouching Giant serendipitously in 1987 and nicknamed it Malin 1. That no more have turned up may simply reflect that Malin 1 has an anomalously high M_{HI}/L_B ($\sim 5 \times$ solar), and therefore N_{HI}. Most such Crouching Giants with normal M_{HI}/L_B's (~ 0.3) would be far harder to find.

More puzzling is the measured deficiency of high SB galaxies. I can only surmise that in the optical they may be largely opaque—in which case they will turn up in surveys at 300μ and beyond (Jura 1980, Disney *et al.* 1989, Davies and Burstein 1994).

The future of this field probably lies with deep blind 21-cm surveys made possible by recent developments in technology (Staveley-Smith *et al.* 1996). We are embarking on such a survey with multi-beams at Parkes and Jodrell in 1997. The Doppler effect prevents foreground HI from blinding us to the extra-galactic variety and should enable us to reach down to 10^{18} cm^{-2} in limited areas, and therefore to corresponding SB's of 29 Bμ or dimmer.

I have tried to show how dramatic SB selection effects could be in two well-studied wavelengths, and how reluctant we have been to recognise them. Observers at all wavelengths should be alert to similar pernicious effects.

7. In Conclusion

I have tried to demonstrate, from one very parochial point of view, that we have the majority of the great astronomical discoveries still to make (see Harwit 1981 for a much deeper discussion of this point). Even in optical astronomy our deepest all-sky survey has been made with 1-metre telescopes with 1 hour exposures and 1% efficient detectors. Is a 36 second glance at the Universe all we are going to need?

References

Bothun G.D., Impey C., Malin D. and Mould J. 1987, Astron.J., 94, 23
Briggs F.H. 1990, Astron.J., 100, 999
Davies J.I. *et al.* 1994, Mon.Not.R.astron.Soc., 268, 984
Davies J.I. and Burstein D. 1994, "The Opacity of Spiral Discs," Kluwer
Disney M.J. 1976, Nature, 263, 573
Disney M.J. 1980, "Dwarf Galaxies," ESO Conf Proc, ed Tarenghi and Kjar, 151
Disney, M.J. and Sparks, W.B. 1982, The Observatory, 102, 231
Disney M.J. and Phillipps S. 1983, Mon.Not.R.astron.Soc., 205, 1253
Disney M.J. and Phillipps S. 1985, Mon.Not.R.astron.Soc., 216, 53
Disney M.J. and Sparks W.B. 1982, Observatory, 102, 231
Disney M.J., Davies J.I. and Phillipps S. 1989, Mon.Not.R.astron.Soc., 239, 939
Disney M.J. and Banks G.D. 1996, Procs.Astr.Soc.Australia, submitted.
Freeman K.C. 1970, Astrophys.J., 160, 811.
Harwit M. 1981,"Cosmic Discovery," Harvester Press.
Jones J.B. and Disney M.J. 1997, "HST & the Deep Universe," Procs Conf (submitted)
Jura M. 1980, ApJ, 238, 337
Phillipps S., Disney M.J. and Davies J.I. 1991, Mon.Not.R.astron.Soc., 242, 235
Shostack G.S. 1977, Astron.Astrophys., 54, 919
Staveley-Smith L. *et al.* 1996, Procs.Astr.Soc.Australia, submitted.

RADIO SURVEYS

J.J. CONDON
National Radio Astronomy Observatory, Charlottesville

1. Introduction

Radio surveys have an important role in astronomy, one that has changed with technology and scientific requirements. Most objects studied by radio astronomers today are the unexpected discoveries of early surveys. The survey "discovery" phase began with Jansky's detection of Galactic radio emission and Reber's 160 MHz maps showing that this emission is nonthermal. Surveys made just after World War II revealed strong discrete sources which were later identified with supernova remnants, radio galaxies, and quasars. Pulsars were discovered during a sky survey for scintillating sources. BL Lac objects were recognized in early high-frequency surveys. The first gravitationally lensed quasar appeared in the extensive Jodrell Bank 960 MHz survey, and the first measurement of gravitational radiation came from the binary pulsar serendipidously found in a pulsar survey.

2. Cosmological Evolution

The strong cosmological evolution evident in the first samples containing hundreds of extragalactic sources drove the second phase of radio surveying. Most radio sources in flux-limited samples are extragalactic, and evolution dominates their redshift distribution at all flux cutoffs $S_0 < 1$ Jy. Consequently, samples of radio sources are quite different from the local samples of bright galaxies found at optical or infrared wavelengths:
(1) Nearby sources are rare. Less than 1% of the 5×10^4 northern-hemisphere radio sources with $S \geq 25$ mJy at 4.85 GHz are associated with the 10^4 UGC galaxies larger than $\theta = 1'$, most of which are within about 100 Mpc (Condon *et al.* 1991). Similarly, there is $< 1\%$ overlap of this radio sample with the *IRAS* Faint Source Catalog, Version 2 (Moshir *et al.* 1992)

galaxies (Condon et al. 1995). Unfortunately, disjoint samples discourage multiwavelength cooperation. Optical and near-infrared astronomers generally prefer the nearby galaxies which they can observe well, while radio astronomers concentrate on luminous AGN at the edge of the universe.

(2) The majority of radio galaxies and quasars are found around their median redshift $\langle z \rangle \approx 0.8$, which is nearly independent of S_0 (Condon 1989). In the nearly Euclidean optical universe, faint objects are statistically more distant and have smaller angular sizes, but are intrinsically similar to bright ones. In the hollow-shell radio universe, reducing S_0 does not yield a "deeper" sample; it adds sources with lower absolute luminosities. Characteristic flux densities and angular sizes exist because features in the radio luminosity and linear-size functions are mapped directly onto flux-density and angular-size distributions. For example, there is a transition from classical radio galaxies and quasars to a mixture of starburst galaxies, Seyferts, and "normal" galaxies below $S \approx 1$ mJy at 1.4 GHz.

3. Extragalactic Astronomy

All-sky continuum surveys capable of detecting $\sim 10^5$ sources became practical with multichannel HEMT receivers in the 1980's. The NRAO 7-beam 4.85 GHz receiver was used on the Green Bank 91 m telescope to make the GB6 survey of about 75,000 sources (Gregory et al. 1996) stronger than 18 mJy in the northern hemisphere and on the Parkes 64 m to make the PMN surveys of the southern sky (Wright et al. 1996). Astronomical applications of such surveys include: (1) Obtaining the first statistically useful ($N > 10^2$) samples of nearby radio sources (e.g., Condon et al. 1991, 1995), (2) discovering intrinsically rare objects such as gravitational lenses, and (3) probing the large-scale structure of the universe at redshifts $z \sim 1$.

4. New Large Continuum Surveys

Large aperture-synthesis surveys capable of detecting $\sim 10^6$ sources are possible with modern computing power. Three are under way: the Westerbork Northern Sky Survey (WENSS), the VLA B-configuration Faint Images of the Radio Sky at Twenty Centimeters (FIRST) survey, and the NRAO VLA Sky Survey (NVSS) being made with the D and DnC configurations. A fourth, covering $\delta < -30°$, has been proposed (Large et al. 1994) for the Molonglo Observatory Synthesis Telescope. They all have significantly higher sensitivity, resolution, and position accuracy than existing single-dish surveys, and the WENSS and NVSS are the first source surveys sensitive to polarization as well as total intensity.

Surveys which can resolve many extragalactic sources face conflicting demands for completeness and position accuracy. Surveys are not flux

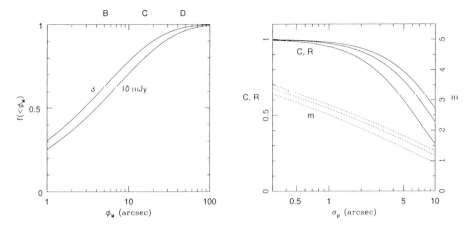

Figure 1. The cumulative fraction $f(< \phi_M)$ of faint sources with angular size ϕ_M is compared with the VLA B-, C-, and D-configuration resolutions in the left panel. The right panel indicates the completness and reliablity (for $C \approx R$ and search radius $r_s = m\sigma_p$) for optical identifications of sources having rms position uncertainty σ_p with objects brighter than $J = 22.5$ near the Galactic pole (top), at $|b| = 30°$, $|l| = 90°$ (middle), and in the Galactic bulge $|b| \leq 30°$, $|l| \leq 45°$ (bottom).

density (Jy) limited—they are surface-brightness (Jy beam^{-1} or K) limited. For example, the average face-on disk surface brightness of spiral galaxies is ≈ 1 K at 1.4 GHz, so this surface-brightness sensitivity is needed to detect complete samples of spiral galaxies above *any* flux limit S_0. Figure 1a shows the cumulative fractions of extragalactic sources versus angular size at 1.4 GHz flux densities $S = 3$ and 10 mJy. Surveys with resolution $\ll 1'$ miss faint resolved sources. On the other hand, noise limits the rms position uncertainty σ_p of faint sources to $\sigma_p \geq \sigma\theta/(2S_M)$, where σ is the rms map noise, θ is the FWHM resolution, and S_M is the source peak flux density. At the detection limit $S_M \approx 5\sigma$ and $\sigma_p \approx \theta/10$. The desire to make complete and reliable identifications with optically faint objects favors low σ_p and hence low θ. For example, $\sigma_p \approx 2\rlap{.}''5$ is needed to identify radio sources with $J = 22.5$ (the POSS II limit) objects at galactic latitude $|b| = 30°$ with 90% completeness and reliability (Figure 1b), so $\theta \leq 25''$ is needed to make such identifications at the survey limit. The WENSS and NVSS are optimized for completeness and photometric accuracy, FIRST for high position accuracy.

4.1. WENSS

The WENSS (de Bruyn *et al.* 1994) will cover the $\Omega = 3.14$ sr north of $\delta = +30°$ at 325 MHz and about one-third of this area at 610 MHz. Its FWHM resolution is $55'' \times 55''\text{cosec}\delta$ at 325 MHz and $30'' \times 30''\text{cosec}\delta$

at 610 MHz. The 5σ detection limit is about 15 mJy beam^{-1} at both frequencies, and about 3×10^5 sources are expected. The sky is being covered with a mosaic of fields separated by half the primary beamwidth. Every field is observed at 18 different hour angles in each of six array configurations over a period of six weeks. Thus the (u,v)-plane coverage and sensitivity to extended structures is excellent, and the data sample short-term variability at low frequencies. Finally, images in all four Stokes parameters (I, Q, U, and V) will yield polarization information for both discrete sources and diffuse Galactic emission.

Scientific goals of the WENSS include: (1) Selecting sources with extremely steep spectra, such as luminous radio galaxies at high redshifts, relic radio sources in clusters of galaxies, and pulsars. (2) Selecting a large sample of flat-spectrum sources to be used in a search for gravitational lenses. (3) Selecting and studying compact steep-spectrum (CSS) sources. In addition, the WENSS should be effective at finding giant (> 1 Mpc) radio galaxies, low-brightness disk and halo emission from nearby spiral galaxies, and cross-identifying objects found in other wavelength regions.

4.2. FIRST

The 1.4 GHz FIRST is a high-resolution total-intensity survey of the north Galactic cap. It was designed to identify faint galaxies and quasars found by the Sloan Digital Sky Survey (Gunn 1995), detect significant numbers of faint starburst galaxies, and resolve many extended extragalactic sources. The $5''\!.4$ resolution ensures that the rms position errors are $< 1''$ even at the survey sensitivity limit, ≈ 1 mJy beam$^{-1} \approx 20$ K. A catalog of fitted components from the first 1550 deg^2 observed (≈ 75 sources deg^{-2}) is now available on-line (Becker *et al.* 1995). Unlike the WENSS, both FIRST and NVSS cover each field with a single snapshot, so the (u,v) coverage is poor and the data must be processed carefully to maximize dynamic range.

4.3. NVSS

The 1.4 GHz NVSS is an "all-sky" survey, covering the $\Omega = 10.3$ sr with $\delta \geq -40°$ by a mosaic of 217,446 snapshot observations in the compact D and DnC configurations of the VLA. The principal data products will be (1) a set of 2326 $4° \times 4°$ "cubes" having three planes containing the Stokes I, Q, and U images with $45''$ FHWM resolution and (2) a catalog of almost 2×10^6 discrete sources brighter than 2.5 mJy beam$^{-1} \approx 0.8$ K (≈ 50 sources deg^{-2}). The rms position errors are $< 1''$ for strong sources, $< 2''\!.5$ for the $\approx 10^6$ sources stronger than 5 mJy, and $\approx 5''$ for the faintest detectable sources. The principal scientific goal of the NVSS is to encourage multiwavelength research by providing complete and reliable samples of sources, especially nearby ones, for use by all astronomers. The NVSS should detect most bright galaxies (*e.g.*, UGC galaxies), most sources in the *IRAS* Faint Source Catalog, and most classical radio galaxies and quasars (*e.g.*, M87 at $z \approx 2$). To guarantee equal access, the NVSS team members have agreed to use only the electronically released results for their own research. Image cubes covering about half of the survey area are currently available on-line, along with a catalog of $\approx 9 \times 10^5$ sources (Condon *et al.* 1996). The NVSS observations will be essentially complete by 1996 October 1.

5. Future Radio Surveys

Imagine a phase space whose axes are survey parameters—sensitivity, frequency, polarization, time resolution, *etc.* (cf. Harwit 1981). Astronomical phenomena are scattered throughout the phase space, some still awaiting discovery. Which regions remain to be explored, which will be scientifically interesting, and how can they be surveyed?

5.1. SPECTROSCOPIC SURVEYS

The first large-scale (survey volume $\approx 10^7$ Mpc$^3 \gg 10^3$ Mpc$^3 \approx$ galaxy correlation volume) spectroscopic survey for galaxies and extragalactic H I clouds was recently proposed (Stavely-Smith *et al.* 1996). A 13-beam receiver on the Parkes 64 m telescope will cover the southern sky to a depth of $140h^{-1}$ Mpc with $14'$ angular resolution. The 5σ sensitivity of 20 mJy in one 14 km s^{-1} channel corresponds to an H I mass detection limit $\approx 10^6 M_\odot (D/\mathrm{Mpc})^2$ for a galaxy with 200 km s^{-1} velocity width. This survey will rediscover known gas-rich galaxies (dwarf galaxies within about 30 Mpc, normal Sc galaxies within about 45 Mpc, and giant gas-rich galaxies such as Malin-1 out to 140 Mpc) and provide data on their H I mass function, space distribution, and group/cluster dynamics. It should discover new galaxies with low optical surface brightness and any isolated H I clouds. If numerous, they may contain a large portion of all H I in the universe.

The longitude range $75° < l < 145°$ of the Galactic plane is being imaged in H I with $\approx 1'$ resolution and in continuum at 408 and 1420 MHz (Normandeau *et al.* 1996). Single-dish data are used to fill the hole in the DRAO (u,v) plane and restore very extended structures.

5.2. $\nu \gg 5$ GHZ CONTINUUM SURVEYS

Continuum surveys now span 30 MHz to 5 GHz. Most known blazars were found in the higher-frequency surveys sensitive to compact sources with flat radio spectra. Surveys at higher frequencies have been considered in the hope that they might find a "new population" of sources with peaked or inverted spectra, but they are technically difficult. The beam solid angle of a telescope scales as ν^{-2} and system noise generally increases with frequency, so the the time needed to survey a fixed area of sky to a given flux limit rises very rapidly above 5 GHz. For example, the 100 m Green Bank Telescope (GBT) will be capable of making an all-sky survey at $\nu = 15$ GHz with the resolution and sensitivity of the 1.4 GHz NVSS, but it would take five years of continuous observing even with a 7-beam receiver!

This region of phase space may also be empty. Chauvanistic radio astronomers see the IRAS $\lambda = 60$ μm survey as a $\nu = 5000$ GHz radio survey complete to $S = 0.28$ Jy over most of the sky. The hypothetical blazars with spectral peaks between 5 and 5000 GHz should appear in cross-identifications with radio sources stronger than 25 mJy at 5 GHz. Unfortunately, all IRAS blazars in the northern hemisphere are stronger than 250 mJy (or weaker than 25 mJy) at 5 GHz (Condon et al. 1995). Their absence in the decade above the radio flux-density limit suggests that few nonthermal sources peak at short cm wavelengths. The only significant new population of "radio" sources found by IRAS is dusty galaxies.

Dusty galaxies at high redshifts remain the best candidates for surveys at the shortest radio wavelengths, in the $\lambda \approx 0.8$ mm atmospheric window. At frequencies below the blackbody peak, the spectra of dusty galaxies and quasars are so steep ($S \sim \nu^3$) that flux density is nearly independent of redshift above $z \approx 1$. Distant galaxies with luminosities near the knee of the evolved luminosity function pile up around $S \approx 1$ mJy, and the source counts near this critical flux density may exceed the Euclidean extrapolation by two orders of magnitude. Blain and Longair (1996) showed that the SCUBA multibeam bolometer mounted on the JCMT could survey about 0.1 deg^{-2} to this level and detect up to 50 galaxies, enough to provide useful constraints on the early evolution of star-forming galaxies.

5.3. SYNOPTIC SURVEYS

Strongly variable and transient radio sources are produced by X-ray binaries, black holes, gamma-sources, radio stars, novae, and other exotic objects in our Galaxy. Some have been discovered serendipidously and a few systematically (Gregory and Taylor 1986), but most remain unrecognized against the background of extragalactic radio sources. Known Galactic variables tend to be weak, have flat or inverted radio spectra, variability time scales ranging from hours to days, and low duty cycles. A systematic survey of this population requires repeated observations with high sensitivity and resolution (to avoid Galactic and extragalactic confusion) at a short

μm, and 32 × 32 for 40–120 μm. Only for wavelengths longer than 120 μm is the technology limited to arrays of fewer than 100 detectors.
- The ability to work from space with cold telescopes is critical to sensitivity in the infrared. Escaping the Earth's atmosphere allows two huge advantages: freedom from atmospheric obscuration that otherwise limits observations to a few translucent windows; and reduced thermal background that results in observations to a given sensitivity on similarly-sized telescopes being a million-times faster from space than from the ground. A number of satellites have demonstrated the advantages of space-borne IR telescopes, including: IRAS, the Cosmic Background Explorer (COBE), the Infrared Space Observatory (ISO), and the Japanese Infrared Telescope in Space (IRTS).
- Finally, advances in high performance computing are critical to large scale surveys. All-sky surveys like 2MASS and its European counterpart DENIS produce up to 20 GByte of data per day and over 10 Terabytes during a complete survey. Powerful workstations with inexpensive disks and millions of floating point operations per second are a *sine qua non*. Disseminating the results of these surveys efficiently will take advantage of the ever-expanding global networks.

2. Overview of Infrared Surveys

2.1. A SURVEY FIGURE OF MERIT

With all these technological advances, the infrared is now approaching the all-sky sensitivity already achieved at other wavelengths. Figure 1 shows survey detection limits over the complete electromagnetic spectrum, from the radio to gamma-rays. A few, very-deep pencil beam surveys from ISO, HST, and SIRTF are also shown. Table 1 lists the properties of some past, on-going, or planned surveys. While the radio and optical surveys (particularly the Sloan Digital Sky Survey (SDSS)) lead other wavelength bands in overall sensitivity, the infrared is catching up.

Of particular interest is the volume that a particular survey explores. This figure of merit tells how well a particular survey can both find rare objects and detect large numbers of objects of various classes to determine their spatial distribution and other statistical properties. The survey volume is proportional to the survey solid angle and, in Euclidean space, to the cube of limiting distance. Since the limiting distance is proportional to the square root of the limiting flux density, we can write:

$$Relative\,Volume = \frac{\Omega}{\Omega(IRAS)} \left(\frac{\nu F_\nu}{\nu F_\nu(IRAS)}\right)^{-1.5}$$

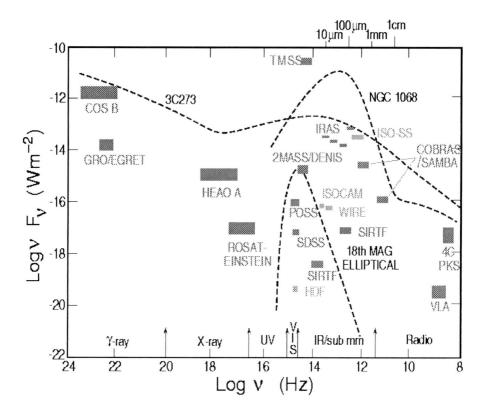

Figure 1. A comparison of surveys across the entire electro-magnetic spectrum shows the greatest sensitivity in energy per logarithmic interval (νF_ν) has been achieved in the optical and radio. Surveys plotted include: the VLA First radio survey, the planned COBRAS/SAMBA all-sky survey in the submillimeter, the various IRAS bands from the Faint Source Survey, the Wide-Field Infrared Explorer(WIRE), the 2MASS and DENIS surveys in the near-IR, the original Two Micron Sky Survey (TMSS), the Palomar Observatory Sky Survey (POSS) and the Sloan Digital Sky Survey (SDSS) in the visible, the ROSAT and Einstein Surveys in the X-ray region, and the EGRET survey from the Gamma Ray Observatory. A few very deep surveys with small solid angle coverage are shown in grey, *e.g.*, the ISO Serendipity Survey, the Hubble Deep Field and its potential NICMOS follow-up, various ISO and SIRTF deep surveys.

where we have normalized the volume relative to the ∼96% sky coverage of the IRAS Survey.

Examination of Table 1 shows that the Sloan Digital Sky Survey (SDSS) will probe the largest volume of space at the high-energy end of the infrared (0.7–0.9 μm). However, the 2MASS and DENIS surveys, SIRTF, the Wide-Field Infrared Explorer (WIRE; Shupe *et al.*, this volume, p. 118), and ESA's recently selected COBRAS/SAMBA mission will all explore important wavelength bands with great sensitivity.

TABLE 1. Selected Infrared Point Source Surveys

Survey Name	Wavelength (μm)	Sensitivity (mJy)	Area (sq.deg.)	Relative Volume (FSC-60=1)
SDSS	0.7	0.0018	10,300	1.7×10^4
COBRAS/SAMBA	350	20	41,250	650
SIRTF	3	0.0014	2	42
2MASS	2.2	1	41,250	29
2MASS/DENIS	1.25	1	41,250	12.4
WIRE	25	1	400	11
HST-NICMOS	1.6	3×10^{-5}	2.5×10^{-3}	6.9
DENIS	2.2	2	20,600	5.1
IRAS PSC/FSC	60	250	39,600	1.0
SIRTF	70	0.5	2	0.6
IRAS PSC/FSC	25	200	39,600	0.38
IRAS PSC	100	1,000	39,600	0.27
IRAS FSC	12	150	39,600	0.19
ISO-CAM Parallel	6.7	1	33	0.12
ISO-PHOT Serendipity	180	2,000	4,125	0.024
ISO-CAM Deep	15	0.3	1	0.08
ISO-CAM Deep	6.7	0.05	0.2	0.07

2.2. GROUND-BASED NEAR-IR SURVEYS

Two near-IR sky surveys are presently underway. The 2MASS survey will start in mid-1997 to survey the entire sky to sensitivity limits of 15.8, 15.1, and 14.3 mag at wavelengths of 1.25 (J), 1.65 (H) and 2.2 μm (K_s). The European DENIS project is already surveying the Southern sky at I, J and K down to comparable limits. These surveys will detect ~300–500 million stars (K<14.3 mag) and 1 million galaxies (K< 13.5 mag) and should be complete by the year 2000. One of the important aspects of the near-IR surveys is that a long wavelength baseline is necessary to identify interesting objects in these vast source catalogs. The near-IR colors of most objects cover only a small range of values. The addition of one or more optical bands (from simultaneous I-band measurements for DENIS and from POSS-I, POSS-II or SDSS for 2MASS) breaks this degeneracy, enabling the identification of interesting galaxies, quasars, and late type stars (Figure 2). Association of near-IR sources with optical, radio, and X-ray identifications will be straight-forward with the < 1″ positional accuracy of these catalogs. Detection of proper motions relative to POSS I will give a 40–50 year baseline to identify nearby stars.

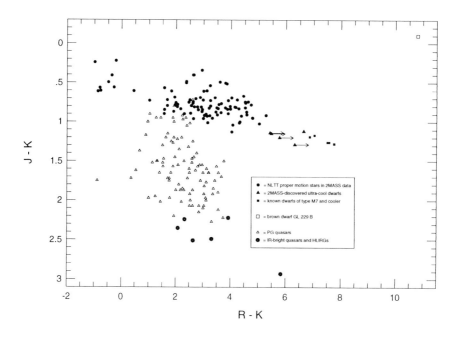

Figure 2. A optical-near-IR color-color diagram shows a wide variety of astronomical objects can be identified through their colors.

2.3. SPECTROSCOPIC SURVEYS FOLLOW SPATIAL SURVEYS

Once an initial reconnaissance has been carried out in a new wavelength, spectroscopic surveys become important to characterize the nature of particular objects and to study the physical conditions in the diffuse medium. The infrared is only now obtaining these critical data. The IRAS-Low Resolution Spectrometer measured the brightest ~5,000 stars from 7 to 18 μm. The COBE Far Infrared Spectrometer (FIRAS) made all-sky spectral maps at low spectral and spatial resolution and measured the global properties of a few key spectral lines in the ISM of our Galaxy, *e.g.*, [CII] at 157 μm and [NII] at 205 μm (Wright *et al.* 1991). But the greatest step forward in this area is taking place now with the flight of ISO. Perhaps the greatest legacy from ISO will be spectra obtained with its Short and Long Wavelength Spectrometers (SWS and LWS), the ISOPHOT-Spectrometer, and low spectral resolution images obtained with ISOCAM. Together these instruments provide unparalleled access to the 2 to 200 μm spectra of all types of astronomical objects at a variety of spatial (3″ to 3′) and spectral ($R \sim 50$ to 30,000) resolutions. These data are just becoming coming available, but the wealth of information is spectacular:

- Lines previously obscured by the Earth's atmosphere, such as thermally excited transitions of H_2O and H_2, as well as lines of condensed volatiles such as CO_2 and H_2O ices have been detected toward late type stars and embedded young stars.
- Maps have been made in various gas lines and dust features in regions showing a broad range of density and photon energy density. These maps will make possible realistic physical models of gas excitation, dust and chemical abundances in obscured regions.
- Fine-structure lines of elements, e.g., Neon, at various ionization stages are extremely useful diagnostics of the density, temperature and exciting spectra. Because they can reach into highly obscured nuclei of galaxies, they are invaluable for addressing the nature of the power source, starburst or black hole, in those nuclei.

The Japanese IRTS mission systematically mapped about 10% of the sky at modest spectral resolution with a $8'$ beam and will complement ISO spectroscopy of specific sources. IRTS will provide spectra of thousands of stars and detailed information on the distribution of various gas phase species, PAHs and other grains throughout the ISM, e.g., Onaka et al. (1996).

A number of future infrared missions will carry out spectroscopic surveys, including:

- The Submillimeter Wavelength Satellite (SWAS) will be launched in 1997 and will take ~ 500 GHz spectra in specific lines of H_2O, O_2, CO, and C, toward molecular clouds and galaxies to help understand the distribution of these fundamental species.
- SIRTF will take modest-resolution spectra of thousands of faint galactic and extra-galactic sources from 5 to 40 μm.
- SOFIA will obtain spectra with particular emphasis on high spectral resolution measurements in the submillimeter.
- FIRST will provide complete spectral coverage of selected areas with 1 km s^{-1} resolution for $100 > \lambda > 800$ μm.

3. Representative Science Projects for Next Generation Surveys

Some of the most important questions in astrophysics will be addressed by the spatial and spectral surveys, planned or now underway. The following list is hardly exhaustive, but represents the breadth of science possible with surveys expected within the next decade.

3.1. THE LOW-MASS END OF THE MAIN SEQUENCE

One of the first uses of the near-IR surveys will be to determine the shape of the low-mass portion of the stellar luminosity function and to search for isolated brown dwarf stars. It is difficult to answer these questions in the optical due to the faintness and redness of these stars, and to the uncertain conversion from magnitude to mass. Existing visible data on the lowest mass stars are ambiguous and stars of low mass could still account for a significant amount of the mass in the solar neighborhood (Mera *et al.* 1996).

Data from the 2MASS prototype camera have already increased the number of stars known to be later than M8 by 30% and identified the lowest mass field star, >M10.5V (Kirkpatrick *et al.* 1996). Figure 3 shows the numbers of M dwarf stars that might be expected in the 2MASS survey. It is interesting to note that the one brown dwarf detected to date, GL 229 B has the near-IR colors of an A0 star due to deep CH_4 absorption at 1.2 and 1.6 μm (Oppenheimer *et al.* 1996), but when the visible band color is added, the remarkable properties (R−K$\sim$ 10 mag!), of this object stand out in the upper right corner of Figure 2. The 2MASS magnitude limits correspond to distance limits of 5–10 pc for an object like GL 229 B. Assuming a flat mass function beyond the terminus of the main sequence we find that there might be as many as 600 brown dwarfs in the mass-range between 0.08 and 0.02 $M_\odot$ in the full-sky 2MASS survey.

The wavelength of peak emission from objects less massive or older than GL 229 B shifts further into the thermal infrared where sensitive observations are possible only from space. The 5–18 μm camera on ISO, ISOCAM, will search for brown dwarfs in a variety of environments, including:

- Targeted surveys in Ophiuchus and the Hyades to look for young, high luminosity objects.
- The Parallel Mode survey at 6.7 μm will examine $\sim$33 sq. deg. to 1 mJy (100 times deeper than IRAS) to look for field brown dwarfs.
- A shallow, wide-area survey will examine a few sq. deg. to 1 mJy
- A number of deep, high latitude surveys at 6.7 and 15 μm will look for very faint brown dwarfs over 0.2–1 sq. degree to 0.05–0.3 mJy.
- Deep maps of external galaxies will look for brown dwarf halos.

The ultimate mission to search for elusive field brown dwarfs will be the WIRE mission which will be launched in 1998. WIRE will cover 400 square degrees with the mJy sensitivity necessary to detect brown dwarfs over the entire mass-age plane. WIRE will either find field brown dwarfs or set a definitive limit to their local space density.

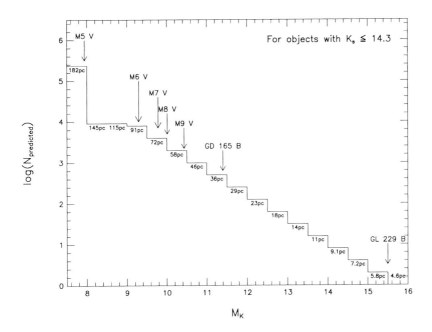

Figure 3. The number of M dwarf stars of different types expected in the 2MASS survey.

3.2. THE STRUCTURE OF THE GALAXY AND THE ISM

The Galactic plane beyond a distance of 1 kpc is almost completely obscured at visible wavelengths. Infrared studies from COBE and IRAS suggest that the Galaxy has a bar (Weinberg *et al.* 1992; Blitz and Spergel 1991). 2MASS and DENIS will find AGB stars ($K_s < 10$ mag) across entire galaxy to test this assertion and to determine the horizontal and vertical scale lengths of many late-type stars. ISO and the infrared instrument on the Mid-Course Space Experiment (MSX; Price, this volume, p. 115) will make ~ 10 μm maps of selected areas of the plane with far higher spatial resolution and sensitivity than was possible with IRAS.

The large scale structure of the ISM has already been mapped by IRAS and COBE. Recently, all the 60 and 100 μm IRAS data within $\pm 4°$ of the plane were processed with the High Resolution (HIRES) algorithm (Aumann *et al.* 1990; Cao *et al.* 1996) to reveal new structures in the dust emission. The IRAS HIRES maps have $\sim 1'$ resolution and can be compared directly with radio data in HI (from DRAO) and CO (from

FCRAO) lines to yield a complete picture of the energetics and kinematics of the gas (atomic and molecular) and the dust.

ISO has revealed the presence of compact regions within molecular clouds having $F_\nu(200\,\mu m) > 50 F_\nu(60\,\mu m$ (Laureijs et al. 1996; Bogun et al. 1996). Complete spectral energy distributions of cloud cores from a combination of IRAS, ISO, and IRTS observations will be used to unravel the contributions of different dust populations and heating mechanisms. A complete survey of the cold component of the ISM must await the launch of the COBRAS/SAMBA satellite which will make all-sky maps with $5'$ resolution in the submillimeter.

3.3. THE STRUCTURE OF THE LOCAL UNIVERSE

Deep, near-IR pencil-beam surveys from ground-based telescopes have established the K luminosity function of galaxies to K$\sim$24 mag and have probed galaxy evolution at redshifts $z < 3$ (Cowie et al. 1994; Djorgovski et al. 1995). Small K-corrections and negligible effects of dust make conversion of source counts into a luminosity function straight-forward for comparison with evolutionary models. Although the effects of evolution do not appear to be as pronounced in the K counts as at optical wavelengths (Djorgovski et al. 1995), some evolution is required for an $\Omega \sim 1$ Universe.

Variations in the galaxy counts at the bright end (K<15 mag) are found in different parts of the sky, implying the existence of structures on the scale of 300 h^{-1} Mpc (Huang et al. 1996). Investigating these variations is a fundamental goal of the 2MASS and DENIS surveys. The near-IR surveys will be more uniform in their coverage of galaxy types than IRAS which was biased toward star-forming spirals. Preliminary analysis of 2MASS prototype camera data indicates that ellipticals and spirals are detected almost equally well out to redshifts of $z \sim 0.1$ (Figure 4).

3.4. A UNIFORM CENSUS OF QUASARS

The lack of a near-IR sky survey leaves unanswered important questions about quasars and other energetic objects: Is there a missing population of dust-embedded quasars that has been missed in optical surveys? Is there a missing link between IRAS ultra-luminous objects and UV-selected quasars? The broad range of optical-IR colors for radio-selected quasars suggests that up to $A_V \sim 5--10$ mag of dust might be present in a parent population of quasars and that only a small, un-extincted part of that population may have been found in optical searches (Webster et al. 1995). A number of authors have contested this view based on optical follow-up of red, radio-selected quasar-candidates (Wall; this volume, p. 191) and of X-ray selected quasars (Boyle and diMatteo 1995). These arguments

INFRARED SURVEYS

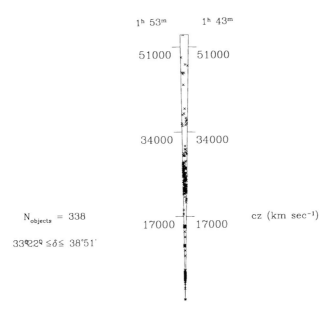

Figure 4. A pencil beam survey showing 2MASS galaxies detected with the prototype camera toward Abell 262. Galaxies are seen far beyond the cluster redshift of 3,000 km s^{-1} (Huchra and Schnieder, private communication).

are necessarily indirect and deep red-sensitive surveys such as SDSS and 2MASS are required for a definitive answer. Figure 5 (Beichman *et al.* 1997) shows how the surface density of a given population of quasars might vary at different wavelengths for different amounts of dust. Extrapolation of optical source counts to the near-IR using quasar colors (Elvis *et al.* 1994) suggests that 2MASS will find ~ 0.3 quasar per sq. deg. at $K_s \sim 14.5$ mag in the absence of dust. Finding significantly more quasars than this value might imply that extinction has hidden a significant part of the true quasar population. Although optical surveys are biased even by small amounts of dust, the SDSS will be sufficiently sensitive over a broad enough area that it should find all but the most deeply reddened objects. The combination of these two surveys will unambiguously determine the population of nearby quasars.

There is as yet insufficient data from 2MASS or DENIS to determine the near-IR quasar density. However, the first 2MASS QSO has been identified through its red optical-IR colors. Optical follow-up of objects in a 0.15 sq. deg. region led to the discovery of a relatively nearby, low luminosity quasar at $z = 0.147$ (Beichman *et al.* 1997).

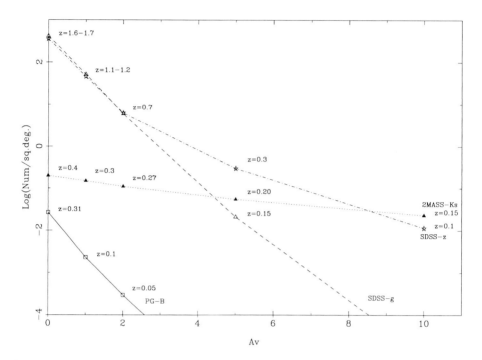

Figure 5. The surface density of quasars expected in different surveys as a function of internal quasar extinction. Surveys shown include 2 bands from the SDSS, 2MASS at K_s, and a representative blue survey with the POSS.

3.5. THE SEARCH FOR PROTO-GALAXIES

When did the first galaxies originate? Did galaxies form via collapse of larger objects, or through agglomeration of smaller clumps? The answers to these questions will be found through observations of objects at $z \sim 5$ and demand surveys in two separate IR wavelengths regimes:

- Observations in the near-IR to find the emission from redshifted UV-Optical light from young galaxies.
- Observations in the far-IR and sub-millimeter to find the emission from dust heated by UV-Optical light in young galaxies.

Which spectral region is most important depends on the unknown rate of dust formation in proto-galaxies. While the Hubble Deep Field (HDF) represents the first look at individual objects in the early Universe at $V \sim 30$ mag (!), objects at very high redshift will be much brighter in the near-IR than in the optical. It will be fascinating to see what the NICMOS observations of the HDF reveal when that instrument is launched early next year. At wavelengths longer than those probed by NICMOS, mid-IR surveys will look for starlight and very hot dust in young galaxies. ISOCAM surveys are presently underway at 6.7 and 15 μm, and WIRE surveys are planned at 12 and 25 μm, which will reach to $z \sim 1$. SIRTF surveys at 3–8 μm will be sensitive to L_* galaxies out to $z > 5$.

If dust plays a strong role in the appearance of proto-galaxies, then far-infrared and submillimeter observations are essential to study galaxy formation. Some interpretations of the COBE FIRAS data suggest the existence of a significant extra-galactic background (Abergel et al. 1996) due to distant, young galaxies. SIRTF will help resolve this question with 70 μm surveys planned to reach ~ 1 mJy over a few sq. deg. The CO-BRAS/SAMBA mission will make a definitive all-sky survey to $\sim$20 mJy at a number of far-IR and submm wavelengths. These space observations will be complemented by millimeter and sub-millimeter array telescopes on the ground which are also sensitive to the dust continuum emission from high redshift sources.

Acknowledgements

It is a pleasure to thank my 2MASS colleagues Mike Skrutskie, Tom Chester, Roc Cutri, Carol Lonsdale, John Huchra, Steve Schneider, and Will Pughe for discussions of 2MASS prototype camera results. I am indebted to Davy Kirkpatrick for his valuable insights into the properties of late type stars in the 2MASS data. George Helou and Martin Harwit graciously described some of the exciting new results from ISO.

This work was supported by IPAC which is operated by JPL and Caltech under contract with NASA.

References

Abergel, A. et al. 1996. Astron. Astrophys., 308, L5.
Aumann, H.H., Fowler, J.W. and Melnyk, M. 1990. Astron. J., 99, 1674.
Beichman et al. 1997. in preparation.
Blitz, L. and Spergel, D.N. 1991. Astrophys. J., 379, 631.
Bogun, S. et al. 1996. Astron. Astrophys., in press.
Boyle, B. J. and diMatteo, T. 1995. Mon.Not. R. astron. Soc., 277, L63.
Cao, Y., Terebey, S. Prince, T.A. and Beichman, C.A. 1996. Astrophys. J., in press.
Cowie, L.L., Gardner, J.P., Hu, E.M., Songaila, A., Hodapp, K.W. and Wainscoat, R.J. 1994. Astrophys. J., 434, 114.
Djorgovski, G. et al. 1995. Astrophys. J., 438, L13.
Elvis, M., Wilkes, B.J., Mcdowell J.C., Green, R.F., Bechtold, J., et al. 1994. Astrophys. J. Suppl., 95, 1.
Huang, J.-S., Cowie, L. L., Gardner, J.P., Hu, E.M., Songaila, A. and Wainscot, R. J. 1996. preprint.
Kirkpatrick, D. , Beichman, C.A., and Skrutskie, M. 1996. Astrophys. J., in press.
Laureijs et al. 1996. Astron. Astrophys., in press.

Mera, D., Charbier, G., and Baraffe, I., 1996. Astrophys. J., 459, L87.
Onaka, T., Yamamura, I., Tanabé, T., Roellig, T. L. and Yuen, L., 1996. preprint.
Skrutskie *et al.* 1996. in "The Impact of Large Scale Near-Infrared Surveys".
Webster, R. L., Francis, P.J., Peterson, B.A., Drinkwater, M.J. and Masci, F.J., 1995. Nature, 375, 469.
Weinberg, M.D., 1992. Astrophys. J., 384, 81.
Wright, E. L. *et al.* 1991. Astrophys. J., 381, 200.

PHOTOGRAPHIC SKY SURVEYS

Days of Future Passed

I.N. REID
Palomar Observatory

1. Introduction

Photographic surveys have played a vital role in virtually every area of astronomical research over the last 50 years. Indeed, one can make a strong case that Schmidt telescopes in general, and the Palomar 48-inch Schmidt in particular, have made the most substantial contribution to our understanding of the Universe—at least at optical wavelengths. The all-sky atlases compiled since the initiation of the POSS I O/E survey in 1949 will continue to provide fundamental reference catalogues and potent research tools for many years to come (for our children's children's children), particularly when combined with the second epoch surveys currently under way. However, as far as undertaking new, large-scale sky surveys are concerned, it is clear that we have reached the end of an era. Alternative detector technology and new instrumentation (*i.e.*, CCD mosaics) can now match and, indeed, exceed the capabilities of the photographic Schmidt telescopes in undertaking effective wide-field photometric surveys.

In this review, I report on the progress being made in completing the all-hemisphere photographic surveys currently being undertaken from Siding Spring with the UK Schmidt and from Palomar with the Oschin Schmidt. I also provide a summary of the availability of machine scans and/or object catalogues derived from the various surveys. Photographic survey work is not confined to these atlases, however, and section 3 provides discussion of other more specialised projects currently being undertaken. Finally, I consider the future possibilities for photographic Schmidt surveys.

2. The All-Sky Surveys

Atlas-style photographic surveys covering both the northern and southern celestial hemispheres are underway at the Oschin Schmidt telescope

B.J. McLean et al. (eds.), New Horizons from Multi-Wavelength Sky Surveys, 41–48.

on Palomar mountain, California, and the UK Schmidt telescope at Coonabarabran, New South Wales. There have been numerous reviews summarising the properties of the atlases produced from these telescopes, most recently by Morgan (1995), so I only cover recent progress on the incomplete surveys in the present talk.

Both Schmidts are 1.2-metre (48-inch) aperture telescopes and both are currently working on completing surveys using IIIaJ (blue), IIIaF (red) and IVN (near infrared: $\lambda_{eff} \sim 8000 \text{\AA}$) emulsions. Each telescope has an unvignetted field of view of ~ 3 degrees radius, a plate scale of $67\rlap{.}''14$ mm^{-1} and is taking plates on a grid of field centres with 5-degree spacing. The seventy-two fields at 0° declination are being covered from both sites, allowing direct comparison of the sensitivity of the two surveys. Overall, the agreement is excellent—the dispersion in limiting magnitude between the northern and southern plate material is comparable with the internal dispersion of either atlas alone. Thus, once suitable calibrated, it will be possible to combine data from these two enterprises to provide the deep, uniform, all-sky catalogues required for investigations such as studies of large-scale structure in galaxy clustering and irregularities in our own Galaxy.

TABLE 1. Sky surveys currently in progress

Survey	Declination range	Emulsion	Filter	Accepted	Total	Comp.
SERC EJ	$0 \geq \delta \geq -15$	IIIaJ	GG395	288	288	100 %
SERC ER	$0 \geq \delta \geq -15$	IIIaF	OG590	257	288	89 %
SERC I	$\delta \leq 0$	IVN	RG715	665	894	74 %
AAO R	$\delta \leq -20$	IIIaF	OG590	508	606	84 %
POSS II J	≥ 0	IIIaJ	GG385	808	894	90 %
POSS II F	≥ 0	IIIaF	RG610	785	894	88 %
POSS II I	≥ 0	IVN	RG9	558	894	62 %
USNO J	≥ 0	IIIaJ	GG385	808	894	90 %

Table 1 summarises the status of plate-taking in the surveys currently approaching completion. In the south the SERC Equatorial J survey has been completed since the Bandung meeting, while significant progress has also been made in obtaining accepted plates for the SERC equatorial and the southern AAO R-band surveys. Finally, the I-band IVN survey is an extension of the original Milky Way survey (spanning 163 fields) to cover the entire southern hemisphere. The completion of both this survey and the POSS II I-band survey were originally threatened by possible difficulties in obtaining plates from Kodak. Fortunately, Kodak has developed a synthetic gelatin and IVN plate availability is no longer a problem. Besides the survey observations, at least 50 % of the time at the UK Schmidt is devoted to smaller scale individual research projects, both photographic and with the FLAIR multi-object spectrograph (Watson 1995).

The Oschin Schmidt is devoted full-time to POSS II, so we have made somewhat greater progress towards the completion of the northern second-epoch sky surveys. (The USNO J survey consists of 3-minute, unhypered II-IaJ exposures designed to provide an intermediate grid of reference stars to tie astrometric measurements of faint objects to the HIPPARCOS/TYCHO catalogues.) However, I should emphasise that the fields remaining in all three passbands are *not* distributed in a uniform manner over the celestial sphere, but are concentrated between right ascensions of 6 hours and 14 hours. This is not a coincidence—it is, in fact, known as winter. An inspection of the plate logs from Palomar shows that while we were able to take 30–40 plates a month between November and March during the early years of the survey, when California was experiencing a severe drought, an average of only 5–10 plates per month is more characteristic of recent years. This concentration is something which other (terrestial) surveys should bear in mind. One can, perhaps, even codify this as Reid's two rules of sky surveys: first, the survey will take at least twice (and usually thrice) as long to complete as your most pessimistic estimate; and, second, that the weather will be best when you are least prepared to take advantage of it, and will deteriorate throughout the course of the survey.

As described in Reid *et al.* (1991), Palomar now lies at the crossroads of several well-used aircraft routes, and, as a result, between 10 and 15 % of the plates taken have aeroplane trails. Aesthetically, these trails are displeasing, but, with a width of $\sim$ 1 arcminute at most, each trail covers a mere 0.03 % of the area of a Schmidt field, and the main aim of POSS II is to provide a scientifically-useful modern survey of the northern sky, not a picture album. Had we but world enough and time, we would aim to replace all of these plates, together with all other plates with small-scale non-uniformities. Unfortunately, we do not have that luxury, while at my back I hear, Sloan Digital Sky Survey hurrying near. Thus, while we will obtain replacement plates for as many as fields as possible, some will remain in the final film atlas issued by the European Southern Observatory (1100 fields issued to date).

Few surveys have the luxury of being able to assume that instrumental parameters remain constant throughout the entire course of observations—and that is certainly not the case for the Oschin Schmidt. Modifications to the system (telescope, plate-holder, guider, *etc.*) have been made throughout the course of the survey, with the most recent being the installation of a motorised polar-axis adjustment at the beginning of the present year. Some of these modifications can influence astronomical analysis. In particular, a vacuum system, similar to that in use at the UK Schmidt, was installed in the plate-holder in April, 1987 and the plate mandrells re-ground to the correct radius of curvature later that year (see Reid *et al.* 1991 for an explanation of why they were originally ground to match the wrong radius of curvature). Since the vacuum system forces the glass plate to conform more closely to the surface of the supporting mandrell, there is a significant difference in the systematic astrometric residuals for plates taken before (*e.g.*, Quick V survey) and after this modification. This obviously has to be borne in mind when computing average global plate solutions.

Most of the available sky surveys have been scanned and processed to produce object catalogues by one or more of the various plate-scanning engines described by Lasker (1995). Table 2 lists those currently available.

TABLE 2.

Machine	POSS I	POSS II	Southern	Availability
APM	O, E		J, R	WWW
APS	O, E			WWW
COSMOS			J	WWW/NRL
DSS	E	J, F, N	J, EJ, R, ER	WWW
PMM	O, E, Wh	UJ, J, F, N	J, EJ, R_{ESO}, ER	CD-ROM
SKYCAT/DPOSS		J, F, N		

APM: Matched O/E image catalogues for POSS I fields above $b = 20°$ are currently available from the RGO/Cambridge WWW site, and a similar high-latitude catalogue is under construction.

COSMOS : The ROE COSMOS scans of the SERC J survey are on-line at AAO, MPI and NRL, with the last set accessible over the world-wide web. Scans of the other southern surveys and POSS II films are planned using the next-generation machine, SUPERCOSMOS.

APS: The Minnesota group has produced a POSS I blue/red matched image catalogue, listing co-ordinates, magnitudes, colours and morphology. A complementary image database (0.33 arcsecond pixels) is currently partially complete, while scans of Luyten's second-epoch (1962–1970) Palomar E-plates will be incorporated in a proper motion database (see Cornuelle, this volume, p. 467).

DSS: STScI is currently adding scans of second-epoch POSS II and southern J, F survey plates to their on-line Digital Sky Survey. Scans of the Palomar Quick V survey are also available.

PMM: Monet and co-workers at Flagstaff have completed scans of not only all the available POSS I, POSS II and southern atlas plates, but also of the short-exposure UJ plates and the Palomar/Whiteoak extension. A preliminary version of the resultant catalogue is available, while a full catalogue (including proper motions to $-48°$) is expected by mid-1997 (Canzian, this volume, p. 422).

SKYCAT: This is a separate analysis of the STScI POSS II scans being undertaken by a Caltech team led by Djorgovski (see Djorgovski, this volume, p. 424).

One point to emphasise is that the utility of these catalogues is crucially dependent on the accuracy of the photometric calibration. Various studies have shown how galaxian large-scale structure analyses can be compromised by inadequate calibration. A bare minimum requirement is one standard sequence extending to the plate limit in each field, and multiple sequences cannot hurt. GSPC-II and the work at Palomar by de Carvalho *et al.* will go some ways towards meeting this requirement, while (in 5 years time) SDSS will also have a substantial impact in the north, but obtaining reliable calibration must stand as the highest priority in large-scale survey work.

3. Specialised Photographic Projects

Besides the encyclopaedic all-sky surveys, various groups have been conducting smaller-scale efforts directed towards more specific aims. While there is insufficient space in these proceedings to cover all such projects, it would be misleading as a question of balance, to ignore them entirely.

UK Schmidt: As mentioned above, at least 50 % of the time on this telescope is devoted to non-survey projects. Multicolour (UB$_J$VRI) plates have been used to search for high redshift QSOs (McMahon), bright QSOs (Hewett *et al.*), low-mass stars (Hawkins, Jones, Thackrah) and blue stellar objects (the Edinburgh-Cape survey), as well as used to map the neighbouring LMC & SMC (Hatzidimitriou). The AAO staff have also been at the forefront in the use of fine-grain Techpan film, which offers significant gains over IIIaF emulsion (see Parker, this volume, p. 179)

Hamburg: The Calar Alto (formerly Hamburg) Schmidt is being used undertake an objective prism survey of the northern sky above $b = 20°$, with the plates (and complementary direct plates) scanned using the Hamburg PDS (Hagen *et al.* 1995). While the main aim is compilation of a QSO catalogue, the survey has also proved invaluable as a source of hot subdwarfs and for identifying ROSAT sources.

Paris: Several intermediate-scale projects are being undertaken with the MAMA scanning machine, based primarily on plates from the ESO

Schmidt. These include the DUO (Bulge) and EROS (LMC) gravitational lensing projects, surveys for long-period variables, Galactic structure analyses as well as studies of large-scale galaxy clustering.

Yale: The Lick Northern Proper Motion survey established a grid of astrometric standards to 18th magnitude in the northern hemisphere (these plates are slated for measurement on the PMM in the near future). This survey is being extended to the southern hemisphere by Yale.

Extensive photographic projects are also being conducted at Kiso, Asiago, ESO and Muenster.

4. New Horizons

The first survey undertaken in a particular wavelength régime requires no ulterior motive—something interesting and new will turn up (even a null result can be interesting). Subsequent surveys, however, demand more specific justification. As a rule of thumb, I suggest that a minimum requirement is an improvement of at least a factor of three (preferably 10) in measuring a scientifically interesting parameter. Marginal improvements generally only give marginal results. POSS II satisfies this criterion by providing an order of magnitude ($\times 10$) more depth at B and I, and a factor of 4 higher accuracy in μ over previous surveys.

Accepting this threshold criterion, what does the future hold for large-scale photographic surveys? Not a great deal. The main advantage offered by Schmidt telescope photography is surveying a large solid angle to relatively faint magnitude limits in a relatively short time. We can define the photometric efficiency of the system as

$$E_p \;=\; \Omega \times T_{mag}$$

where T_{mag} is the time taken to reach a given apparent magnitude with a given signal-to-noise—a parameter which depends on telescope aperture, the number and efficiency of optical elements, detector quantum efficiency and overheads such as mounting/dismounting plates and reading out a CCD. If we compare photography on a 1.2-metre Schmidt against CCD observations on a 2.5-metre conventional telescope, there is a factor of four difference in aperture and, at least at R and I, a factor of > 15 in DQE. CCD readout time is negligible if one operates in scanning mode along a great circle (which places stringent, but not excessive requirements on the telescope drive). The unvignetted field of the Schmidt is ~ 29 square degrees, so a $30' \times 30'$ CCD imager can match the Schmidt in photometric efficiency.

Astrometrically, the large-scale rigidity of the glass substrate supporting the emulsion on a plate (plus the fact that one is working at room temperature at all times) gives photography a substantial advantage over CCD mosaics. However, once one all-sky survey is completed, the main purpose of subsequent surveys (astrometrically) is to determine proper motions. The surveys listed in Table 1 provide second-epoch data for the whole sky and, since proper motion is linear with time and we have baselines of 20 and 40+ years in South and North respectively, there appears to be no urgent necessity for a third epoch survey until at least 2045. By that time I would expect photography to have been completed replaced by space missions such as SIM and GAIA, ground-based CCD transit surveys, *etc.*

However, while the justification for a new all-sky photographic survey is at best tenuous, that does not mean that Schmidt photography will not remain a useful tool for smaller-scale projects. The uniformity of photographic emulsions makes them well suited for studying low surface-brightness features (Malin 1994), while the wide field covered on a single exposure remains a distinct advantage in many programmes, such as supernova searches (the POSS II team, mainly J. Mueller, has discovered 57 supernovae in the course of the survey—an order of magnitude more than most dedicated SN surveys), comet and asteroid surveys, A photographic Schmidt is always more efficient if the requirement is a wide-field and a *bright* limiting magnitude. These smaller-scale, scientifically-focused projects, using modern emulsions, should represent the only future application of astronomical photography.

Acknowledgements

On behalf of those associated with both POSS II and the UK Schmidt, I would like to thank Eastman Kodak for their continued support of astronomical photography, particularly the development of the synthetic gelatins required to allow continued production of IIIa and IVN emulsions. The Second Palomar Observatory Sky Survey is funded by the Eastman Kodak Company, the National Geographic Society, the Samuel Oschin Foundation, the Alfred Sloan Foundation, the National Science Foundation grants AST84-08225, AST87-19465, AST90-23115 and AST93-18984, and the National Aeronautics and Space Administration grants NGL 05002140 and NAGW 1710.

References

Hagen,H.-J., Groote, D., Engels, D., Reimers, D. (1995), Astron.Astrophys.Suppl., **111**, 195

Lasker, B.M. (1995) "The Future Utilisation of Schmidt Telescopes," Astron.Soc.Pacific Conference Series, 84, p. 177 (ed. J. Chapman, R. Cannon, S. Harrison & B. Hidayat)

Malin, D. (1994) IAU Symposium 161, "Astronomy from Wide Field Imaging," p. 567 (ed. H.T. MacGillivray *et al.*: Kluwer Academic Publishers, Dordrect)

Morgan, D.H. (1995) "The Future Utilisation of Schmidt Telescopes," Astron.Soc.Pacific Conference Series, 84, p. 137 (ed. J. Chapman, R. Cannon, S. Harrison & B. Hidayat)

Reid, I.N. *et al.* (1991) Publ.Astron.Soc.Pacific, 103, 661

Watson, F.G. (1995) "The Future Utilisation of Schmidt Telescopes," Astron.Soc.Pacific Conference Series, 84, p. 71 (ed. J. Chapman, R. Cannon, S. Harrison & B. Hidayat)

ELECTRONIC OPTICAL SKY SURVEYS

T.A. MCKAY
University of Michigan Department of Physics

1. Introduction

The introduction of of Charge Coupled Devices (CCDs) in the middle 1970s provided astronomy with nearly perfect (linear, high-sensitivity, low-noise, high dynamic-range, digital) optical detectors. Unfortunately, restrictions imposed by CCD production and cost has typically limited their use to observations of relatively small fields. Recently a combination of technical advances have made practical the application of CCDs to survey science. CCD mosaic cameras, which help overcome the size restrictions imposed by CCD manufacture, allow electronic access to a larger fraction of the available focal plane. Multi-fiber spectrographs, which couple the low-noise, high QE performance of CCDs with the ability to observe spectra for many objects at once, have improved the spectroscopic efficiency of telescopes by factors approaching half a million. An improved understanding of image distortion gives us telescopes on which we expect sub-arcsecond images a large fraction of the time. Finally, and perhaps most important, the performance of computer hardware continues to advance, to the point where analysis of multi-terabyte datasets, while still daunting, is at least conceivable.

Optical sky surveys occupy a special place in the astrophysical survey world. Due both to the information content of optical emission and to the great power of the global collection of optical instrumentation, no object is generally considered "understood" until it's optical counterpart is found. These electronic optical survey projects, providing accurate optical data for much of the sky, will thus have a major impact on surveys at all other wavelengths.

We discuss the status and prospects for several subclasses of surveys; time domain surveys, redshift surveys, and deep, wide-field imaging surveys. This is followed by a relatively detailed description of the Sloan Digital Sky Survey, which includes elements of all three. We conclude with some general considerations about the difficulties these surveys face, and what prospects they present for the future of astronomy. This discussion will concentrate on the instruments which enable surveys to take place. Many of these instruments will be used for a variety of survey projects, and it is their parameters which define what is possible.

2. Time Domain Surveys

The study of variable objects has taken an unsurpassed leap in the last ten years, driven by the ability of CCDs to provide accurate photometry of many objects at once. The achievements of the microlensing experiments represent the state of the art in these photometric monitoring programs. The scale of these experiments is set by the need to monitor millions of objects on a regular basis for several years. The MACHO project, for example, has been monitoring about 22 million stars in two colors for 4 years. These experiments have observed several hundred microlensing events, impressively vindicating the effort put into them. That said, the most remarkable new results of these projects are the variable star catalogs. The MACHO project alone has detected and heavily sampled at least 80,000 variable stars in every known class, and several unknown ones. This is to be compared with the 20–30,000 variable stars known across the entire sky in the early 90's. The power of the increased sensitivity and accuracy of CCDs is demonstrated clearly here.

The new science being probed by the time domain surveys includes measures of the microlensing optical depth to the LMC (Alcock *et al.* 1996), a surprisingly high optical depth to the galactic bulge, the previously mentioned detailed studies of many classes of variable stars, new and challenging constraints on H_0 (Saha *et al.* 1996, Ferrarese *et al.* 1996), a deeper understanding of the details of Supernova light curves (Riess, Press, and Kirschner 1995), and the discovery of many type IA supernova at high redshift (Perlmutter *et al.* 1996). Prospects for the future in the time domain are directly connected to the construction of large area CCD mosaics intended for use on new telescopes.

3. Redshift Surveys

Redshift surveys are in many ways the progenitors of all electronic surveys. Starting from target lists selected as carefully as possible, they obtain homogeneous data sets designed for statistical study. The field of redshift surveys has been extremely active for the last 15 years (Strauss 1996, Colless 1996). Significant recent advances include the introduction and rapid advance of multi-fiber spectrographs, and the use of homogenous CCD photometry for target selection. Probably the premier wide area survey of the moment is the Las Campanas Redshift Survey, which has recently reported photometry and redshifts for about 26,000 galaxies in six 1.5 degree strips with a mean redshift of about 0.1 (Schectman et al. 1996). Also extraordinary are the narrow but deep redshift surveys like CFRS (Lilly 1996) and DEEP (Koo et al. 1996). In these surveys we're beginning to see samples of galaxies with median redshifts of 0.8!

The progress rapid progress in wide area redshift surveys will be continued in the near future by two new projects; the 2DF (Taylor 1996) and the SDSS. The "Two-degree field" project will utilize a pair of robotically positioned 400 fiber spectrographs in combination with a focal reducer at prime focus of the AAT 4m. This instrument will be used to measure redshifts of 250,000 galaxies in a total of about 1700 square degrees. Galaxy targets will be selected from the APM scanned plates. Details of the SDSS are outlined below; essentially it will record spectra for 10^6 galaxies over 10^4 square degrees selected throughout from 5 color CCD data.

Redshift surveys are performed for two reasons; to study the intrinsic properties of galaxies, and to study the large-scale structure implied by their locations. Primary results on intrinsic properties include the galaxy luminosity function and the first really solid indications of galaxy evolution (Ellis et al. 1996, Lilly et al. 1996). Large-scale structure surveys continue to find unexpected structure at scales close comparable to the survey size, and have for several years been combined with independent distance indicators to study peculiar velocities and flows (Lauer and Postman 1994). As several large new telescopes come on line, studies of galaxy evolution will be done in which many galaxies at redshifts beyond one will be studied spectroscopically. Meanwhile the two new wide area surveys will expand the explored volume of the local universe by nearly a factor of 100.

4. Wide-Field Imaging Surveys

CCD Imaging surveys of hundreds of square degrees of sky have only recently become practicable. They are enabled by the increased availability of large, easily mosaiced CCDs, along with the data acquisition and computing power required to handle the data. Perhaps the most exciting development in this area is the imminent availability of a number of new imaging detectors, each capable of performing deep imaging surveys on a photographic scale. The canonical new instrument has a field of view of several tens of arcminutes, instrumented with thinned CCDs, on a telescope of several meters aperture. The remarkable thing is that there are more than 10 of these instruments either online or planned for the near future.

A good example of a camera already in operation is the BTC (Walker 1996). This camera places four thin CCDs at the prime focus of the CTIO 4m to obtain a 30' field of view. Spanning the range of cameras in development are a 21' FOV prime focus imager for the SUBARU telescope and the 3 degree FOV SDSS imager. Data from these new cameras flows fast; a single image from the MMT 16K×16K camera will be half a gigabyte!

TABLE 1. New Wide-field Imaging Instruments

Name	Pixels	FOV □°	Telescope	Status
BATC	2K×2K Thin	0.90	0.9m BAO	Operational
QUEST	4K×4K Thick	1.3	1.0m CIDA	In construction
LMT	2K×2K Thick	0.35	2.6m LMT	Operational
BTC	4K×4K Thin	0.24	4.0m CTIO	Operational
UH8K	8K×8K Thick	0.23	3.6m CFHT	Operational
Kiso Camera	5K×8K Thick	—	4.2m WHT	Operational
UH	8K×12K Thin	0.37	3.6m CFHT	In construction
UW/APO	8K×8K Thin	0.13	3.5m ARC	In construction
SDSS imager	$\simeq$12K×12K Thin	1.8	2.5m SDSS	In construction
MMTCAM	16K×16K Thin	0.16	6.5m MMT	In design
SubPrime	10K×8K Thin	0.14	8.3m Subaru	In design

These instruments will be used for a broad range of science programs. Early deep CCD surveys discovered an unexpectedly large population of faint blue galaxies, which many of these instruments will use to measure weak gravitational lensing. Others will search for Kuiper belt and other faint nearby objects, or be used for the supernova search projects mentioned earlier. A special opportunity exists here; for many of these projects, particularly time domain surveys and deep co-added imaging, can utilize exactly the same data. It is hoped that collaborations can be designed to take advantage of the full range of science possible with deep survey data, so the time and effort required to produce them can be optimally used.

The particular advantages of CCDs have begun to make themselves felt, especially in the areas of time domain and redshift surveys. Having spent some time surveying the field, we will now concentrate on the SDSS, and use it to illustrate some of what is possible with new electronic surveys.

5. The Sloan Digital Sky Survey

The Sloan Digital Sky Survey will produce a digital image of the entire northern galactic cap (10^4 sq.degrees). Data will be obtained with 0.4″

resolution, in 5 colors, to 23rd magnitude in r'. From this imaging catalog, approximately 10^6 galaxies and 10^5 quasars will be selected. Spectra with a resolving power of about 2000 will be taken for all of these objects from about 400 nm to 910nm. In a complementary survey, a single $3\times100°$ strip in the southern galactic cap will be imaged about 35 times over a five year period, providing both a large time domain survey and a deep co-added imaging survey to about 25th magnitude. Imaging and spectroscopy will be done on the same telescope in an interleaved fashion; only photometric nights with sub-arcsecond seeing will be used for imaging. The SDSS (Gunn and Knapp 1993, see also http://www.astro.princeton.edu/GBOOK for details) is a collaboration of seven US and one Japanese institution, including over 200 scientists and engineers. Major support has been provided by the Alfred P. Sloan Foundation.

To conduct the survey we have constructed a new 2.5m telescope at the Apache Point Observatory in southern New Mexico. The telescope features a novel modified Ritchy-Chretien design, which provides a flat, undistorted, 3 degree field of view, a focal plane about 27" in diameter. Median seeing at APO is 0.8"; all the imaging will be done with sub-arcsecond seeing. The imaging camera for the survey, the largest CCD mosaic yet constructed, is being assembled at Princeton. There are six columns in the camera, each consisting of 5 SITe 2048×2048 pixel CCDs. Each of the five CCDs in a column views the sky through a different filter, allowing essentially simultaneous measurement in five colors. The camera will be operated in powered TDI mode, with 55 second effective exposures. A single pass through an area yields a "strip," a second interleaved pass fills the complete 2.6 degree wide "stripe." There are 54 CCDs with 145 million pixels in the imager, and it produces data at 8.2 Mbytes per second; 240 Gbytes in an 8 hour night.

The ultimate usefulness of the survey depends sensitively on the quality of the photometry. As the SDSS will observe most of the sky only once, we must carefully monitor all elements which affect photometry. An automated 24" "Monitor Telescope" has been constructed to measure atmospheric extinction and establish survey zeropoints on an hourly basis. We will also directly observe cloud cover with a $10\mu m$ imager. Careful astrometry is also needed, with limits set by our need to properly position the fibers for obtaining spectra. The required astrometric accuracy of 200 milliarcseconds should be immediately achievable, and recalibration with HIPPARCOS data may allow us to improve this to 50 milliarseconds.

In order to select targets and conduct spectroscopy within a month of obtaining imaging data, analysis of each 240 gigabyte night must be completed within one week. A custom data processing system has been developed which will perform the analysis. Much of this software and all of the hardware is in place, and we expect to be able to handle survey data essentially in real time by the time the survey begins.

Targets for SDSS spectroscopy will be selected from the SDSS imaging data. Galaxies will be chosen primarily by a magnitude cut; yielding a median redshift of about 0.1. Quasars targets will be selected on the basis of their non-stellar colors; an area where accurate five color data is exquisitely useful. The spectra will be taken with a pair of dual channel spectrographs fed by 640 fibers. Ten manually plugged fiber cartridges will be prepared for each night; 6400 spectra on a good night. The spectrographs provide a resolution sufficient to measure typical galaxy velocities to ± 20 km sec^{-1}.

The SDSS promises to provide an unprecedentedly complete and detailed picture of the local universe. The spectroscopic survey will support large-scale structure studies in a fully sampled volume about 100 times that previously available, and will permit direct LSS studies with quasars for the first time. The combined imaging and spectroscopy will provide a measure of the galaxy luminosity function in five colors; luminosity/density and luminosity/morphology relations will be there to explore. The five color data will allow estimation of photometric redshifts for 50 million more galaxies with an estimated error of about $\delta Z = 0.02$. The entire Northern Galactic Cap will have a 5 color calibration to a few percent; every field you look at will have objects accurately calibrated to 20th magnitude. And of course the most important discoveries will be those we haven't thought of.

All major telescope components are complete and the instruments are being delivered to the mountain this fall. We should begin taking data early next year, and begin the survey proper early in 1998. Data from the SDSS will be released to the public in as timely a fashion as is practical; data from the first two years will be released two years after they are taken, with the remainder of the survey released two years after it is completed. The "data" will consist first of complete object catalogs from all the imaging observations, a total of about 50 million galaxies, a million quasars, and 70 million stars and a spectroscopic catalog will including not only redshifts, but complete spectra. A second level of data release will consist of "atlas" images of every detected object. Finally the imaging observations will be collected into a single image of the full survey area.

6. Conclusions

Electronic surveys on a scale comparable to photographic surveys are now possible, and are being undertaken at a furious pace. This new generation of surveys is beginning to probe the universe in the spatial, time, and spectroscopic domains on a greater scale, and especially with greater accuracy, than ever before. Most of the technical impediments to conducting these surveys have been solved. We can construct large mosaics of high-performance CCDs, build spectrographs capable of measuring more than 500 spectra at once, and cajole our telescopes into providing sub-arcsecond images over large fields. A problem which all the projects will face, and which we have perhaps not yet satisfactorily solved, is the automated analysis and easy distribution of the immense quantity of data they electronic surveys will produce. Another point to note is that the most powerful and successful of these surveys all utilize dedicated equipment designed specifically for the purpose. Serious surveys require a serious commitment, and cannot generally be conducted as side projects.

Major discoveries made by electronic surveys so far include the faint blue galaxy population, many examples of weak and strong gravitational lensing and microlensing, the first real measures of large-scale structure and flows, and a more detailed quantitative understanding of SNe.

Just as the Palomar survey provided targets for 4m class telescopes for 40 years, surveys like the SDSS will provide the basic survey material required for the selection of targets by the new generation of 8–10m telescopes coming on line now. They will also provide the optical counterpart identification for many of non-optical surveys reviewed in these proceedings. These surveys, combined with the ready accessability of the immense quantities of data they will produce, hold the promise of again revolutionizing our view of our environs. It is an exciting time, a step forward in survey science which will probably not occur again for a generation.

References

Alcock, C., *et al.* 1996 Astrophys.J. Submitted
Colless, M., 1996 Proc. of the Athens conf. on 'Wide Field Spectroscopy,' in press
Ellis, R., *et al.* 1996 Mon.Not.R.astron.Soc. 280, 235
Ferrarese, L., *et al.* 1996 Astrophys.J. 464, 568
Gunn, J., and Knapp, G., 1993 in "Sky Surveys," ASP Conf. 43, 267
Koo, D., *et al.* 1996 Astrophys.J. Accepted
Lauer, T., and Postman, M., 1994 Astrophys.J. 425, 418L
Lilly, S., *et al.* 1996 Astrophys.J. 460, 1L
Perlmutter, S., *et al.* 1996 Astrophys.J. Submitted.
Riess, A., Press, W., and Kirschner, R., 1995, Bull.Am.Astron.Soc. 187, 1712
Saha, A., *et al.* 1996 Astrophys.J. 466, 55
Schectman, S., *et al.* 1996 Astrophys.J. 470, 172.
Strauss, M., 1996 Proceedings of Jeruselum Winter School, in press
Taylor, K., 1996 this volume, 135
Walker, A., 1996 this volume, 129

ULTRAVIOLET SKY SURVEYS

N. BROSCH
*Dept. of Astronomy and Astrophysics
and the Wise Observatory, Tel Aviv University*

1. Introduction

Among all spectral bands, the ultraviolet has long been neglected, despite the advantage of small space experiments: the sky is very dark, thus detection of faint objects does not compete against an enhanced background (O'Connell 1987) and the telescope construction techniques are very similar (at least longward of ~ 50 nm) to those of optical astronomy.

The short history of UV astronomy can be divided into two eras, until the flight of TD-1 and since the availability of the TD-1 all-sky survey. Very little was accomplished in terms of general sky surveys during this second era. The UV domain may be divided into the "regular" ultraviolet, from shortward of the spectral region observable from ground-based observatories (~ 320 nm) to below the Lyman break at ~ 80 nm, and the region from the Lyman break to the fuzzy beginning of the X-ray domain, arbitrarily defined as ~ 6 nm≈ 200 eV). The first segment is called "UV" and the second "extreme UV" (EUV). Observational techniques used in the EUV are more similar to those in X-ray astronomy, whereas the UV is more like the optical.

The units used here are "monochromatic magnitudes," defined as:

$$m(\lambda) = -2.5 \log[f(\lambda)] - 21.175 \tag{1}$$

where f(λ) is the source flux density in erg sec^{-1} cm^{-2} Å^{-1}, at wavelength λ. The background brightness is described in "photon units" (c.u.=count units) which count the photon flux in a spectral band, per cm^2, per steradian, and per Å. At 150 nm, 1 c.u.=1.32 10^{-11} erg cm^{-2} sec^{-1} Å^{-1} steradian^{-1}, or 1.32 10^{-13} W m^{-2} nm^{-1} steradian^{-1}, or 32.6 mag arcsec^{-2}.

Although only few missions performed full sky surveys in the UV or EUV, many scanned or imaged restricted sky regions and provided information about the deeper UV sky. O'Connell (1991) reviewed UV imaging experiments and their results updated to 1990.

In parallel with the development of UV astronomy, first steps were taken to study the EUV sky. Detections in the EUV range are hampered by the opacity of the interstellar medium (ISM). From 91.2 nm shortward to about 10 nm the opacity is high, because of the photoelectric cross-section of H^0, and to a lesser extent of He^0 (below 50.4 nm) and He^{+1} (below 22.8 nm).

The first studies in the EUV range were with rocket-flown instruments (Henry et al. 1975 a, b, c), which measured a few very bright sources and established calibrators. The earliest observations below Lyman α were by Belyaev et al. (1971). The culmination was the EUV instrument flown on the Apollo-Soyuz mission in 1975, when four EUV point sources were discovered (Lampton et al. 1976, Margon et al. 1976, Haisch et al. 1977, Margon et al. 1978). The Voyager spacecraft explored the EUV sky with their Ultraviolet Spectrometers (UVS: Sandel, Shemansky and Broadfoot 1979). For a number years the two Voyager spacecraft were the most distant astronomical observatories (Holberg 1990, 1991).

2. The TD-1 Era

Modern UV astronomy began with the first UV all-sky survey by the ESRO **TD-1** satellite, described by Boksenberg et al. (1973). The all-sky catalog of UV sources was published by Thompson et al. (1978) with 31,215 stars with S/N>10 in all four TD-1 bands. An unpublished version, with lower S/N, has 58,012 objects. The TD-1 S2/68 experiment is a benchmark against which all other sky surveys are and will be measured.

After TD-1, the various UV and EUV efforts can be characterized as either imagers or spectrometers. Among the imagers, some were orbiters and others were on short-duration flights. Some major missions were ANS and IUE. **ANS** was described by Van Duinen et al. (1975), Wesselius et al. (1982), and de Boer (1982). One of the greatest successes of any orbiting astronomical instrument was the **IUE** observatory (Boggess et al. 1978). The IUE data are a valuable resource, mainly after the final reprocessing of all the low-dispersion spectra into the final Uniform Low-Dispersion Archive (ULDA).

The NRL experiment **S201** was described by Page et al. (1982) and operated automatically on the Moon during the Apollo 16 mission in April 1972. Ten 20° diameter fields were observed, the experiment covering ~7% of the sky. The results were discussed by Carruthers and Page (1976, 1983, 1984a, 1984b). Two experiments flew on sounding rockets. **GUV** flew on 21 February 1987 and is described by Onaka et al. (1989). The GUV observations were re-analyzed by Kodaira et al. (1990). The **UIT prototype** flew on a number of rocket flights and with different focal plane assemblies. Bohlin et al. (1982) described the instrument and its observations of the Orion nebula. Other observations are described by Smith and Cornett (1982), Smith *et al.* (1987), and Bohlin *et al.* (1990).

UV observations from balloon-borne telescopes at 40+ km were performed by a collaboration between the Observatoire de Geneve and the Laboratoire d'Astrophysique Spatiale of Marseille. The stabilized gondola (Huguenin and Magnan 1978) carried telescopes tuned for imaging observations in a bandpass centered at ~200 nm and ~15 nm wide. **SCAP-2000** was described by Laget (1980), Donas *et al.* (1981), and Milliard *et al.* (1983). The results were published by Donas *et al.* (1987) and Buat *et al.* (1987, 1989). **FOCA** is a 39 cm diameter telescope (Milliard *et al.* 1991) which surveyed some 70 sq. degrees of the sky to ~19 mag. Results were reported by Laget *et al.* (1991a, 1991b), Vuillemin *et al.* (1991), Courtes *et al.* (1993), Buat *et al.* (1994), Bersier *et al.* (1994), Reichen *et al.* (1994), Donas *et al.* (1995), and Petit *et al.* (1996). Galaxy counts and color distributions for objects in the range 15.0–18.5 mag were published by Milliard *et al.* (1992) and used to predict UV galaxy counts (Armand and Milliard 1994).

The **Wide-Field UV Camera** flew in December 1983 on the Space Shuttle and produced some very wide-field UV images (Courtes *et al.* 1984). The NRL group headed by Carruthers flew a number of far-UV wide-field imagers on rockets (Carruthers *et al.* 1980). These flights used the Mark II **FUVCAM** (Carruthers *et al.* 1993, 1994) which flew on the Space Shuttle in spring 1991. The results were published in a series of papers dealing with individual fields (Schmidt and Carruthers 1993a, 1993b, 1995). The 40 cm telescope **GLAZAR** operated on the Mir space station (Tovmassian *et al.* 1988, 1991a, 1991b) and results were reported by Tovmassian *et al.* (1993a, 1993b, 1994, 1996a).

FAUST is the Fusee Astronomique pour l'Ultraviolet Spatiale, or the Far Ultraviolet Space Telescope, first described by Deharveng *et al.* (1979). On SPACELAB-1 in 1983 FAUST did not obtain significant data because of high on-orbit background (Bixler *et al.* 1984). During the second flight, on board the Shuttle Atlantis in March 1992 (Bowyer *et al.* 1993), FAUST observed 22 regions ~ 8° in diameter and produced a catalog of 4,698 UV sources (Bowyer *et al.* 1994a). Selected results from the FAUST imagery are by Deharveng *et al.* (1994), Haikala *et al.* (1995), and Courtes *et al.* (1995). A program of systematic investigation of FAUST images takes place the Tel Aviv University and include optical observations from the Wise Observatory. To date, we analyzed completely four FAUST fields, the North Galactic Pole (Brosch *et al.* 1996a) and three fields covering Virgo (Brosch *et al.* 1996b).

The Ultraviolet Imaging Telescope (**UIT**) was described by Stecher *et al.* (1992). It flew on the Space Shuttle during the ASTRO-1 (December

1990) and ASTRO-2 flights (March 1995). First results were published in a dedicated publication (1992 Astrophys. J. Lett. **395**). Other results are in Hill *et al.* (1993, 1994, 1995a, 1995b, 1996). The UIT source catalog (Smith *et al.* 1996) covers 16 sq.degrees of the sky and contains 2244 objects from 48 pointings in the ASTRO-1 flight.

Attempts to measure the diffuse UV background consist of many observations with wide-field instruments. Notable among these are the two Shuttle-borne UVX instruments from JHU and Berkeley (Murthy *et al.* 1989; Hurwitz *et al.* 1989). In addition, observations done for other purposes were used to derive the UV background, *e.g.*, Waller *et al.* (1995).

3. Modern EUV observations

The EUV sky was explored by the **EUV Wide Field Camera** on the ROSAT X-ray all-sky survey satellite described by Pounds and Wells (1991). The first EUV all-sky survey was during 1990–1991 (Pye 1995). Initial results were reported by Pounds *et al.* (1993) as the WFC Bright Source Catalog (BSC). The reprocessed data make up the 2RE catalog (Pye *et al.* 1995) with 479 EUV sources. In 2RE 52% of the sources are active F, G, K, and M stars, 29% are hot white dwarfs, and less than 2% are AGNs.

The region bordering the EUV and the X-rays was explored by the **ALEXIS** spacecraft (Priedhorsky 1991). The first sky maps were produced on 4 November 1994. The ALEXIS team calculated that ~10% of the brightest EUVE sources (see below) should be detectable. Most sources are WDs; the catalog from the first three years of operation will probably contain ≤50 sources.

The EUV sky was investigated by the Extreme Ultraviolet Explorer (**EUVE**) spacecraft (Bowyer and Malina 1991). EUVE mapped the sky in four spectral bands, from 7 to 70 nm (18 to 170 eV). The first results were published as "The First EUVE Source Catalog" (Bowyer *et al.* 1994) with 410 sources. The Second EUVE Source Catalog (2EUVE) has recently been published (Bowyer *et al.* 1996). The majority of the identified sources in 2EUVE (55%) are G, K, and M stars.

A new catalog, to ~60% of the thresholds of the second EUVE catalog, has been produced by Lampton *et al.* (1996) with 534 coincident sources between the EUVE 10 nm list and the ROSAT all-sky survey sources detected in the broadband event window (0.1–2 keV), of which 166 were not previously discovered. Of these, 105 have been identified and 77% of them are late-type stars. White dwarfs and early-type stars make up only 14% of the sources, and there are no extragalactic objects at all.

4. Comparison of Survey Missions

The various missions surveying the UV sky can be compared in terms of a "power" parameter θ, introduced by Terebizh (1986), used by Lipovetsky (1992) in a comparison of optical surveys, and slightly modified here:

$$\theta = \frac{\Omega}{4\pi} 10^{0.6(m_L - 10)} \qquad (2)$$

where Ω is the sky area covered by the survey (4π for TD-1) and m_L is the limiting magnitude of the survey.

The various parameters relevant to the missions discussed here are collected in Tabel 1. An all-sky survey to $m_L \approx 8.5$ (such as TD-1) has the same "survey power" as one HST WFPC-2 image exposed to show $m_{UV}=21$ objects. Because of this, and because not all surveys cover the entire sky, it may be more useful to look at another estimator, the density of sources detected (or which are expected to be detected) by a certain experiment. This estimator indicates that the field of UV astronomy retains its vitality; the source density increases exponentially with time.

5. The Resultant Sky Picture

The combined results yield a picture in which most of the stars detected by TD-1, FAUST, SCAP and FOCA are early-type B, A and F. However, most of the stars included in the UIT catalog are probably late-type (G and later). In the FAUST fields where the reduction and identification processes are complete, we find almost equal fractions of A–F stars (70 and 75%). Except for the fields studied at Tel Aviv (Brosch *et al.* 1995, 1996), most surveys used exclusively correlations with existing catalogs to identify sources. These sometimes mis-identify objects, as some likely early-type stars are just below the catalog thresholds.

The UV information on galaxies is very sparse and a statistically complete sample of a few 1000's galaxies is lacking. In the absence of very deep surveys in more than a single spectral band, our information about a significant number of galaxies originates from SCAP-2000 (Donas *et al.* 1987) and FOCA (Milliard *et al.* 1992). These measurements consist of integrated photometry at 200 nm of a few hundred galaxies. In the range 16.5–18.5 mag galaxies dominate the source counts at high |b|. These have B=18–20 and [2000-V]≈ -1.5. Using the "field" galaxy luminosity function in the UV from Deharveng *et al.* (1994) the differential number density is:

$$\log N(m) = 0.625 \times m_{200} - 9.5 \qquad (3)$$

Studies by UIT and FAUST emphasize the importance of the dust in understanding the UV emission. Bilenko and Brosch (1996) analyzed the

TABLE 1. UV and EUV survey missions

Mission	Year	Ω (ster)	m_L	θ	$\lambda\lambda$ (nm)	$N_{sources}$	Notes
TD-1	1968–73	4π	8.8	0.19	150–280	31,215	1
S201	1972	0.96	11	0.30	125–160	6,266	
WF-UVCAM	1983	1.02	9.3	0.03	193	?	
SCAP-2000	1985	1.88	13.5	18.9	200	241	2
GUV	1987	$5\ 10^{-3}$	14.5	0.2	156	52	3
GSFC CAM	1987+	0.03	16.3	14.4	242	$\sim$200	4
FOCA	1990+	0.02	19	377	200	$\sim$4,000	5
UIT-1	1990	$3.8\ 10^{-4}$	17	0.48	$\sim$270	2,244	6
GLAZAR	1990	$4.4\ 10^{-3}$	8.7	$6\ 10^{-4}$	164	489	
FUVCAM	1991	0.09	10	$7.5\ 10^{-3}$	133, 178	1,252	7
FAUST	1992	0.33	13.5	3.3	165	4,698	
UIT 1+2	1990, 95	$1.3\ 10^{-3}$	19	26	152–270	6,000 ?	8
HST WFPC	1990+	$3.9\ 10^{-4}$	21	123	120–300	50,000 ?	9
MSX UVISI	1997+	4π ?	13.9	218	180–300	?	
GIMI	1997+	4π	13.6	136	155	$2.5\ 10^5$	10
TAUVEX	1998+	0.06	19	11,700	135–270	10^6	11
WFC	1992 ?	4π	-	-	10, 16	479	
ALEXIS	1994+	4π	-	-	13–19	50?	
EUVE	1992 ?	4π	-	-	7–70	734	12

Notes to Table 1:
1: The unpublished extended version has 58,012 sources.
2: 92 stars (Laget 1980) and 149 galaxies (Donas *et al.* 1987).
3: Pointed phase.
4: Virgo observation.
5: Estimated.
6: UIT Catalog.
7: Only the Sag and Sco fields (Shuttle flights) included.
8: Assumes 66 pointings for ASTRO 1 and 100 for ASTRO 2.
9: Assumes 1000 observations with HST with UV filters on WFPC-2.
10: Assumes 2× stars per magnitude w.r.t. TD-1.
11: Assumes 5000 independent pointings to end-of-life.
12: Number of sources in the 2nd EUVE catalog.

TD-1 catalog and a version of the Hipparcos Input Catalog transformed to the TD-1 bands and showed that the UV exinction is very patchy, with different extinction gradients on scales $< 10°$. Tovmassian *et al.* (1996b) used GLAZAR observations of a 12 degree^2 area in Crux to establish that the dust distribution is very patchy, with most of the space relatively clear of dust. The EUVE catalogs (Bowyer *et al.* 1994, 1996; Lampton *et al.*

1996) confirm the previously known features of the local ISM (a "tunnel" to CMa with very low HI column density to 200 pc and close to the Galactic plane, a cavity connected with the Gum Nebula in Vela, a shorter 100 pc tunnel to 36 Lynx, and the very clear region in the direction of the Lockman hole).

The accurate measurement of the UV sky background (UVB), with the expectation that it could set meaningful cosmological limits, has been the goal of many rocket, orbital, and deep space experiments. Observational results were summarized by Henry (1982), Bowyer (1990), Bowyer (1991), and Henry (1991). The various origins of the UVB can be separated into "galactic" and "high latitude." The latter is an ∼uniform pedestal, onto which the former is added in various amounts depending on the direction of observation. The "galactic" component can be ∼one order of magnitude more than the "high latitude" one. Most is probably light scattered off dust particles in the ISM and the rest is from the gaseous component of the ISM (HII two photon emission and H_2 fluorescense in molecular clouds). The "high latitude" component is also mostly galactic, light scattering off dust clouds at high $|b|$.

The extragalactic component of the UVB (eUVB) can be at most 100–400 c.u. (Murthy and Henry 1995). Whenever $N(HI) > 2 \ 10^{20}$ cm^{-2}, the main contributor is dust-scattered starlight. The low level eUVB is probably integrated light of galaxies (Armand et al. 1994), or Milky Way light scattered off dust grains in the Galactic halo (Hurwitz et al. 1991), or intergalactic Lyα clouds contributing their recombination radiation (Henry 1991).

The low UVB away from orbital and galactic contaminants has recently been confirmed by ASTRO-1 UIT images (Waller et al. 1995). After correcting for orbital background and zodiacal light, and after accounting for scattered Galactic light by ISM cirrus clouds (from the IRAS 100 μm emission), the extrapolated UV-to-FIR correlation to negligible FIR emission indicates [eUVB]≈200±100 c.u.

The shorter wavelength UVB has been observed with the Voyager UVS down to 50 nm (Holberg 1986). A very deep EUVE spectroscopic observation of a large region on the ecliptic has recently been reported (Jelinsky et al. 1995) but the only emission lines observed were He I and He II (58.4, 53.7, and 30.4 nm), which originate from scatted Sunlight by the geocoronal and/or interplanetary medium and no continuum was detected.

The "true" eUVB can be evaluated from the FOCA data (Milliard et al. 1992). The galaxy counts, for $15.0 \geq m_{200} \geq 18.5$, extrapolated to $m_{200}=20.0$, give a contribution of ∼100 c.u.'s from only UV galaxies. Milliard (1996, private communication) studied the nature of sources in the

A2111 FOCA field. The majority are emission-line galaxies, only 16% in A2111 and the rest are in the foreground or background up to z≈0.7.

The detection and identification of faint UV galaxies may help resolve the issue of a real eUVB. Recent indications are that the eUVB is negligible. Waller *et al.* (1995) estimated 200±100 c.u. in the UIT near-UV band. Unpublished results from an analysis of 17 years of Voyager UVS spectra (Murthy *et al.* 1996, in preparation) indicate that at ∼100 nm the UVB is ≤100 c.u. (1σ). However, when adopting the FOCA galaxy counts and extrapolating to m_{UV}=20 and to ∼1000Å using the SEDs of starburst galaxies from Kinney *et al.* (1996), the contribution due to sources unresolved by UVS violates the Murthy *et al.* limit. An extrapolation to m_{UV}=23 violates also the UIT constraint. It is therefore necessary to investigate the faint end of the UV galaxy distribution to understand the nature and reality of the eUVB.

6. The Future

Two UV missions are approved, funded, built, and integrated into their carrier spacecraft. These are the UVISI on MSX, and GIMI on ARGUS, which will produce full or partial UV sky surveys. MSX was launched on 24 April 1996 and the operation of UVISI is expected to start in 1997. ARGUS has been slightly delayed and will be launched in 1997.

The narrow field UV imager of **UVISI** (Heffernan *et al.* 1996) is sensitive to sources which produce 2 photons cm^{-2} sec^{-1}, *i.e.*, m_L ≈13.9 (monochromatic, at 240 nm). It is not clear how much of the sky will MSX survey. However, **GIMI** on ARGOS has as a declared goal the production of a full sky survey in three UV bands. The most recent description of GIMI (Carruthers and Seeley 1996) indicates m_L ≈13.6.

TAUVEX, **T**el **A**viv University **UV** **Ex**plorer, (Brosch *et al.* 1994) is the most advanced attempt to design, build and operate a flexible instrument for observations in the entire UV band. TAUVEX images the same 0$\overset{d}{.}$9 FOV with three co-aligned telescopes with an image quality of about 10″. It is part of the scientific complement of the Spectrum X-γ spacecraft. The projected performance is detection of objects 19 mag and brighter with S/N>10 in three bands ∼40 nm wide, after a four hour pointing. At high |b|, each pointing is expected to result in the detection of some tens of QSOs and AGNs (mainly low-z objects) and some hundreds of galaxies and stars. A three-year operation will cover ∼5% of the sky to m_{UV} ≈19 mag.

The deepest observations of the UV sky should be made far from the Earth's geocorona, away from the Sun, and out of the ecliptic. Multi-purpose missions to the outer planets could be used for astronomy during their cruise phase. Much cheaper options are UV observations from long-

GAMMA RAY SKY SURVEYS

N. GEHRELS
NASA/Goddard Space Flight Center

1. Introduction

Prior to the current Compton Gamma Ray Observatory (*Compton*) mission, no comprehensive all-sky gamma-ray surveys had been performed. There were, however, some surveys performed over limited energy bands and/or over portions of the sky. These include the HEAO-A4 hard X-ray survey and the COS-B and SAS-2 high-energy gamma-ray surveys. The early work forms a basis for understanding and appreciating the *Compton* results, and so is reviewed in Section 2.

The Compton Observatory was launched in April 1991 and spent the first 18 months performing an all-sky survey in the 1 MeV–30 GeV band. The results of this survey are given in Section 3. Plans for survey missions beyond *Compton* are presented in Section 4.

2. Pre-Compton Surveys

The pre-*Compton* surveys naturally divide into hard X-rays, high-energy gamma rays and gamma-ray bursts. There were no surveys in the medium energy range (0.1–30 MeV) prior to *Compton*.

2.1. HARD X-RAYS

The HEAO-A4 instrument performed an all-sky survey in 1977–79 (Levine *et al.* 1984). It was complete to a flux threshold at $\sim 8 \times 10^{-3}$ ph cm^{-2} s^{-1} ($\sim$15 mCrab) in the 13–80 keV band. The total number of sources detected was 70 of which $\sim 10\%$ were extragalactic. The galactic sources were almost all identified with known X-ray binaries or pulsars. The extragalactic sources were largely active galactic nuclei (AGNs) plus clusters.

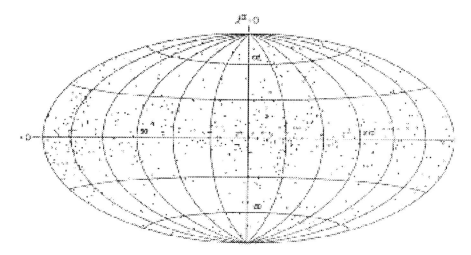

Figure 1. Distribution in galactic coordinates of individual gamma rays at > 50 MeV observed by OSO-3. From Kraushaar *et al.* (1972).

A diffuse galactic continuum radiation was detected in the hard X-ray range by OSO-7, HEAO-1 and balloon instruments (see Gehrels and Tueller 1993 and references therein). Known as the "galactic ridge," it is centered on the galactic plane and has a width of approximately 5° in latitude and ±40° in longitude.

A diffuse extragalactic radiation had also been observed at hard X-rays in the 1960's and 70's by Ranger 3, Apollo 16/17, HEAO-1, and balloons (see Gruber 1992 and references therein).

2.2. HIGH ENERGY GAMMA RAYS

The first high energy (> 50 MeV) gamma-ray observation was in fact an all-sky survey. It was performed by OSO-3 in 1967–68 (Kraushaar *et al.* 1972) and detected only 621 celestial gamma rays. The distribution on the sky of these individual photons is shown in Figure 1. This early "map" already shows evidence for two main features now known to be characteristic of the high energy sky, namely a concentration along the galactic plane of a diffuse galactic radiation and an isotropic cosmic diffuse radiation.

The COS-B and SAS-2 missions followed on OSO-3 and early balloon instruments to perform the first good survey of the high energy sky. The coverage of the sky was not complete for either mission, but did include all of the galactic plane.

The COS-B map of the galactic plane is shown in Figure 2 (Mayer-Hasselwander *et al.* 1982). Strong emission is seen concentrated along the plane that was modeled as gamma rays produced by cosmic ray electrons

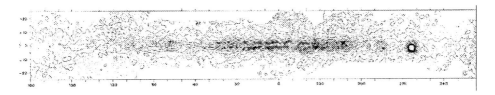

Figure 2. Contour map of the galactic plane emission at > 70 MeV observed by COS-B. From Mayer-Hasselwander *et al.* 1982.

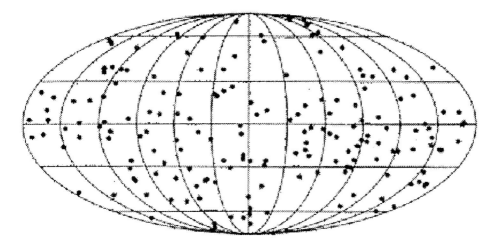

Figure 3. Map of 180 pre-*Compton* gamma-ray bursts in galactic coordinates. From Hurley (1992).

and protons interacting with the ISM. Also detected along the plane were ~25 flux concentrations consistent with point sources of emission (Swanenberg *et al.* 1981). Two of these in the >100 MeV sky. Some of the other sources were thought to be ISM clouds or gas concentrations. Many of the sources were unidentified. The one extragalactic object seen by COS-B was 3C273.

2.3. GAMMA-RAY BURSTS

There were many different gamma-ray burst detectors flown prior to *Compton*. The missions included Vela, Konus, HEAO-A4, Sigma, Apex, Lilas, PVO, ISEE-3 and Ginga. Several hundred bursts were detected between 1968 and 1991. However, few of the instruments gave good positional information for the bursts and none provided positions for a large number of bursts. A map of 180 pre-*Compton* bursts from several missions was compiled by Hurley (1992) and is shown in Figure 3. The map is consistent with an isotropic distribution of burst sources on the sky.

3. Compton Observatory Sky Surveys

Compton has four instruments on board with parameters summarized in Table 1. It was launched into a 28° inclination, 450 km altitude orbit on 1991 April 5. One of the prime scientific objectives of the mission was to carry out an all-sky survey in the 1 MeV to 30 GeV range with the COMPTEL and EGRET wide-field instruments. This was accomplished between 1991 May and 1992 November. Since that time, *Compton* has been observing selected fields. Because of the wide-field nature of COMPTEL and EGRET, this additional time has tremendously deepened the exposure in many regions of the sky. The maximum exposures are in the 10^9 cm^2 s range.

Throughout the *Compton* mission, BATSE has been detecting gamma-ray bursts from the whole sky. In addition, the instrument monitors the sky for bright steady sources using the Earth occultation method.

Results from the *Compton* sky surveys are presented in the sections below. For more information about the observatory, see Shrader and Gehrels (1995) and the Web site
    http://cossc.gsfc.nasa.gov/cossc/cossc.html.

TABLE 1. Compton Observatory Instrument Parameters

Name	Key Capabilities	Energy Range	Field-of-view
BATSE[1]	gamma-ray bursts, all-sky monitor	15 keV–1.2 MeV	all sky (not occulted by Earth)
OSSE[2]	narrow-field spectrometer	0.1–10 MeV	4° × 11°
COMPTEL[3]	wide-field imager and spectrometer	1–30 MeV	64° (1 sr)
EGRET[4]	wide-field imager and spectrometer	20 MeV–30 GeV	45° (0.6 sr)

[1] BATSE = Burst and Transient Source Experiment (PI: G.J. Fishman)
[2] OSSE = Oriented Scintillation Spectrometer Experiment (PI: J.D. Kurfess)
[3] COMPTEL = Compton Imaging Telescope (PI: V. Schönfelder)
[4] EGRET = Energetic Gamma-Ray Experiment Telescope (co-PIs: C.E. Fichtel, K. Pinkau)

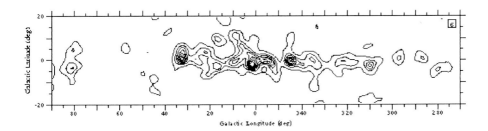

Figure 4. Map from COMPTEL observations of the 1.8 MeV line of ^{26}Al during 1991–93. From Diehl et al. 1995a.

3.1. EGRET SURVEY

A contour map of the >100 MeV sky as measured by EGRET is shown in Plate 1. As with the COS-B map, bright emission is seen along the galactic plane, but now the sensitivity and angular resolution are much improved. The emission is well fitted by models of cosmic rays interacting with the ISM (Hunter et al. 1996). The dominant component below 150 MeV is bremsstrahlung of cosmic ray electrons and above 150 MeV is nucleon-nucleon interaction of cosmic ray protons with interstellar gas.

Along the plane EGRET sees more than 50 point sources. Six of these are pulsars: Crab, Vela, Geminga, PSR 1706-44, PSR 1055-52, PSR 1951+32 and possibly PSR 0656+14 (Thompson et al. 1994; Nolan et al. 1996; Ramanamurthy et al. 1996). The remaining plane sources are largely unidentified. This is a remaining and increasing mystery of the gamma-ray sky since COS-B.

Off the galactic plane, EGRET detects more than 40 AGN compared to the one (3C273) seen by COS-B. All of them are in the blazar class consisting of BL Lacs and flat spectrum radio quasars (*e.g.*, Montigny et al. 1994; Dermer and Gehrels 1995). They are typically variable and have photon spectral indices of approximately -2. The emission is thought to be due to photons Compton upscattered by relativistic jets of electrons and positions emanating from the central engine.

3.2. COMPTEL SURVEY

The COMPTEL instrument has performed the first medium energy (1–30 MeV) sky survey. The results in the upper portion of this band are qualitatively similar to the EGRET map. At lower energies, COMPTEL has made important observations of nuclear line emission.

Figure 4 shows the COMPTEL map along the galactic plane of the 1.809 MeV line from ^{26}Al (Diehl et al. 1995a). This radioisotope is spread into the ISM by supernovae, novae and winds of massive stars. It has a half life of 7×10^5 years and is therefore a tracer of the sites of nucleosynthesis in the Galaxy over the last million years. The clumpiness of the emission is not well understood. A hot spot at $l = 265°$ seems to be associated with the Vela supernova remnant (Diehl et al. 1995b). Other concentrations coincide with positions of the tangents to the arms of the Galaxy (Prantzos 1991; Chen et al. 1995) and therefore may be the sum of many supernova remnants and massive stars. COMPTEL also detects gamma-ray line emission from ^{44}Ti in Cas A (Iyudin et al. 1994) and nuclear excitation from cosmic ray interactions in the Orion region (Bloemen et al. 1994).

3.3. BATSE AND OSSE SURVEYS

Although there are no all-sky maps to show yet from BATSE and OSSE (excluding gamma-ray bursts which are discussed below), both instruments are contributing significantly to our understanding of the hard X-ray and gamma-ray sky. The BATSE team is developing software to generate all-sky maps using Earth occultation imaging. The maps will be complete to about 10 mCrab every month. OSSE has been doing some scanning surveys over limited portions of the sky. An all-sky survey is probably not feasible, but a galactic plane scan may be done.

A few characteristics of the hard X-ray/low-energy gamma ray sky as determine by BATSE, OSSE and previous satellite and balloon instruments are as follows.

1) There is a diffuse emission along the galactic plane (Section 2.1).

2) Some 100 steady or variable point sources are known in the plane. Most are X-ray binaries or pulsars.

3) Bright transient sources appear along the galactic plane every few weeks, with typical "on" times of days to weeks. They include X-ray novae, superluminal sources (GRS 1915+105 and GRO J1655-40) and Be binary pulsars.

4) There is a bright ($\sim 10^{-3}$ ph cm^{-2} s^{-1}) gamma-ray line emission at 511 keV from positron annihilation that forms a diffuse glow of $\sim 10°$ extent around the galactic center (possibly with imbedded hot spots or point sources).

5) Approximately 30 AGNs are seen, most of which are Seyferts.

3.4. GAMMA-RAY BURSTS OBSERVED BY BATSE

The primary scientific objective of BATSE is to observe gamma-ray bursts. BATSE is the first large gamma-ray burst instrument ever flown and has the capability to position each burst to few degree accuracy. Thus, BATSE has provided the first uniform and deep sky survey for bursts.

The BATSE burst map accumulated from 1991 April through 1995 November is shown in Figure 5. The burst distribution is statistically consistent with being isotropic. No physically meaningful sub-sample of the bursts has been found to deviate from isotropy.

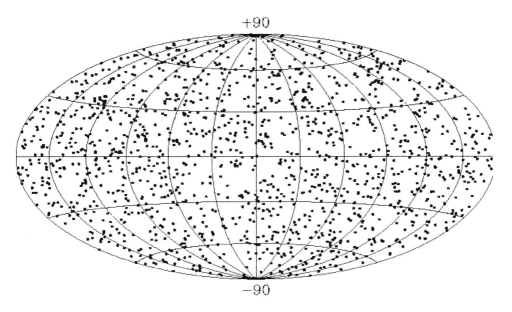

Figure 5. BATSE gamma-ray burst map from 4.5 years of observations. Courtesy of the BATSE team.

TABLE 2. Future Gamma Ray Survey Missions

Mission	Energy (MeV)	WWW URL (preceded by http://)
*INTEGRAL	0.02–10	astro.estec.esa.nl/SA-general/Projects/Integral/integral.html
GLAST	10–300,000	www-glast.stanford.edu
EXIST	0.005–0.6	hea-www.harvard.edu/EXIST/EXIST.html
*HETE	0.006–1	nis-www.lanl.gov/nis-projects/hete/
BASIS	0.01–0.15	lheawww.gsfc.nasa.gov/docs/gamcosray/legr/BASIS/basis.html
ETA	0.05–0.5	
BLAST	0.01–0.15	osse-www.nrl.navy.mil/blast.htm

*mission under development

4. Future Missions

Several future missions are currently being developed or planned that will perform excellent new surveys of the gamma-ray sky. Web sites and energy ranges for many of these missions are listed in Table 2. INTEGRAL is an approved ESA mission that will perform pointed gamma-ray observations, including a survey of the galactic plane, starting in 2001. EXIST is a proposed hard X-ray all-sky survey mission. GLAST is a proposed follow-on mission to EGRET. HETE is a gamma-ray burst mission that will be launched in late 1996. BASIS, ETA and BLAST are proposed gamma-ray burst missions.

References

Bloemen, H., et al. 1994, Astron.Astrophys., 281, L5.
Chen, W., Gehrels., N., and Diehl, R. 1995, Astrophys.J., 440, L57.
Dermer, C. D., and Gehrels, N. 1995 Astrophys.J., 447, 103.
Diehl, R., et al. 1995a, Astron.Astrophys., 298, 445.
Diehl, R., et al. 1995b, Astron.Astrophys., 298, L25.
Gehrels, N., and Tueller, J. 1993, Astrophys.J., 407, 597.
Gruber, D. E. 1992, in "The X-Ray Background," eds. X. Barcons & A. C. Fabian (Cambridge: Cambridge University Press) p. 44.
Hunter, S., et al. 1996, Astrophys.J., accepted.
Hurley, K. 1992, in "Gamma Ray Bursts," eds. W. S. Paciesas & G. J. Fishman (New York: AIP) p. 3.
Iyudin, A. F., et al. 1994, Astron.Astrophys., 284, L1.
Kraushaar, W. L., et al. 1972, Astrophys.J., 177, 341.
Levine, A. M., et al. 1984, Astrophys.J.Suppl., 54, 581.
Mayer-Hasselwander, H. A., et al. 1982, Astron.Astrophys., 105, 164.
Montigny, C. von, et al. 1994, Astrophys.J., 440, 525.
Nolan P. L., et al. 1996, Astron.Astrophys.Supp., in press.
Prantzos, N. 1991, in "Gamma Ray Line Astrophysics," eds. Ph. Durouchoux & N. Prantzos (New York: AIP) p.129.
Ramanamurthy, P. V., et al. 1996, Astrophys.J., 458, 755.
Shrader, C. R., and Gehrels, N. 1995, Publ.Astron.Soc.Pacific, 107, 606.
Swanenburg, B. N., et al. 1981, Astrophys.J., 243, L69.
Thompson, D. J., et al. 1994, Astrophys.J., 436, 229.

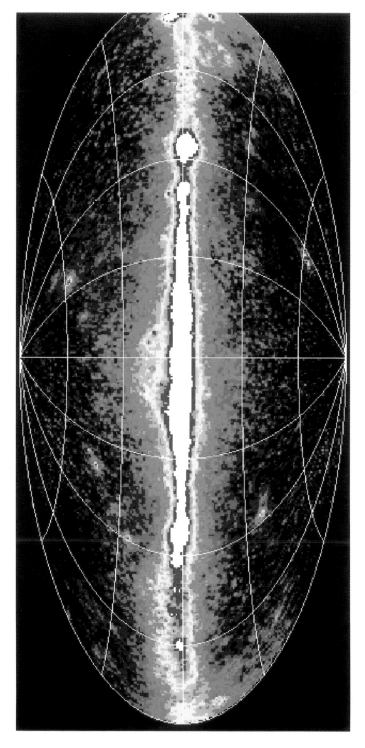

Figure 6. All-sky EGRET map at > 100 MeV energies observed from 1991 to 1995. Courtesy EGRET Team.

PROSPECTS FOR FUTURE ASTROMETRIC MISSIONS

P.K. SEIDELMANN
U.S. Naval Observatory

Abstract. Hipparcos and the Hubble Space Telescope have demonstrated the astrometric capabilities in space. SIM and GAIA are being studied for future missions. There have been many proposals for new astrometry missions from many different countries, but most of them have not been funded.

The best possibility for a mission within the next five years would be through a collaborative effort for a small, fast, cheap spacecraft which could be a precursor for future, larger, more accurate missions, which are under study.

1. Introduction

In 1997 there will be available the Hipparcos Astrometric Satellite star catalog of 120,000 stars with accuracies of 1 milliarcsecond (mas) and proper motions of 1.1 mas year^{-1} for stars brighter than 9th magnitude. In addition, the Tycho Catalog will include one million stars with accuracies of 30 mas. The Hubble Space Telescope can do only a limited amount of astrometry due to the competition for observing time.

Ground-based astrometric, optical observations over wide angles will be made at mas accuracies with optical interferometers, such as the Navy Prototype Optical Interferometer, for several thousand stars. CCD observations of small fields can be made with 30–50 mas relative accuracies.

There are many proposals for future astrometric missions. They can be divided into scanning and pointed missions. The scanning missions, like Hipparcos, can observe many stars repeatedly and achieve good accuracy. The pointed missions measure relative positions more accurately, but cannot make as many observations and cannot observe as many stars. The technological improvements are based on the use of CCD detectors, optical interferometers, metrology systems, and increased data rates. The Space Interferometry Mission (SIM) in the U.S. and the Global Astrometric Interferometer for Astrophysics (GAIA) in Europe have some financial support for studies.

2. Why Future Astrometric Space Missions?

The real prospects for future astrometric missions must be tied to the reasons for such missions. The observables are positions, proper motions, parallaxes, photometry and images. New missions can achieve more accuracy and more stars. There are the traditional astrometric applications of navigation, guidance, and reference frames. In addition, space activities have added space surveillance and autonomous space navigation as applications for accurate astrometric data. However, this probably will not justify a future astrometric mission. So consider the purpose in terms of NASA's four themes, where astrometry has a fundamental scientific role:

2.1. STRUCTURE AND EVOLUTION OF THE UNIVERSE

- Calibrate the cosmological distance scale by measuring absolute parallaxes of Cepheids and RR Lyrae.
- Detemine positions, proper motions, and absolute parallaxes to 10% accuracy or better (as well as apparent magnitudes and spectral energy distributions) of a complete sample of stars brighter than 14th magnitude within 2.5 kiloparsecs of the Sun.
- Calibrate the absolute luminosities of solar neighborhood stars, including population I and II stars, enabling studies of stellar evolution.
- Determine transverse velocities of a complete sample of stars within 2 kpc of the Sun to 14th magnitude. From this, the mass distribution and gravitational surface mass density in the disk near the Sun can be determined. This relates directly to the dark matter implied by dynamical studies of globular clusters and rotation of galaxies (the "Oort problem").
- Detect astrometric perturbations of lensed sources during a microlensing event, and directly image a MACHO object shortly after a lensing event.
- Determine parallaxes and relative positions of binaries to determine masses, including unusual systems such as those containing white dwarfs and black holes, by analysis of positions of photocenter emission and use of multiple colors.
- Calibrate the distances to open star clusters and determine the absolute color magnitude diagrams of newly formed star clusters.

– Determine accurate reference frames which will ensure identification of sources in radio, optical, infrared and X-ray wavelengths.

2.2. SEARCH FOR ORIGINS AND PLANETARY SYSTEMS.

– Identify candidate stars for brown dwarfs and planets from inconsistent proper motion values.
– Detect astrometrically planets and brown dwarfs from non-linear proper motions.
– Image brown dwarfs and planets directly using interferometric nulling.
– Image the scattered light from the disks and exo-zodiacal dust surrounding young stars and main sequence stars like Beta Pictoris using interferometric nulling.
– Calibrate the cosmological distance scale which is critical to the origins of the universe question.

2.3. SOLAR SYSTEM EXPLORATION.

– Determine the accurate relationships between the radio, optical and dynamical reference frames which are necessary for solar system exploration.
– Make accurate positional observations of asteroids which will contribute to determination of asteroid masses.
– Make accurate observations of small bodies to improve the ephemerides for mission objectives.
– Make accurate observations of positions of near Earth objects which will improve knowledge of the motions of these objects.

2.4. SUN EARTH CONNECTION.

– Image other Suns and determine their radius, mass, luminosity and distance.

3. Possible Future Astrometric Missions

At this time two missions seem to be the most likely for future launch and are being actively studied. The Space Interferometer Mission (SIM) (OSI 1996) is a U.S. spacecraft with pointed Michelson interferometers of variable baselines and a nulling interferometer backend. It can observe about 5000 stars down to 20th magnitude to an accuracy of 0.004 mas. Plans are to launch SIM by 2005. It is a precursor for the ExNPS mission and will test technology for that future mission.

The Global Astrometric Interferometry for Astrophysics (GAIA) (Lindegren and Perryman 1995) is a European Space Agency study for a stack of three scanning Fizeau interferometers. It can observe 50,000,000 stars at an accuracy of 0.010 mas brighter than 15th magnitude. It is currently competing for a cornerstone mission with a possible launch date of 2015.

TABLE 1. Small scanning astrometric missions.

	FAME	LIGHT	DIVA
Purpose	Astrometry Photometry	Galactic Halo Tracer	Astrometry Photometry
Technique	Fixed Angle Scanning	Scanning Fizeau Interferometer	1/10 size Gaia
Num. of Stars	10,000,000	10,000,000	100,000
Accuracy	0.05 mas	0.1 mas	0.8 mas
Mag. Limit	15	15	10.5
Launch Date	2001	2007	2005
Status	Proposal	Proposal	Proposal

There are currently three proposals (Table 1) for a scanning instrument similar to Hipparcos, but using CCD detectors and higher data rates, so that more stars and accuracy can be achieved: FAME from USA (Seidelmann et al. 1995), LIGHT from Japan (Yoshizawa et al. 1997), and DIVA from Germany (Bastian et al. 1997). One such mission would fill the needs for a large all sky, very accurate star catalog down to about 15th magnitude. In addition to excellent science, such as calibrating the distance scale, this project would provide both a second epoch for Hipparcos and an independent set of short-time-period proper motions. The comparison between the sets of short-period proper motions and long-period proper motions should identify discrepancies and thus most likely candidates for planetary systems and brown dwarfs. This type of project is currently the most needed, unfunded astrometric spacecraft.

Two pointed spacecraft have been proposed (Table 2), POINTS (Chandler and Reasenberg 1990) and a scaled down POINTS called Newcomb (Johnston et al. 1995). These have been supplemented by the SIM project. A number of Russian spacecraft have been proposed in the last decade. The three most actively being discussed at the present time are AIST or Struve (Yershov et al. 1995; Chubey et al. 1995; Kopylov et al. 1995), Lomonosov (Nesterov et al. 1990) and Zenith (Table 3).

TABLE 2. Pointed astrometric missions.

	NEWCOMB	POINTS
Purpose	Astrometry	Astrometry
Technique	Stacked, small Michelson Interferometer	Rigid Michelson Interferometer
Num. of Stars	3,000	5,000
Accuracy	0.1 mas	0.001 mas
Mag. Limit	15	12
Launch Date	?	?
Status	Unfunded	Unfunded

TABLE 3. Russian astrometric missions.

	AIST/STRUVE	LOMONOSOV	ZENITH
Purpose	Astrometry Photometry	Astrmetry Photometry	Astrometry
Technique	2 Schmidt telescopes, Fixed angle scanning	1 m Mirror 90° separation, rotating	Michelson Interferometer, pointing
Num. of Stars	500,000	400,000	3,000
Accuracy	0.3 mas	2–10 mas	0.1 mas
Mag. Limit	18	10	12
Launch Date	2001	2003	?
Status	Study	Study	?

There are also discussions about observations from the Moon (Mission to the Moon 1992). These proposals try to take advantage of the large surface, absence of interference, distance to Earth, or lunar characteristics (Table 4). All of these ideas appear to be far in the future.

4. Conclusion

The Hipparcos satellite has proven to be a great success and demonstrated the capability to do astrometry in space. Since its design, there have been significant technical developments that make it possible to build a satel-

TABLE 4. Proposed lunar based observations.

TECHNIQUE	PURPOSE
VLF Array of antennas	Extragalactic
Optical Interferometer	Small angel astrometry
Earth-Moon VLBI	Astrometry
Complex Optical Interferometer	Imaging
1 M Transit Telescope with CCDs	Macho detection

lite which can observe many more stars much more accurately. There is excellent scientific justification for future astrometric satellites.

As a result many satellites have been proposed for astrometric, photometric and imaging purposes. At this time the SIM and GAIA proposals appear to be the most likely to be launched. These two satellites complement each other in their capabilities and the differences in their launch dates. There appears to be good scientific justification for a scanning satellite with 50 microarcseconds accuracies for stars brighter than 9 magnitude launched in the first years of the twenty first century.

References

Bastian, U.; Hog, E.; Mandel, H.; Quirrenback, A.; Roser, S.; Schalinski, C.; Schilback, E.; Seifert, W.; Wanger, S.; and Wicenec, A. (1997) "DIVA, An Interferometric Minisatellite for Astrometry and Photometry" Astronomische Nachrichten (in preparation)

Chander, J.F. and Reasenberg, R. D. 1990, "POINTS: A Global Reference Frame Opportunity," Inertial Coordinate System on the Sky, J. H. Lieske and V. K. Abalakin eds, Kluwer Academic Publishers, Dordrecht, 217–228.

Chubey, M.S.; Paskkov, V.S.; Kopylov, I.M.; Kirian, T. R.; Nickiforov, V. V.; Markelov, S.V; Ryadchenko, V.P. (1995) "On the Registration System of the AIST-Project" Astronomical and Astrophysical Objectives of Sub-Milliarcsecond Optical Astrometry, E. Hog and P. K. Seidelmann eds, Kluwer Academic Publishers, Dordrecht, 323–326.

Eichhorn, H. (1974) "Astronomy of Star Positions" Frederick Ungar Publishing Co, New York

Hindsley, R. B. and Harrington, R.S. (1994) "The U.S. Naval Observatory Catalog of Positions of Infrared Stellar Sources" Astron.J. 107, 280–6

Hog, E. (1995) "A New Era of Global Astrometry II. a 10 Microarcsecond Mission" Astronomical and Astrophysical Objectives of Sub-Milliarcsecond Optical Astrometry, E. Hog and P. K. Seidelmann eds, Kluwer Academic Publishers, Dordrecht, 317–322.

Johnston, K. J.; Seidelmann, P.K.; Reasenberg, R. D.; Babcock, R.; Phillips, J.D. (1995) "Newcomb Astrometric Satellite" Astronomical and Astrophysical Objectives of Sub-Milliarsecond Optical Astrometry, E. Hog and P. K. Seidelmann eds, Kluwer Academic Publishers, Dordrecht, 331–334.

Kopylov, I.M.; Gorshanov, D.L.; Chubey, M.S (1995) "Photometry Facilities of the AIST Space Project" Astronomical and Astrophysical Objectives of Sub-Milliarcsecond Optical Astrometry, E. Hog and P. K. Seidelmann eds, Kluwer Academic Publishers, Dordrecht, 327–330.

Lindegren, L.; Perryman, M.A.C. (1995) "A Small Interferometer in space for Global Astrometry: The GAIA Concept" Astronomical and Astrophysical Objectives of Sub-Milliarcsecond Optical Astrometry, E. Hog and P. K. Seidelmann eds, Kluwer Academic Publishers, Dordrecht, 337–344.

"Mission to the Moon," European Space Agency, esa SP-1150, June 1992.

Nesterov, V. V. Ovchinnikov, A. A., Cherepashchuk, A. M. and Sheffer E. K. 1990 "The Lomonosov Project for Space Astrometry," Inertial Coordinate System on the Sky, J. H. Lieske and V. K. Abalakin eds, Kluwer Academic Publishers, Dordrecht, 355–360.

OSI, Orbiting Stellar Interferometer, NASA Science and Technology Review, March 1996, JPL.

Seidelmann, P. K.; Johnston, K. J.; Urban, S.; Germain, M.; Corbin, T.; Shao, M.; Yu, J.; Fanson, J.,; Rickard, L. J.; Weiler, K., and Davinic, N. (1995) "A Fizeau Optical Interferometer Astrometric Satellite" in Future Possibilities for Astrometry in Space, a Joint RGO-ESA Workshop, Cambridge UK 19–21 June 1995 ESA SP-379

Volonte, S. (1995) "Astronomy from a Lunar Base," Astronomical and Astrophysical Objectives of Sub-Milliarcsecond Optical Astrometry, E. Hog and P. K. Seidelmann eds, Kluwer Academic Publishers, Dordrecht, 347–350.

Yoshizawa, M.; Sato, K.; Nishikawa, J.; and Fukushima, T. (1997). Two Astrometric Projects: MIRA (Mitaka optical and infrared interferometer Array) and LIGHT (Light Interferometer Satellite for the Studies of Galactic Halo Tracers); Proceedings of IAU Colloquium 165, Wytrzyszczak, Lieskie, and Mignard eds, Kluwer Academic Publishers.

Yershov, V. N.; Chubey, M.S.; Il'in, A.E.; Kopylov, I.M.; Gorshavov, D.L.; Kanayev, I.I.; and Kirian, T.R. (1995) "Struve Space Astrometric System. Scientific Grounds of the Project," (in Russian).

"Future Possibilities for Astrometry in Space" A Joint RGO-ESA Workshop, Cambridge UK 19–21 June 1995 ESA SP-379

Part 2. Survey Projects

WIDE-FIELD LOW-FREQUENCY IMAGING WITH THE VLA

N.E. KASSIM, D.S. BRIGGS AND R.S. FOSTER
Naval Research Laboratory, Washington

1. Introduction

The 330 MHz observing system at the VLA is a potentially powerful survey system. It can map fields many degrees in size quickly and at high sensitivity. However imaging characteristics unique to this data pose excessive computational burdens on conventional mapping systems. Hence while data acquisition is quick and efficient, data reduction has been difficult and slow. Here we describe how powerful new scalable processing algorithms have been used to generate the first full resolution 330 MHz images, demonstrating also that lower resolution survey work is now a tractable problem.

2. Solution to The Wide-Field Imaging Problem: DRAGON

Imaging at low frequencies is complicated because the VLA is a non-coplanar array (Cornwell and Perly 1992). At commonly used centimeter wavelengths, 2-D approximations to the celestial sphere work well. However for the large fields of view and enhanced source count densities encountered at 330 MHz these approximations break down and introduce phase errors which severely limit sensitivity. Tim Cornwell of the NRAO has implemented a full solution to this problem via the polyhedron algorithm DRAGON. We have recently ported DRAGON to the SGI Power Challenge Array at the Army Research Laboratory.

2.1. LOW RESOLUTION IMAGES REQUIRE CONTIGUOUS FACETING

Low resolution VLA data acquired in the B (< 12 km), C (< 3 km) and D (< 1 km) configurations can be handled by contiguous faceting. Here the 3-D image volume is broken up into a set of contiguous 2-D facets. The advantage over a full 3-D transform is a significant savings in computational expense. Images produced in this way are now achieving thermal and confusion limited sensitives on a routine basis. See Frail, Kassim, and Weiler (1994) and Kassim & Frail (1996) for examples. Common "B array DRAGONs" take several days to run on conventional workstations but only a few hours on the Power Challenge Array.

2.2. HIGH RESOLUTION IMAGING WITH TARGETED FACETING.

Full pixellation of the primary beam in the A array (< 36 km) is impossible even with the most powerful contemporary machines. Hence we have implemented Cornwell's suggested technique of targeted faceting in which small outlier fields are placed only on the hundreds of small-diameter background sources which must be deconvolved, avoiding pixellation of large regions of empty sky. A smaller contiguously faceted region is centered on the target source. The location of the outlier fields is determined from a tapered resolution image or an image from a smaller VLA configuration. This technique has been demonstrated on the supernova remnant W49B, producing the first full resolution 330 MHz image from the VLA generated using proper 3-D algorithms. This image would have taken months to produce on conventional workstations but was produced in only a few days on the Power Challenge Array.

3. Summary

We have implemented algorithms to achieve thermal noise limited sensitivity, wide-field 330 MHz images at the full resolution of the VLA. We have successfully implemented this code on scaled processing computers and generated images in manageable lengths of time. This capability allows data reduction to keep up with data acquisition which could be quick and efficient for surveying purposes. Hence the capability of undertaking sensitive, high resolution large sky surveys with this system is realized. As a look forward we estimate that only ~ 40 hours of telescope time would be required to survey the first quadrant of the Galactic plane with a latitude coverage of $\pm 3°$ at 1 mJy sensitivity. Science goals would define the required resolution.

References

Cornwell, T.J., and Perley, R.A., 1992, Astron.Astrophys. 261, 353. Frail, D.A., Kassim, N.E. and Weiler, K.W.,1994, Astron.J. 107, 1120. Kassim. N.E. and Frail, D.A., 1996, Mon.Not.R.astron.Soc. in press

THE MIYUN 232 MHZ GENERAL CATALOGUE:

A reliability and positional uncertainties study

B. PENG, W. YUAN, R. NAN AND Y. YAN
Beijing Astronomical Observatory, China

1. Reliability Study

A meter-wave sky survey of the region north of declination $+30°$, excluding only four fields straddling the Galactic plane (each $8° \times 8°$ in angular size), has been conducted with the Miyun Synthesis Radio Telescope operating at 232 MHz. The reduced Miyun General Catalogue contains 34,426 radio sources in total. To distinguish which sources were not observed by other sky surveys, we first identify sources in the MGC with those in the 6C (151 MHz) catalogue, using a matching radius of 100 arcseconds. The remaining MGC objects were then matched with the B2/B3 (408 MHz), 4C (178 MHz), Texas (365 MHz, 1400 MHz) and 87GB (4.85 GHz) catalogues. 6850 MGC sources were found to lack any radio counterpart in these reference catalogues. It means 19.9% of sources listed in the MGC are probably new.

The "new" sources found were divided according to their flux densities into 5 classes, for each class, we denote the number of members by N_m and the subset of members not observed by others(here taking only the 6C and B2/B3 surveys) due to sky coverage limitation, N_{scl}. To be an indicator for source reliability, $R_{reli} = 1 - \frac{N_m - N_{scl}}{N_m}$ has been calculated for each class and listed in Table 1. The higher the R_{reli}, the higher the reliability.

2. Estimation of Positional Uncertainty

We assume that the positional errors of a point source are approximately proportional to its signal to noise ratio taking into account the declination dependence. This implies one measurement for N sources with the same flux density of S can be equivalent to N measurements for a source with this flux density. The apparent positional errors $\sigma_{app}(S)$ can be calculated by $\sigma_{app}(S) = \sqrt{\sum_{i=1}^{N} \frac{\Delta x_i^2(S)}{N-1}}$, where Δx_i refers to a positional difference of

TABLE 1. Reliability investigation results

Source class	Flux range Jy	Source members N_m	N_{scl}	R_{reli} %
CL1	$S \geq 1.5$	108	104	96
CL2	$1.5 > S \geq 1.0$	310	296	95
CL3	$1.0 > S \geq 0.5$	1631	1426	87
CL4	$0.5 > S \geq 0.3$	1923	1339	70
CL5	$S < 0.3$	2878	1815	63

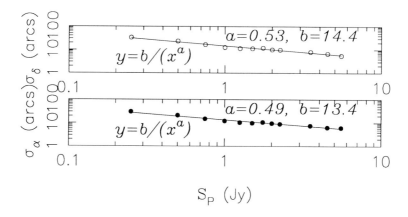

Figure 1. Resultant position errors of the MGC fitted by a least-squares method

a source between its measured position x_i and real(reference) position x_r. The positional uncertainty from a reference catalogue(the 6C, Baldwin et al. 1985; Hales et al. 1991) should be considered in calculating the true uncertainty.

Assuming that the sources have roughly the same flux densities when their relative flux errors are less than 10%, the positional errors in the MGC as a function of flux densities are plotted in logarithmic cooridinates as cycles, fitted by a least-squares straight line ($y = \frac{b}{S^a}$) in Figure 1. The results for α and δ in *arcsec* are $\sigma_\alpha = 13.4/S_P^{0.49}$, $\sigma_\delta = 14.4/S_P^{0.53} csc\delta$. The coefficients are all -0.98 for the best fits to the uncertainties.

References

Baldwin J.E., et al. 1985 Mon. Not. R. astron. Soc., 217, 717.
Hales, S.E.G., et al. 1991 Mon. Not. R. astron. Soc., 251, 46.

KILOMETER-SQUARE AREA RADIO SYNTHESIS TELESCOPE

KARST project

B. PENG AND R. NAN
Beijing Astronomical Observatory, China

1. Introduction

One way to realize the Large radio Telescope with a collecting area approaching one square kilometer, continuously covering a frequency range between 200 MHz and 2 GHz, is to construct a passive spherical reflector array of about 30 individual unit telescopes, each $\sim$ 300 m diameter(Butcher 1995). Valleys amid the hills of southwest China would be ideal for such LT concept. We refer to this effort as the Kilometer-square Area Radio Synthesis Telescope project. Site surveying and Radio Interference monitoring looks promising. Engineering considerations are summarized.

2. The Site, RIF Monitoring and Engineering Considerations

A large number of karst depressions, more than 400, have been found in the south Guizhou. There are less than 5 days of snowfall, and no ice build-up at the sites. Statistical results (Figure 1) by using remote sensing, geographic information system techniques and on-the-spot observation provide suitable sites.

A series of measurements at various sites in Guizhou have been carried out to check on their suitability, from the point of view of interference for realizing the KARST. The first measurements were made at 8 karst depression sites in Nov. 1994 in Pingtang and Puding counties. In addition we monitored a site in the center of Guiyang and at the Urumqi astronomical station for comparison. Additional measurements were made in March 1995 in an attempt to understand distance effects. The RIF monitoring results (Peng *et al.* 1996) look quite promising. Remoteness together with the futility of industrial development potentially benefit the RIF environment.

A backup structure for the main spherical reflector could be composed of ring-radial beams of concrete (Wang 1996). A synthetic design with

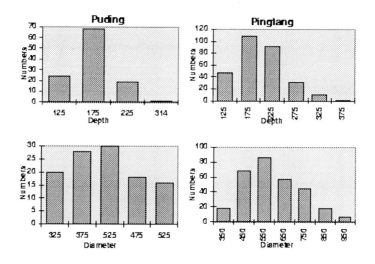

Figure 1. Statistical results at Pingtang and Puding counties

mechanics and electronics is proposed (Duan 1996), in which the line feed is supported by six suspended cables connected to a mechanical servo system controlled by a computer, and the ends of the line feed move along the two spherical surfaces. Laser techniques would be employed to accurately detect the position of the line feed in time. Line feeds (Mao & Jiao 1996), dual reflectors with a phased array (Guo 1996), and hybrid feeds (Xiong & Xie 1996) consisting of a co-spheric array and a line feed which diminishes the length of a single line feed required and reduces the blockage effect were it a single array feed have been discussed at an elementary level for the spherical reflectors. Several problems such as mutual coupling effects, overlapping and spillover are expected to be studied in more detail.

References

Butcher H., 1995. Astron NFRA. Issue no.9 page 5-7.
Duan B., Zhao Y., Wang J. and Xu G., 1996. Proc. of the 3rd Meeting of the Large Telescope Working Group and of a Workshop on Spherical Radio Telescope, eds: Richard Strom, Bo Peng and Rendong Nan 85-102.
Guo Y., Shu X. and Yan, H., 1996. Proc. of the LTWG-3 & W-SRT. eds: Richard Strom, Bo Peng and Rendong Nan 175-177
Mao Y. and Y. Jiao Y., 1996. Proc. of the LTWG-3 & W-SRT. 81-84
Peng B., Strom R., Nan R., Nie Y., Piao T., Kang L., Yan Y. and Wu S., 1996. Proc. of the LTWG-3 & W-SRT. 144-151
Wang J., Xu G., Duan B., and Li H., 1996. Proc. of the LTWG-3 & W-SRT. 158-159
Xiong J. & Xie S., 1996. Proc. of the LTWG-3 & W-SRT. 116-122

RADIO CONTINUUM SURVEYS ABOVE 2 GHZ

J.L. JONAS
Department of Physics and Electronics, Rhodes University

1. Current and Future Surveys

The Rhodes/HartRAO 2326 MHz radio continuum survey was observed at HartRAO over a 13 year period (Jonas *et al.* 1985, Mountfort *et al.* 1987, Jonas & Baart 1995, Jonas *et al.* 1996). This is the highest frequency and highest angular resolution (HPBW=20′) all-sky radio continuum survey made using a ground-based telescope. The 1σ noise level is better than 30 mK. A high-pass filtered image of the data is shown in Figure 1.

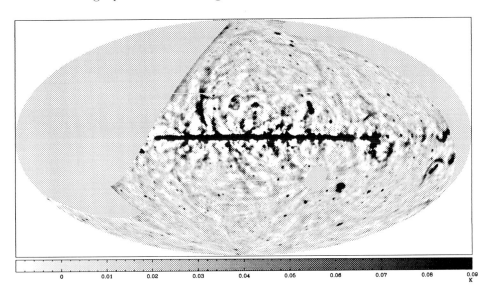

Figure 1. A high-contrast image of the Rhodes/HartRAO 2326 MHz radio continuum survey after processing with a median high-pass filter which removed structures with a scale size greater than 3°.

The scientific competence of ground-based, single-dish radio continuum survey data at GHz frequencies is constrained by the following environmental realities and experimental difficulties: (a) Contamination from ground radiation and atmospheric absorption makes absolute temperature measurements difficult and contributes to baseline uncertainties. (b) The increasing number of sources of radio frequency interference (terrestrial and satellite) are making ground-based radio continuum measurements impossible in certain bands. (c) Receivers with exceptional gain and noise temperature stability are required for large area surveys. (d) Temperature scale calibration is difficult for large dish antennae. These factors lead to systematic errors in the survey data, specifically "scanning effects," false large-scale structures and temperature calibration errors.

We have constructed a small, portable 2.3 GHz horn telescope with which we intend to make a low-resolution (HPBW=15°) survey of the southern sky. The results from this experiment will be used to improve the calibration of the main survey.

We have made observations of three regions of the Galactic plane at 8400 MHz using the HartRAO telescope (*e.g.*, du Plessis *et al.* 1995) to determine the feasibility of an extensive survey of the southern Galactic plane. The results of these observations were encouraging, and when the HartRAO 8400 MHz receiver has been upgraded to support dual polarization we will pursue this new survey project.

2. COBE DMR Comparisons

We have cross-correlated the 2.3 GHz survey data (convolved to 7° HPBW) with the 4-year COBE DMR data (Bennett *et al.* 1996) in an attempt to quantify the Galactic foreground contribution at large angular scales. A preliminary investigation yields an RMS contribution of $13\mu K$ at 53 GHz, which corresponds to a temperature spectral index of -2.7. This amplitude is somewhat larger than the Galactic synchrotron residual estimated by Kogut *et al.* (1996).

References

Bennett,C.L., *et al.* 1996, Astrophys.J. 464, L1
du Plessis,I., *et al.* 1995, Astrophys.J. 453, 746
Jonas,J.L., de Jager,G. & Baart,E.E., 1985, Astron.Astrophys.Supp. 62, 105
Jonas,J.L. & Baart,E.E., 1995, Astron.Astrophys.Supp. 230, 351
Jonas,J.L., Baart,E.E. & Nicolson,G.D., 1996, Mon.Not.R.astron.Soc. in preparation
Kogut,A., *et al.* 1996, Astrophys.J. 464, L5
Mountfort,P.I., *et al.*, 1987, Mon.Not.R.astron.Soc. 226, 917

THE MPIFR RADIO CONTINUUM SURVEYS AND THEIR WWW DISTRIBUTION

E. FÜRST, W. REICH, P. REICH,
B. UYANIKER AND R. WIELEBINSKI
Max-Planck-Institut für Radioastronomie, Bonn

1. Introduction

The observation of an area of $120° \times 56°$ centered on RA=8^h, DEC=$20°$ at 408 MHz was the first astronomical use of the MPIfR 100-m telescope (1970) and was designed to compile a complete sky survey using also data from Jodrell Bank and Parkes (Haslam et al., 1982). The observation of the northern sky at 1420 MHz started in 1972 using the Stockert 25-m telescope and was finished in 1976 (Reich and Reich 1986). This survey has been completed to an all sky survey using data from Villa Elisa (Argentina). The two surveys are absolutely calibrated. The angular resolutions are $0.8°$ and $0.59°$, respectively. A number of surveys of the Galactic plane have been made with the 100-m telescope at arc minute angular resolution. Surveys at 2695 MHz ($|b| \leq 5°$) (Reich et al. 1990, Fürst et al. 1990) and at 1410 MHz ($|b| \leq 4°$) (Reich et al. 1990) are public.

At medium Galactic latitudes (up to $|b| = 20°$) the emission consists mainly of faint extended ridges or arcs superimposed on the still dominating, about 10 times stronger, diffuse Galactic emission. They have never been investigated in a systematic way although they provide important clues for the understanding of the "disk-halo connection". This region is covered by new observations at 1400 MHz with the 100-m telescope.

2. A New 1400 MHz Survey with the 100-m Telescope

The new survey covers the accessible Galactic area for the latitude range $4° \leq |b| \leq 20°$ in total power and linear polarization. The survey will complement the running 21cm VLA survey (Condon et al.) for the missing extended predominantly Galactic component. The sensitivity is about

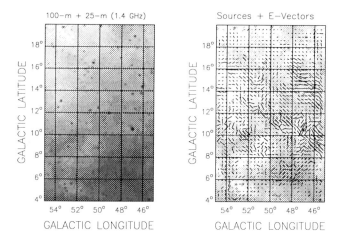

Figure 1. Part of the new 1400 MHz Survey. Left panel: data on an absolute scale. Right panel: The diffuse Galactic emission removed, polarization E-vectors superimposed.

15 mK r.m.s. in T_B, close to the confusion limit. The data will be adjusted to the absolutely calibrated 1420 MHz northern sky survey made with the Stockert 25-m telescope. It is also planned to combine these data with the VLA data to obtain a survey at an angular resolution of about one arc minute. The polarization U and Q will be adjusted using the polarization data obtained with the Dwingeloo 25-m telescope (Brouw and Spoelstra 1976), wherever it is possible.

2.1. FIRST RESULTS

About 15% of the survey has been completed. The first example is shown in Figure 1. The left panel shows the new 100-m data. Large-scale structures down to 3° have been removed in the right panel where polarization vectors are superimposed. A large number of emission structures are visible up to $b = 20°$. The polarized emission seems to conincide with these structures, but part of it may be associated with the removed diffuse emission. To distinguish between local and distant structures, the comparison with HI surveys will be important. For a distance of 1 kpc the observed emission structures at $b \approx 15°$ are about 250 pc above the Galactic plane. A substantial energy is required to push these structures out of the plane.

The second example (Figure 2) demonstrates the combination of the 100-m and the VLA data. The left panel (single dish data) shows the general decline of the Galactic disk emission away from the plane and a faint partial shell of about 200 mK T_B brightness temperature. This emission is not visible in VLA data (middle panel). This data shows a great number of compact sources. The combination of both data sets is shown in the

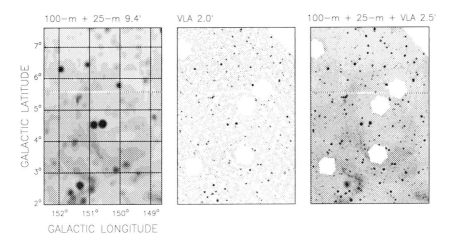

Figure 2. Left panel: 1400 MHz data on an absolute scale. Middle panel: The VLA 1400 MHz data. Right panel: Both data sets combined.

right panel. After subtraction of the compact sources from the combined map flux densities and spectral indices of diffuse emission structures can be accurately determined.

3. WWW

All the radio continuum surveys mentioned in this report are or will be available to the community via the WWW. A special retrieval software is offered at http://www.mpifr-bonn.mpg.de/survey.html. It is planned to include pulsar and spectral line data later.

References

Brouw, W.N., Spoelstra, T.A.Th. (1976) Linear Polarization of the Galactic Background at frequencies between 408 MHz and 1411 MHz. Reductions Astron.Astrophys.Suppl. 26, 129.

Fürst, E., Reich, W., Reich, P., Reif, K. (1990) A radio continuum survey of the Galactic Plane at 11cm wavelength. II. The area $76° \leq l \leq 240°$, $-5° \leq b \leq 5°$ Astron.Astrophys.Suppl. 85,691.

Haslam, C.G.T., Salter, C.J., Stoffel, H., Wilson, W.E. (1982) A 408 MHz All-Sky Continuum Survey. II. The Atlas of Contour Maps, Astron.Astrophys.Suppl. 47, 1.

Reich, P., Reich, W. (1986) A Radio Continuum Survey of the northern Sky at 1420 MHz - Part II Astron.Astrophys.Suppl. 63, 205.

Reich, W., Reich, P., Fürst, E. (1990) The Effelsberg 21cm radio continuum survey of the Galactic plane between $l = 357°$ and $l = 95.5°$ Astron.Astrophys.Suppl. 83, 539.

Reich, W., Fürst, E., Reich, P., Reif, K. (1990) A radio continuum survey of the Galactic Plane at 11cm wavelength. II. The area $358° \leq l \leq 76°$, $-5° \leq b \leq 5°$ Astron.Astrophys.Suppl. 85, 633.

THE UTR-2 VERY LOW FREQUENCY SKY SURVEY AND ITS MAIN RESULTS

K.P. SOKOLOV
Institute of Radio Astronomy,
Ukrainian National Academy of Sciences

1. Introduction

During the past decade there has been a dramatic increase in the amount of high-frequency ($\nu > 1000$ MHz) data currently underlying the studies of bright compact sources ($T_b \sim 10^{11-12} K$, $l \leq 1$ kpc) with flat spectra. But in order to determine physical conditions inside extragalactic radio sources at different stages of their evolution the studies of old extended ($l \sim 100$ kpc) sources with low surface brightness and steep spectra which constitute the dominant radio source population at very low frequencies ($\nu \ll 100$ MHz) are also needed. These sources are known to represent the final stage in the evolution of extragalactic objects.

On the other hand, from the literature data it follows that extended sources observed at very low frequencies form one of the most astrophysically significant class of objects for studies of the general properties of extragalactic radio source space distribution, because: (a) these sources constitute the so-called "parent population" of extragalactic objects whose observed parameters are not biased by relativistic effects; (b) these sources are characterized by the least scatter in their luminosities (Chambers *et al.*, 1988); (c) these sources have a lifetime $t \sim 3 \cdot 10^9$ years (Cordey, 1986) which is comparable with the period passed from the suggested epoch of radio source initial formation in early Universe.

Nevertheless these radio sources represent one of the least studied classes of extragalactic objects. The main goal of the UTR-2 Very Low Frequency Sky Survey is to provide the database on radio sources radiating at very low frequencies for further studies of their physics and statistical properties.

2. Discussion

Since 1973 the very low-frequency radio telescope UTR-2 has been used for radio sky surveying at six frequencies within the range 10 to 25 MHz. The UTR-2 Sky Survey has covered now two regions of the sky within the declination zones $-13°$ to $20°$ and $40°$ to $60°$ (Braude et al., 1994). The total solid angel of these regions is about 4 sr. The maximum sensitivity of the survey is 20 Jy at 25 MHz.

The radio source catalogue of the UTR-2 Sky Survey contains estimates of coordinates and flux densities of more than 2000 sources at 10, 12.6, 14.7, 16.7, 20 and 25 MHz. About 20 per cent of the sources observed can not be identified with objects from other higher-frequency surveys. The highly reliable and statistically complete samples of extended extragalactic radio sources radiating at very low frequencies have been selected on the basis of the data obtained. It should be noted that sources from the UTR-2 samples are weaker than those from the well-studied 3CR-sample of strong sources. The data obtained have been used to study space distribution of extended extragalactic radio sources and their angular structures.

On the basis of the UTR-2 data the statistical relationship $n(S)$ or source counts at 25 MHz have been obtained and physical reasons responsible for the strong cosmological evolution effects seen at very low frequencies have been considered (Sokolov 1988). The analysis of the 25-MHz source count has confirmed the suggestion mentioned in the literature many times that effects of radio source cosmological evolution should manifest themselves particular strongly at very low frequencies. It has been shown that the 25-MHz source count for strong radio sources is conditioned mainly by the nearby powerful radio galaxies. On the contrary, strong evolution effects seen for weak radio sources are conditioned by quasars and unidentified objects. This result differs in principal from those observed at high frequencies where the relative content of radio galaxies and quasars remains practically constant within the wide flux density intervals where the corresponding source counts are analyzed. The analysis of the 25-MHz source count has also allowed us to suggest the existence of a decrease in the space density of the most distant extended radio sources. In view of similarity in the general pattern of cosmological evolution of radio sources of different morphological types the result obtained for extended sources found at very low frequencies indicates on the existence of a redshift cut-off in the space distribution of the whole class of extragalactic sources.

For further implication of the results obtained the study of the astrophysical significance of the source counts observed in different frequency ranges has been carried out (Sokolov,1990). It has been shown that general properties of radio source spatial distribution manifest themselves most distinctly at very low frequencies since the physical conditions for the existence of a "flux density - distance" correlation, which constitutes a basis for every source count analysis, are best satisfied for extended isotropically radiating objects. On the contrary, this correlation cannot exist in principal for bright compact sources, since their observed flux densities are conditioned mainly by Doppler enhancement rather than distance. In the framework of the study explanations were make of: (a) the frequency dependence of the overall shape of the source counts, (b) the cosmological paradox caused by the qualitatively different character of the observed spatial distribution for compact and extended radio sources, (c) the difficulties of detecting the reality of a redshift cutoff from analysis of high-frequency source counts.

Further qualitative studies of the space distribution of extended sources suggest evaluation of the RLF for weak sources which constitute the maximum region in the 25-MHz source count. To reach this aim and to study the physical nature of unidentified radio sources with very steep spectra the complete sample of 265 weak ($S_{16.7} > 29$ Jy) extragalactic sources with very steep spectra ($\alpha_{178}^{16.7} \geq 1.0$) have been observed with the VLA at 333 and 1435 MHz in C-configuration. Detailed discussion of astrophysical aims of angular structure studies of radio sources found at very low frequencies is given in the paper by Sokolov (1993). Analysis of the VLA images obtained is in progress now. The VLA data allow us to obtain accurate positions of unidentified sources from the sample for further clarification of their physical nature and to carry out a search for: (a) very old or the so-called "fossil" radio sources which represent the final stage in the evolution of extragalactic radio sources; (b) very distant objects, as well as to study interaction of the source extended components with gaseous environment at different z.

References

Braude S. Ya., Sokolov K.P. and Zakharenko S.M. (1994) Decametric Survey of Discrete Sources in the Northern Sky. XI, Astrophys.Sp.Sc.213, 1.

Chambers K.C., Miley G.K. and Breugel W.J.M. (1988) 4C40.36: a radio galaxy at a redshift 2.3, Astrophys.J. 327, L47.

Cordey R.A. (1986) Radio sources in giant E and S0 galaxies Mon.Not.R.astro.Soc. 219,575.

Sokolov K.P. (1988) Determination of the Space Distribution Parameters for Extragalactic Radio Sources observed in Decametric-Wavelength Range, Astron.J.(Russian), 63,236.

Sokolov K.P. (1990) Analysis of the Astrophysical Significance of Radio Source Counts Obtained in Different Frequency Ranges, Australian J. Physics, 43,263.

Sokolov K.P. (1993) On Angular Structure Studies of Very Steep Spectrum Sources Found at Decametric Wavelengths, Sub-Arcsecond Radio Astronomy, Cambridge University Press, pp. 282–283

THE RATAN GALACTIC PLANE SURVEY

A Radio Continuum survey of the Galactic Plane with $343° \geq l^{II} \leq 19°$ and $|b^{II}| < 5.5°$ at 0.96, 3.9 and 11.2 GHZ

S.A. TRUSHKIN
Special Astrophysical Observatory of the Russian AS, RUSSIA

1. Survey Description

Now in radio continuum surveys more than 10,000 radio sources have discovered in the Milky Way plane but the Galactic origin only of a small part of them has been determined. The problem comes from the absence of estimates of source distance and the optical identification even for bright radio sources, and the most of sources have not spectral data at 2–3 frequencies. As followed some hundreds of sources have not classified as thermal or non-thermal. Now we don't know the full number of supernova remnants (SNRs) in the Galaxy. The simple estimates show that a sample of Galactic SNRs is not full as for weak and extended ($> 15'$) as for bright and compact ($< 3'$) SNRs (Trushkin 1993).

The three-frequency radio survey of the Galactic plane has been made at 0.96, 3.9 and 11.2 GHz with the RATAN-600 radio telescope in three sets in 1991–1995 (Trushkin 1996b). The resolution (RA×DEC) is $4' \times 75'$, $1' \times 39'$, $21'' \times 14'$ respectively. The noise level is typically 60, 10 and 100 mJy/beam at 0.96, 3.9 and 11 GHz respectively.

The total area of survey is near 400 degrees in square (0.12 sr). It is one third of the volume of the Galaxy in which the outbursts of supernova are possible. The coordinate accuracy estimated on the comparison with Bonn and Texas surveys is better than $15''$ in R.A. Flux densities and spectral indices of near 40 known supernova remnants were measured (Trushkin 1996a).

The preliminary list includes more than 1500 radio sources. The cross-identification, comparison, spectral index studies have been made with $\sim$ 1800 selected sources from Effelsberg, Nabeyama, VLA, PMN, Texas radio surveys and from the IRAS catalog and another radio catalogs, included in radio astronomical database CATS (Verkhodanov & Trushkin 1994, 1996).

2. Science from the Survey

The search of the supernova remnants and variable radio sources has been made. Using the different criteria from the IRAS sources we searched the thermal sources: planetary nebulae and H II regions. We used ISSA (IPAC) infrared maps for comparison with the survey data. Some supernova remnant candidates have been found in survey area when we used IR/Radio criterion. G11.2-1.1 seems one of good SNR candidates.

The superluminal variable source and X-nova GRO J1655−40 has been detected (in 1991.2) in the survey at flux level 100±10 at 3.9 GHz far before the identification at radio wavelengths. The extended radio envelope has been detected around this source. Its spectrum is non-thermal ($\alpha = -0.6$) and size is near 6'. Its origin could be connected either with associated supernova remnant or with blow-up envelope from central active source (Trushkin 1995).

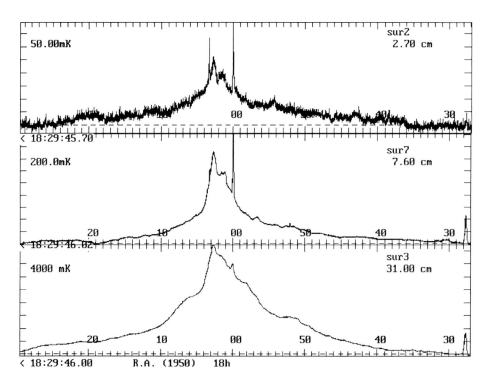

Figure 1. The survey cross-section of the Galaxy (Dec.=−21°40', $|b| < 5.5$) at three wavelengths. The extended galactic background is well seen on these scans.

The interesting point source 1820−239 was identified in the confirmed shell SNR G8.7−5.0. It has a non-thermal spectrum, $\alpha = -0.4$ and size $< 1'$. The sequential sets of the Galactic survey in other epochs allowed to detect the variability of flux from this source on time scale of 3–4 months. Thus 1820−239 seems to be analogous to the Galactic binary systems such as SS433 and Cir X−1.

Then we have made the cross-identification of two catalogs: the Texas and IRAS-point source in the database CATS (Trushkin & Verkhodanov 1995). 97 sources of the total 1208 pairs fall in the survey area. All 63 such sources with $S_{365MHz} > 0.4$ Jy were detected at 3.9 GHz.

In Figure 1 the same cross-cuts of survey on DEC50: $-21°40'$ at three wavelengths. The compact HII region 1800−217 and remnant of SN1604 (Kepler's SN) 1727−215, the latest from optically visible for last 400 years.

The work has been supported by the grant of the Russian Foundation of the Basic Researches N93-02-17086 and the travel RFBR grant N96-02-27170. This report became also possible due to the hospitality of STScI and the SOC for IAU symposium 179.

References

Trushkin, S. A. 1993, in "XXV radio astronomical conference," Pushchino, FIAN, 76
Trushkin, S.A. 1994, Preprint of SAO, 107, 1
Trushkin, S. A. 1995, in "XXVI radio astronomical conference," St-Petersburg, IPA, 114
Trushkin, S.A. 1996a, Bulletin of SAO RAS, 41, 64
Trushkin, S.A. 1996b, Astron. Astrophys. Trans., 11, in press
Trushkin, S.A., & Verkhodanov, O.V. 1995, Bulletin of SAO RAS, 39, 150
Verkhodanov, O. V., & Trushkin, S. A. 1994, Preprint SAO RAS, N106, 1
Verkhodanov, O.V., & Trushkin, S.A. 1996, Baltic Astronomy, in press

DENIS: A DEEP NEAR INFRARED SOUTHERN SKY SURVEY

N. EPCHTEIN
Observatoire de Paris, Meudon, France

1. Aim of the Project and Achieved Performances

The aim of DENIS is to map the southern sky in three near-infrared photometric bands (I, J, K_s) at the 1mJy sensitivity level (14th magnitude in K_s) and arcsecond resolution using the ESO 1 meter telescope at La Silla, Chile. The DENIS operations have started in December 1995 after a *protosurvey* period of one year mainly dedicated to technical tests and pilot surveys of selected regions of high astronomical interest. In September 1996, approximately 3000 square degrees of the sky have been covered, and the survey should be completed in 1999.

The main specifications of the project have been described several times (*e.g.*, Epchtein *et al.* 1994; Copet 1996; Epchtein 1997) and the achieved performances are summarized in Table 1. The dedicated 3 channel camera and its data handling hardware and software have been designed, implemented and tested by a European consortium of laboratories coordinated by the *Département de Recherche Spatiale* of Paris Observatory at Meudon. Important parts of the data acquisition routines were also developed at the University of Innsbruck, Austria.

2. Data Processing

The DENIS data processing is performed routinely in 2 dedicated data analysis centers (DAC) located at Paris (PDAC) and Leiden (LDAC). The PDAC processes and archives the raw images and implements an extended source database. The LDAC tasks (Deul *et al.* 1995) consist in source extraction and photometric calibration, and implementation of a point source database. Cross-identifications with other catalogs will be performed in collaboration with the *Centre de Données Stellaires* in Strasbourg. Data will be released in stages to the astronomical community one year after being

TABLE 1. Main specifications of the DENIS cameras

Channel	I	J	K_s
Central wavelength (μm)	0.8	1.25	2.15
Array manufacturer	Tektronix	Rockwell	Rockwell
Size (nb. of pixels)	1024 × 1024	256 × 256	256 × 256
Pixel size (μm, arcsec)	24, 1	40, 3	40, 3
Array quantum efficiency (aver.)	0.65	0.81	0.61
Read-out noise (e^-)	6.7	38	39
Read-out time (second)	2.98	0.13	0.13
Exposure time (second)	9	10	10
Achieved limiting magnitude (point source 3 σ)	18.5	16.3	14
Magnitude of saturation	9.5	8.5	6.5

archived in the DACs. A first set of data will be widely released in Spring 1997.

3. Science with DENIS

The primary aim of DENIS is to provide astronomers with reference documents (catalogs, atlases, databases) that will be fully available by the end of 2000. In the meantime, scientific programs will benefit of parts of this wealth of new data. Based on the pilot program and first months of observations, several investigations have been performed and have already produced significant scientific results. Some are presented at this Symposium and briefly outlined hereafter.

Multicolour surveys in a spectral domain never explored at this scale and at this level of sensitivity, provide an exceptional opportunity to probe the stellar content of the Galaxy. Lower interstellar extinction and high sensitivity to evolved stars are two major advantages of the near-infrared surveys with respect to optical Schmidt plates. Two main programs are underway in these area, one focuses on the exploration of the galactic disk, and more specifically of the anticenter regions, the other on the bulge. Ruphy et al. (1996) have shown that the DENIS data can be used to determine with a good accuracy the cutoff of the stellar distribution and the scale length (see her paper, this volume, p. 231). Another project, called ISOGAL and managed by A. Omont, combines ISO and DENIS images taken in the galactic bulge. The purpose of this investigation is to probe the bulge populations at 15 μm with ISOCAM in selected areas within the longitude range $-45°$ to $+45°$ at low latitude and to carry out systematic cross-identifications of ISO and DENIS sources. Colour diagrams involving DENIS and ISO data are successfully used to select evolved giant populations and recently formed stars (Pérault et al. 1996)

Search for young stars in nearby molecular complexes is also a domain in which near-infrared surveys will provide worthwhile data. A study of the *Orion A* region has been performed by Copet (1996) which led to the discovery of a large number of new candidates of T Tau stars. A program aimed at finding the infrared counterparts of CS cores is in progress (see his paper, this volume, p. 172). Other interesting regions such as the *Chameleon* cloud are also under investigations, also in connection with ISOCAM observations.

Acknowledgements

The instrument team involves scientists and engineers from Paris Observatory, Paris Institut d'Astrophysique, and Innsbruck University (B. de Batz, P. Fouqué, S. Kimeswenger, F. Lacombe, T. Le Bertre, S. Pau, D. Rouan, J.C. Renault, D. Tiphène). The data analysis centers are managed by G. Simon and J. Borsenberger in Paris and E. Deul in Leiden with the contribution of T. Forveille from Grenoble Observatory. All the contributors to DENIS are warmly thanked. The observations are collected at the European Southern Observatory, La Silla, Chile. The DENIS project is funded partly by the *SCIENCE and the Human Capital and Mobility* plans of the European Commision under grants CT920791 and CT940627, by the European Southern Observatory, and by various national sources of funding.

References

Copet E., 1996, PhD. dissertation, Paris 6, Le relevé infrarouge DENIS: mise en oeuvre de l'instrument et étude de la région de formation d'étoiles d'Orion A

Deul E.R., et al., 1995, Proc. Euroconference on "Near-Infrared Sky Surveys", San Miniato, Italy, P. Persi, W.B. Burton, N. Epchtein, A. Omont (eds.), reprinted from Mem. S. A. It. Vol. 66-3, p. 549

Epchtein N., et al., 1994, Astrophys. Sp. Sc. 217, 3

Epchtein N., 1997, Proc. Euroconference on "The Impact of LargeSscale Near-Infrared Surveys", eds. F. Garzon, N. Epchtein, A. Omont, W.B. Burton, P. Persi, Kluwer Academic Publishers, Dordrecht, in press

Pérault M. et al., 1996, Astron.Astrophys. Lett., in press

Ruphy S., et al., 1996, Astron.Astrophys. Lett. 313, L21

TABLE 1. Summary of Main Areas. These four areas comprise the main survey composed of $40' \times 40'$ rasters. In addition to the cirrus criterion $I_{100} < 1.5$MJy/sr, we restricted ourselves to regions of high visibility $> 25\%$ over the mission lifetime. For low Zodiacal background we required $|\beta| > 40$ and to avoid saturation of the CAM detectors we had to avoid any bright IRAS 12μm sources. (In addition to these are also 6 smaller rasters $20' \times 20'$ centred on well studied areas of the sky or high-z objects)

Area	Rasters	Nominal Coordinates J2000		$\langle I_{100} \rangle$ MJysr^{-1}	Visibility %	β
N1	3 × 2	$16^h10^m01^s$	$+54°30'36''$	1.2	98.0	73
N2	3 × 2	$16^h36^m58^s$	$+41°15'43''$	1.1	58.7	62
N3	3 × 2	$14^h29^m06^s$	$+33°06'00''$	0.9	26.9	45
S1	3 × 3	$00^h34^m44^s$	$-43°28'12''$	1.1	32.4	-43
Lock. 3	1	$13^h34^m36^s$	$+37°54'36''$	0.9	17.3	44
Sculptor	1	$00^h22^m48^s$	$-30°06'30''$	1.3	27.5	−30
TX1436	1	$14^h36^m43^s$	$+15°44'13''$	1.7	22.2	29
4C24.28	1	$13^h48^m15^s$	$+24°15'50''$	1.4	16.8	33
VLA 8	1	$17^h14^m14^s$	$+50°15'24''$	2.0	99.8	73
Phoenix	1	$01^h13^m13^s$	$-45°14'07''$	1.4		36

these objects at much higher redshifts and thus obtain greater understanding of the cosmological evolution of star formation.

If elliptical galaxies underwent a massive burst of star-formation between $2 < z < 5$, they would be observable in the far infrared and may look like F10214 (Elbaz et al., 1992). This survey will provide a powerful discrimination between this and a merging model whose components are individually too faint to detect.

IRAS uncovered a population with enormous far infrared luminosities, $L_{\rm FIR} > 10^{12} L_\odot$. While most of these objects appear to have an AGN it is argued that star formation could provide most of the energy. Interestingly, most of these objects appear to be in interacting systems, suggesting a triggering mechanism. Exploration of these objects at higher redshift will have particular significance for models of AGN/galaxy evolution. Since AGN far infrared emission is relatively insensitive to inclination angle these objects may also constrain certain unification schemes.

F10214 was at the limit of IRAS sensitivity (Rowan-Robinson et al., 1991) and new classes of objects may well be discovered at the limit of the ISO sensitivity. The lensing phenomenon which made F10214 detectable by IRAS may become more prevalent at fainter fluxes, increasing the proportion of interesting objects.

3. Survey Definition

We were restricted to two bands and selected 15 μm (using ISO-CAM) which is sensitive to AGN emission and 90 μm (using ISO-PHOT) which is sensitive to emission from star formation regions. To complement deep CAM surveys (Franceschini *et al.*, 1995) we decided to sacrifice depth at the shorter wavelength for increased area, at the longer wavelength we aimed to reach the confusion limit. This lead us to an integration time of 20s for both with the CAM raster such that each sky position is observed twice. The CAM detector is an array of 32×32 pixels and we use a pixel field of view of $6''$, the PHOT detector we use is a 3×3 array with $43.5''$ pixels.

The satellite is operated in raster mode and a single raster is $40' \times 40'$ (although with PHOT this area is observed in two halves). The main survey is 28 of these rasters grouped in four areas on the sky. Cirrus confusion is a particular problem, so our main selection criterion was low IRAS 100μm intensities, using the maps of (Rowan-Robinson *et al.*, 1991b).

4. Current Status

A CAM test raster and two PHOT half rasters were observed and have now been processed. As of 7th October observations of 4 rasters in S1 and all 6 in N1 have been completed. Subject to scheduling constraints we expect the remainder of the observations to be done by the end of January 1997.

Analysis of the CAM and PHOT test data suggest that we will reach a 5σ limit of around 2mJy at 15 μm and 50mJy at 90 μm .

Our WWW page can be found on `http://artemis.ph.ic.ac.uk/`

References

Elbaz, D. *et al.* (1992) Astron.Astrophys.,265, L29.
Franceschini A., Cesarsky,C Rowan-Robinson, M., (1995) In 'Near-IR Sky Survey' San Maniato (Pisa), Memorie della Societa Astronomica Italiana (in press)
Oliver, S., *et al.* (1995), In "Wide-Field Spectroscopy and the Distant Universe," Maddox, S.J., Aragon-Salamanca, A. eds, World Scientific. p. 264
Rowan-Robinson, M. *et al.* (1991) Nature, 351, 719.
Rowan-Robinson, M. *et al.* (1991) Mon.Not.R.astron.Soc., 249, 729.

ASTRONOMY ON THE MIDCOURSE SPACE EXPERIMENT

S.D. PRICE[1], E.F. TEDESCO[2], M. COHEN[3],
R.G. WALKER[4], R.C. HENRY[5], M. MOSHIR[6],
L.J. PAXTON[7] AND F.C. WITTEBORN[8]
[1]*Phillips Laboratory, Hanscom AFB,*
[2]*Mission Research Corporation,*
[3]*Univ. of California Radioastronomy Laboratory,*
[4]*Jamieson Science & Engineering, Inc.,*
[5]*Johns Hopkins University,*
[6]*Jet Propulsion Laboratory,*
[7]*Applied Physics Laboratory,*
[8]*NASA/Ames Research Center*

Abstract. The Midcourse Space Experiment (MSX) carries a varied suite of sensors for imaging in the ultraviolet to the mid-infrared, hyperspectral imaging in the ultraviolet through visible and infrared spectroscopy with a Michelson interferometer. At comparable sensitivity to the Infrared Astronomical Satellite (IRAS) but with a 40 times smaller point response function, the MSX infrared radiometer is ideal for surveying specific large areas such as those in which IRAS was degraded by confusion or not covered at all. This experiment obtains simultaneous observations over a very wide spectral range, from 0.11 to 25 μm, thus providing unique information on the energetics of such diverse objects as comets and H II regions. Initial observations indicate that the astronomy objectives on this experiment will be achieved.

1. Introduction

The Midcourse Space Experiment (MSX) is a Ballistic Missile Defense Organization mission for research on space-based tracking of Ballistic Missiles. This includes a variety of measurements related to the chemistry and physics of the atmosphere and astronomy. MSX was successfully launched on 24 April 1996 into a $\sim$ 900 km altitude circular orbit. The four to five year mission lifetime is divided into a cryogen phase lasting about a year during which infrared measurements are a priority and the post cryogen phase using the suite of ultraviolet through visible (red) sensors. The infrared astronomy survey objectives are to cover the regions either missed by previous experiments such as the IRAS gaps and the zodiacal cloud near the sun and in the anti-solar direction or where previous experiments were degraded by confusion such as in the Galactic Plane. Additional measurements probe Galactic structure, the physical processes in H II regions and comets, obtain spectroscopy on extended sources and tie a set of secondary stellar calibration standards to the generally adopted primary stars. The post cryogen astronomy goal is to survey the sky in the "solar-blind" UV.

2. Instrumentation

An overview of the spacecraft, instruments and large range of scientific objectives of this mission is given by Mill *et al.* (1994) with more detail provided in Vol. 17, Numbers 1 and 2 of the Johns Hopkins APL Technical Digest (1996). The instruments used for astronomy on MSX are the SPatial InfraRed Imaging Telescope (SPIRIT) III, the four large and small field ultraviolet and visible imaging instruments (Imagers) and the five hyperspectral uv-visible imagers (SPIMs).

SPIRIT III is a 35 cm diameter clear aperture off-axis telescope with two focal plane instruments, five line scanned arrays and a Michelson interferometer. The spectral responses of the line scanned arrays are 6 to 11 μm (designated Band A), 11–13 μm (Band C), 13–15 μm (Band D), 18 to 25 μm (Band E) and a very narrow ($\sim$ 0.25 μm) region at 4.2 μm (Band B). Each array is 8 columns by 192 rows of 18″ pixels with half the columns offset by half a pixel. Only half the columns are active to reduce the data rate to only 5 Mbps. The interferometer is capable of spectral resolutions from 2 to 20 cm^{-1}. Three large detectors, 12′ × 12′, divide up the spectral region between 6 and 28 μm, two of the 6′ by 9′ elements cover the entire 2.7 to 28 μm range while the third is limited to 2.7 to 5 μm.

The narrow field visible and ultraviolet instruments image 1.3° × 1.6° fields with pixels slightly larger than those in SPIRIT III. The wide field imagers have eight times larger pixel with the corresponding increase in field size. Collectively, the imagers span the spectral range from 110 to 900 nm. Five hyperspectral imagers combined cover 110 to 893 nm at a resolving power as high as 200. Uniquely, the SPIMs can provide spectra in 1.5′ increments over a linear spatial extent of 1° or over a 1° by 1° field.

3. Astronomy Experiments on MSX

With Band A having an on orbit, single read sensitivity comparable to the IRAS 12 μm band and a footprint some 40 times smaller, a principal experiment is to survey the Galactic Plane, the Magellanic Clouds and star forming regions in the large molecular clouds in which IRAS was confused. The area within 5° of the Plane will be redundantly covered once and within 3° twice. Deeper probing of Galactic structure, by at least a factor of 20 in sensitivity, is obtained from raster scans of selected regions in the plane. Another priority is to survey areas not previously covered by other infrared experiments. This includes the IRAS gaps and, owing to the superior side-lobe rejection of SPIRIT III, solar elongations between 25° and 30° and greater than 125°. Other infrared experiments include observations of several large galaxies, comets and asteroids and infrared secondary calibration standards. The interferometer obtains spectroscopy on the emission characteristics of the zodiacal dust cloud and the Galactic Plane during the survey scans. Spectroscopy of extended sources will be emphasized during the last half year of the cryogen phase of the mission.

The wide field UV imager has at least ten times smaller pixels and higher sensitivity than TD-1. Thus, the primary post cryogen objective is to survey the sky with this instrument while sampling in the higher resolution and sensitivity narrow field UV imager and SPIMs. The visible imagers will concurrently survey for small, relatively near, low activity comets by looking for OH^- emission from these objects.

Acknowledgements

We are grateful for the work of our Associate Investigators, M. Allen, M. Egan, J. Murthy, R. Shipman, and J. Simpson.

References

Johns Hopkins APL Technical Digest, 1994, Vol. 17, No. 1, 2-116 and No. 2, 134–252.
Mill, J.D., O'Neil, R.R., Price, S., Romick, G.J., Uy, O.M., Gaposchkin, E.M., Light, G.C., Moore, Jr., W.W., Murdock, T.L., & Stair, A.T., 1994, "Midcourse Space Experiment: Introduction to the Spacecraft, Instruments, and Scientific Objectives, Jour. Spacecraft and Rockets, Vol. 31, No. 5. 900–907.

THE WIDE-FIELD INFRARED EXPLORER (WIRE) MISSION

D.L. SHUPE[1], P.B. HACKING[2], T. HERTER[3], T.N. GAUTIER[4],
P. GRAF[5], C.J. LONSDALE[1], G.J. STACEY[3], S.H. MOSELEY[6],
B.T. SOIFER[7], M.W. WERNER[4] AND J.R. HOUCK[3]
[1] *Infrared Processing and Analysis Center, Pasadena, CA*
[2] *Jamieson Science & Engineering, Bethesda, MD*
[3] *Cornell University, Department of Astronomy, Ithaca, NY*
[4] *Jet Propulsion Laboratory, Pasadena, CA*
[5] *Ball Technologies, Boulder, CO*
[6] *Goddard Space Flight Center, Greenbelt, MD*
[7] *California Institute of Technology, Pasadena, CA*

1. Introduction

The Wide-Field Infrared Explorer (WIRE) (Schember *et al.* 1996 and references within) is a small spaceborne telescope specifically designed to study the evolution of starburst galaxies. This powerful astronomical instrument will be capable of detecting typical starburst galaxies at $z \sim 0.5$, ultraluminous infrared galaxies beyond a $z \sim 2$, and luminous protogalaxies beyond $z \sim 5$. The WIRE survey, to be conducted during a four month period in 1998, will cover over 100 deg^2 of high Galactic latitude sky at 25 μm and 12 μm.

WIRE was selected as the fifth in NASA's series of Small Explorer (SMEX) missions in August 1994. It was proposed in 1992 by a teaming partnership of the Jet Propulsion Laboratory (JPL), California Institute of Technology, and the Space Dynamics Laboratory (SDL), Utah State University. Launch is scheduled for September 1998.

2. WIRE Primary Science

The WIRE survey will detect primarily starburst galaxies, which emit most of their energy in the far-infrared. The number of these faint sources at a given flux level depends on their as-yet-unknown evolutionary rate. The objective of WIRE is to answer the following three questions: (1) What fraction of the luminosity of the Universe at a redshift of 0.5 and beyond is due to starburst galaxies? (2) How fast and in what ways are starburst galaxies evolving? (3) Are luminous protogalaxies common at redshifts less than 3?

The WIRE survey will cover over 100 deg^2 of sky and detect sources 200–500 times fainter than the IRAS Faint Source Catalog at 25 μm and 500–2000 times fainter at 12 μm. The resulting catalog, expected to contain at least 30,000 starburst galaxies, will reveal their evolutionary history out to $z \sim 0.5$–1 and the evolutionary history of extremely luminous galaxies beyond $z \sim 5$. This will be the first significant galaxy survey to probe these redshifts at far-infrared wavelengths where extinction effects are small and where most of the luminosity of starburst galaxies, and possibly of the Universe, can be measured. WIRE will measure the $25\mu m - 12\mu m$ color of detected sources, which is a powerful statistical luminosity indicator (Soifer & Neugebauer 1991) as well as an effective means of distinguishing foreground stars.

3. The WIRE Instrument

The WIRE instrument is a cryogenically-cooled 30 cm Ritchey-Chrétien telescope system that illuminates two 128 × 128 Si:As infrared detector arrays. A two-stage solid hydrogen cryostat maintains the optics colder than 19 K and the detector arrays below 7.5 K. The optical system consists of the telescope primary and secondary mirrors, a dichroic beamsplitter, one optical passband filter, and baffles. The two channels of the instrument cover broad bands centered near 12 μm and 25 μm; the 25 μm band is the primary one for detecting starburst galaxies. The pixels are 15.5 arcsec on a side, providing a 33×33 arcmin^2 field of view in each passband. The FWHM beam size in coadded images will be about 26 arcsec at 25 μm and 22 arcsec at 12 μm. The instrument contains no moving parts.

4. Survey and Observing Strategy

The WIRE survey will consist of three parts. The moderate-depth survey is designed to maximize the detection of distant protogalaxies. 60% of the survey time will be spent on this survey, covering hundreds of square degrees, with 15 to 50 minutes total exposure time on each WIRE field. 30% of the survey time will be spent on the deep survey, with a total integration time of several hours per field, set by the point at which confusion noise is equal to instrumental noise. The goal of this survey is to obtain a large sample with the largest lookback time at a given luminosity, which will require covering tens of square degrees to this depth. Finally, the ultra-deep survey will use about 10% of the survey time early in the mission to observe a few WIRE fields for 24 hours or more total exposure time, to measure the confusion distribution.

To reach such large cumulative exposure times, WIRE will use a stare-and-dither observing technique. During a ten minute orbit segment when the target field is near the zenith, the instrument will record several short (44 sec) exposures, each separated by a small dither. This technique will allow accurate subtraction of the background from the data. Target fields will be reacquired on subsequent orbits to accumulate sufficient exposure time and to allow detection of moving or variable sources.

The sensitivity of the deep survey will be limited by source confusion and hence depends on the rate of starburst galaxy evolution. Expected sensitivities are listed in Tables 1 and 2 for two evolution cases. The 12 μm sensitivity will be 1.5 to 3 times lower (in mJy at 12 μm) than the limits tabulated for 25 μm.

TABLE 1. WIRE Sensitivity for $(1+z)^{1.7}$ Density Evolution

Survey	Sky Coverage	25μm Flux limit (5σ)	# of sources
Moderate Depth	400 deg^2	1 mJy	> 90,000
Deep	20 deg^2	0.4 mJy	17,000
Ultra-deep	1 deg^2	0.28 mJy	1,000

TABLE 2. WIRE Sensitivity for No Evolution

Survey	Sky Coverage	25μm Flux limit (5σ)	# of sources
Moderate Depth	145 deg^2	0.65 mJy	> 35,000
Deep	7 deg^2	0.27 mJy	6,000
Ultra-deep	0.5 deg^2	0.22 mJy	500

References

Schember, H., Kemp, J., Ames, H., Hacking, P., Herter, T., Everett, D., Sparr, L., and Fafaul, B., 1996, Infrared Technology and Applications XXII, SPIE Proceedings 2744, in press.

Soifer, B.T., & Neugebauer, G., 1991 Astron.J. 101, 354.

FIRST OBSERVATIONS WITH THE 1.5 M RC TELESCOPE AT MAIDANAK OBSERVATORY

B.P. ARTAMONOV
Sternberg Astronomical Institute, Moscow, Russia

1. Introduction

Since 1975, the Sternberg Astronomical Institute of Moscow University (SAI) has been conducting a search of Middle Asia for good astronomical sites. After investigating the meteorological conditions, temperature fluctuations and seeing quality of different sites, Maidanak Mountain (an isolated summit 150 km south of Samarkand) was chosen. This site has 2000 hours per year of clear observing conditions with a median seeing of about 0.7″ (Artamonov *et al.* 1987, Bugaenko *et al.* 1992). The construction of a 1.5m RC telescope for the Maidanak Observatory was mostly completed when it was nationalized by Uzbekistan in 1993. Tashkent Astronomical Institute is now working with SAI to form the International Maidanak Observatory and to continue with joint observations.

2. The Dome and 1.5 meter RC Telescope

The main instrument at Maidanak Observatory is a 1.5 meter RC telescope (AZT-22) wich was manufactured by the optical firm LOMO (St.Peterburg). This telescope has three configurations (f/8, f/17, f/48). The f/8 configuration with two corrector lenses has a flat field about 1.5 degrees and a resolution of 0.5″. Optical testing with this configuration has shown that the telescope optics are of very high quality and are almost diffraction limited.

To preserve the best quality seeing, the dome was manufactured to have a cold floor to eliminate the problem of heating from the laboratory and control rooms. The dome can be ventilated at up to 25,000 $m^{-3}h^{-1}$. The result is that thermal equilibrium can be attained inside the dome in about

2–3 hours reducing the turbulence in the dome to that of the outside environment.

3. First Observational Results

The first CCD observations were taken in September 1995, in collaboration with an international team (P.Notni, V.Dudinov,V.Bruevich, M.Ibragimov from Germany,Ukraine,Russia,Uzbekistan) in order to test the active ventilator system for improving the dome seeing. During two weeks of observation we estimated a median seeing about $0.7''$ with a range of $0.5 - -1.2''$ which is as good as the site will provide. Exposures of the "Einstein Cross" gravitational lens system Q2237+0305 (V=18) were obtained with an FWHM=$0.5''$ using the Pictor 416 CCD camera. The VRI photometry during 17–23 September 1995 confirmed that the A component again became the brightest and that the colours of all components lie on the reddening line of the two-colour diagram (Vakulik *et al.* 1996).

In April 1996, a University of Pittsburgh CCD camera (TI 800×800 mounted in vacuum in a LN2 dewar) with BVRI filters was installed on the 1.5m telescope. First observations have shown high efficiency with a 5 minute exposure reaching objects of $V = 22^m$ with a s/n=30. Since then, we have continued to study the photometric properties of the telescope/CCD system, and in collaboration with D.Turnshek have begun a QSO and QGN monitoring program.

References

Artamonov B.P., Novikov S.B. and Ovchinnikov A.A., 1987. "Methods to increase efficiency of optical telescopes," Moscow University, p.16

Bugaenko O.I., Dudiniv V.N., Novikov S.B., Ovchinnikov A.A., Popov V.V. and Sinelnokov I.E.,1992. "Results of investigation of Maidanak astroclimate with mirror interferometer and CCD camera," Preprint SAI, No.2

Vakulik V.G., Dudinov V.N., Zheleznyak A.P., Notni P., Shalyapin V.N. and Artamonov B.P. 1996. Astronom. Nachricht (in press)

SOME RESULTS OF THE BATC CCD COLOR SURVEY

J.-S. CHEN
Beijing Astronomical Observatory, China

1. Introduction

The BATC (Beijing-Arizona-Taiwan-Connecticut) CCD color survey started 2 years ago. It is based on observations with the 60/90cm f/3 Schmidt Telescope of the Beijing Astronomical Observatory using a 2k×2k CCD and 15 intermediate band filters covering from 300nm to 1000nm to obtain the spectral energy distribution (SED) of all objects in 500 selected fields down to $m_V = 20$ (Chen 1994). The basic framework of the survey including instrumentation, data acquisition system, archive data system, and the various steps of data reduction has been established. About 60 fields have been observed. Most fields are still short of observations in UV band. We are waiting for a new thinned CCD to improve the quantum efficiency.

The method of using the SED, image structure and astrometric information combined with stellar evolutionary track calculations, stellar population synthesis, and the model SED template data (theoretical and observational) to separate stars/galaxies, to classify the SED of the stellar objects, to select abnormal SED sources (QSO, AGN, CV *etc.*), to discover moving and variable objects and to determine the redshift of galaxies is being developed (Proceedings of Workshop on BAO Schmidt CCD Astronomy,1996).

2. M67 Observations

The studies of M67 field (Fan *et al.* 1996) provide good evidence that we can obtain spectrophotometry from the ultraviolet to 1000nm with an intrinsic accuracy of better than 0.02 mag. for all objects in the nearly 1 square degree field of the CCD, using Oke-Gunn primary standard stars. The CMD shows not only the morphology consistent with previous ones, but define better than most the gap in the main sequence. The stellar track and atmosphere model fits the CMD very well for an age of 4 Gyr and [Fe/H] = -0.10, yielding a reddening of E(B−V) between 0.015 and 0.052 mag and a distance modulus $(m - M)_0 = 9.47 \pm 0.16$ mag. As our data combines deep images, accurate photometry and wide field coverage more than previous survey, we are able to observe both direct and implied evidences of substantial dynamical evolution issues pertaining to this old galactic cluster, such as the mass dependent spatial distribution of "single" stars, binaries and blue stragglers; the two dimensional shape of M67 elongated along an angle of 15° relative to the galactic plane; the volume dependent luminosity function rising from the main sequence turnoff and then flattening out at fainter absolute magnitudes; the leveling off for lower mass stars in the mass function, which may be due to the evaporation of stars through dynamical evolution of this old cluster.

3. Other Observations

An automatic SED classification technique has been applied to this and other fields. The classification is accurate to a subtype of spectral and luminosity class and has been confirmed by a sample of 80 stellar objects observed with slit spectra. The abnormal SED objects can then be selected. Some bright objects were observed spectroscopically and identified as QSOs and metal poor HII galaxies. The SEDs of several known high redshift QSOs in our selected fields shows that our system is very efficient in selecting QSOs.

Other research work in progress using the BATC survey data includes multi-color surface photometry of nearby galaxies and the fields of Abell clusters.

References

Chen, J.-S., 1994, in "Astronomy From Wide-Field Imaging" Proceedings of IAU Symposium no.161, 20.

Fan, X. *et al.* 1996, Astron.J, 112, 628

THE AIST-STRUVE SPACE PROJECT SKY SURVEY

M.S. CHUBEY, I.M. KOPYLOV, D.L. GORSHANOV,
I.I. KANAYEV, V.N. YERSHOV, A.E. IL'IN,
T.R. KIRIAN AND M.G. VYDREVICH
Pulkovo Observatory, Saint-Petersburg, Russia

1. Introduction

The *Struve* space astrometric project (Yershov et al. 1995) is being developed by a consortium of Russian space instrumentation institutes in order to extend the Hipparcos satellite reference system (the project initially named *AIST*, owes its current name to the first director of the Pulkovo Observatory).

The extension of the Hipparcos system means a density of about 100 stars per square degree (at least 4 million stars in the output catalogue). The proper motions of the Hipparcos stars are to be determined with an accuracy of about 0.1 mas year^{-1} due to the epoch difference between two catalogues. The mean accuracy of star positions in the output catalogue is expected to be 0.6 mas which could be achieved by proper design of the satellite (symmetry, smooth rotation, *etc.*), optics and the micrometer. With a properly designed micrometer (with CCD arrays, special processors for image processing and compressing the data flux to the ground station) it will be possible to observe all objects of the sky down to a definite limiting magnitude. This survey technique assumes no input catalogs to be used for observations.

2. Micrometer and Photometric System

One-square degree field of view of the on-board telescopes is planned to be filled with a mosaic of relatively small CCDs (28 detectors of 13×15 mm size each). Some of the CCDs will be covered with color filters (different number of filters for different spectral bands) in order to obtain photometry of stars in the Vilnius 7-band photometric system (Kopylov et al. 1994). This photometric system seems to be optimal for the survey because it provides sufficient information for classification of objects and the determination of their physical properties and chemical composition.

Each CCD array with square $16 \times 16\mu$ pixels will have a rigid adjustable base. The precision of the adjustment is about $\pm 4\mu$, and it is restricted mainly by the CCD flat surface which does not fit strictly the curvature of the spherical focal surfaces of the Schmidt telescope.

3. Limiting Magnitude and Accuracy of Observations

The image is uniformly moving over the field of view because the scanning mode is chosen for the observations. This is one of the most effective ways to survey the sky. The CCDs are planned to be working in the Time Delay and Integration mode, when the accumulated charges are shifted synchronously with the image. The integration time will be approximately 7 seconds for each detector. In total, each transit lasts 35 seconds, and estimations show that the limiting magnitude $V \approx 18^m$ is achievable in the Vilnius photometric bands and $\approx 22^m$ in the wide band $\lambda \in (300, 800$ nm$)$. Computer simulations show the accuracy of the CCD observations will be ≤ 0.4 mas for coordinates of a $V = 14^m$ star image, ≤ 3.0 mas for a $V = 18^m$ star, $\leq$ 4mmag and $\leq$ 35mmag in the photometry of 14 and 18-magnitude stars, respectively.

4. Conclusion

The described project is intermediate between Hipparcos and the planned GAIA project (Lindegren and Perryman 1995). The Struve output catalogue will be less accurate than that of GAIA, but it could be much easier and faster to do then any interferometric project, and it will give very valuable information on the dynamics and evolution of the stars in our galaxy. The accuracy of the Struve catalog will be higher than that of the Hipparcos. The resulting reference system could be maintained at the one milliarcsecond level for at least 40–50 years.

References

I.M.Kopylov, D.L.Gorshanov and M.S.Chubey, 1995. in "Astronomical and Astrophysical Objectives of Sub-Milliarcsecond Optical Astrometry", 327–330. E.Høg & P.K.Seidelmann (eds.), Kluwer Publ.

L.Lindegren, M.A.C.Perryman, 1995. in "Future Possibilities for Astrometry in Space", 23–34, ESA SP-379.

V.N.Yershov, M.S.Chubey, A.E.Ill'in, I.M.Kopylov, D.L.Gorshanov, I.I.Kanayev, and T.R.Kirian, 1995, "Project STRUVE", "Glagol," St-Petersburg, 272 p., (in Russian).

ASPA—SKY PATROL FOR THE FUTURE

C. LA DOUS AND P. KROLL
Sonneberg Observatory, Sonneberg, Germany

1. Introduction

Ever since the turn of the century sky patrol observations—*i.e.*, the routine surveillance of the visible night sky—has been one of the main foundations of variable star research, and thus of astronomy in general. Historically several of the major Observatories all around the world were contributing. Up to this day all these observations are in the form of large-field photographic plates, constituting a total of some 2 million plates which, collectively, contain the history of the light changes of celestial objects (mostly in the northern hemisphere and down to some 13 magnitudes apparent brightness or fainter) during the past 100 years. In modern times the last place left in the world where sky patrols are still being carried out routinely is Sonneberg Observatory.

There seems no doubt that the continuation of sky patrol observations is scientifically essential, not only for variable star research, but reaching from observations of near-Earth objects all the way to cosmological ones, and as direct support to space observations. Nevertheless, the current situation is far from satisfactory as the routine production of photographic plates for astronomical usage was discontinued some time ago; Sonneberg, in central Europe, climatically is not an ideal place for astronomical observations; and finally, the extraction of information from photographic plates is difficult and time consuming.

Until not long ago CCD detectors—which have replaced photographic onces in all astronomy for the last two decades—were too small to fully replace the patrol plates. But in the last few years technology has developed sufficiently for sky patrols on the basis of large CCD-arrays to become possible.

2. The Aims and Essentials of ASPA

The idea of ASPA (*All Sky Patrol Astrophysics*) was born in late 1994 in an attempt to bring sky patrols up to the level of modern technology and to thus fill a badly-needed gap in astronomical observations.

In brief the aims of ASPA are to continue sky patrols using CCD detectors; to construct a world-wide network of fully-automatic telescopes, so-called ASPA-stations; to monitor the entire visible night sky with a time resolution of a few hours; and to make data available on the web within some 24 hours of each observation. In addition, classical photographic observations will be continued until the CCD-based version will be operational. And finally an attempt will be made to scan as many of the old photographic plates as possible and to make them available through the web as well.

For the time being, although the requirements come from astronomy, ASPA is largeley a—rather difficult—technical project, which in turn implies that no observatory on its own is able to put it into practice. Fortunately a collaboration could be started between several departments of the Technical University of Ilmenau, the Optikzentrum in Bochum, and Sonneberg Observatory (all in Germany).

Meanwhile detailed concepts have been prepared for the telescopes (30 cm diameter, 940 mm focal length, each equipped with a large CCD array (7000 × 9000 pixels) covering a 5° × 5° field in the sky, thus providing an angular resolution of around $2''$/pixel; with exposures of some 5 min duration objects down to 18^m–19^m, some 1–2×10^8 objects, and a photometric accuracy of some 3–5% can be reached), the platform (some 20 such telescopes put together), data transmission to the planned data center in Sonneberg Observatory, data reduction and archiving, and easy access to fully reduced data (light curves or individual points) through the WWW. Funding is being applied for with the *Deutsche Forschungsgemeinschaft.*

In the fully-grown version ASPA cannot yet be realized with currently available technology—mainly for reasons of the sheer amount of data to be taken care of. For this reason we will start out with just two telescopes and limited info in the web in excess to the very observations, in order to test all the functions and to replace the photographic sky patrol, to begin with in Sonneberg. As technology evolves and money becomes available, little by little the final goal, as described above, will be approached.

More details and regularly updated information are available at
`http://www.stw.tu-ilmenau.de`.

WIDE FIELD OPTICAL IMAGING AT CTIO

A. R. WALKER
Cerro Tololo Inter-American Observatory, Chile

1. Introduction

The CTIO 4-m and 0.6/0.9-m Schmidt telescopes provide a wide-field CCD imaging capability unequalled in the southern hemisphere. Characteristics of present and future CCD imaging systems for these telescopes are discussed.

2. 4-m Telescope and Prime Focus CCD Imager (PFCCD)

An on-going program of optical, thermal and electronic upgrades have substantially improved the performance of the 4-m telescope. At prime focus, images with FWHM as small as 0.65″ have been recorded, and 0.8–1.0″ is common. The PF corrector (Ingerson 1996 *in preparation*) is a 6-element design incorporating two rotating elements which provide atmospheric dispersion correction. The unvignetted field is 48′ in diameter, with d70 image quality of 0.20″ on-axis, and 0.45″ at the edges, and is useable from 3,400–11,000Å. The PFCCD Imager uses a Grade 0, thinned, quad-amp SITe 2048 CCD, covering a 15′ square field at scale 0.43″ pixel^{-1}. An Arcon CCD controller reads out the CCD in approximately 30 seconds. Two filter wheels can each hold up to five 4×4 inch filters, while a low power, fused silica concave lens compensates for the curvature of the CCD. The PFCCD imager is scheduled for approximately 20% of all 4-m nights, and is the most popular dark time instrument.

3. 4-m Telescope and Big Throughput Camera (BTC)

The BTC, built by A. Tyson (Lucent Technologies) and G. Bernstein (U. Michigan) contains four thinned SITe 2048 CCDs, plus all readout electronics and a Sparc 10 computer with fast disk and tape (DLT) drive. In combination with the 4-m telescope it is the highest throughput CCD imager in the world, and the only one to use thinned CCDs. Field size is four times that of the PFCCD, at the same scale. The BTC mounts at the 4-m PF, behind the ADC corrector. It contains a fast "focal plane" shutter and a filter bolt which holds four 6×6 inch filters. The CCDs live in a large dewar behind lenses which correct for their curvature. Since the CCDs are mounted in standard packages they are not butted together, and there is a "cross-shaped" band approximately 16 mm wide between CCDs. Normal operation is to take many exposures in a shift-and-stare mode, to produce a large, panoramic picture with no gaps, however this is not essential. BTC will be available to visitors as a supported-instrument from February 1997.

4. 4-m Telescope and NOAO Mosaic Imager

The NOAO MOSAIC imager is a joint project between NOAO-Tucson and CTIO. The detector array will initially consist of eight butted Loral 2K×4K's, these CCDs have not been thinned owing to the very low yield of the foundry runs. The Loral's will be replaced with thinned SITe 2K×4K's in late 1997. Both arrays give a field size of 38' square, at a scale of 0.27 arcsec pixel^{-1}. The imager is read out by four Arcon CCD controllers, read time is 80 seconds for the Loral CCDs which each have only one operative amplifier. A Sparc 20 is used for instrument control, with a very powerful Sparc Ultra for image processing and display. It will be possible to display simultaneously, using two separate screens, a binned version of the full 8K×8K image and any 1K×1K section at full resolution. A large filter track holds 14 filters, and there is a fast "focal plane" shutter, plus two guide cameras. The instrument will be operated at KPNO in 1997 and at CTIO from late 1998.

5. 0.6/0.9-m Curtis Schmidt Telescope

Over the past four years the CCD imaging capability at the Newtonian focus of the Schmidt telescope has been steadily improved. The detector is a STIS (Tek) 2048, which gives a field 68' square at a scale of 2.0 arcsec pixel^{-1}. This is a front-illuminated CCD, coated with Metachrome to provide some UV and blue sensitivity, and is read through all four amplifiers by an Arcon CCD controller. A five position filter bolt holds 4×4 inch filters, both this and the focus are under computer control. A planned detector upgrade (1998?) is to install a "mini-mosaic," consisting of 2 butted 2K×4K SITe CCDs. This would double the QE, increase the field size to 85' square, and provide better sampling of 1.25 arcsec/pixel.

LAMOST PROJECT

Y. CHU[1] AND Y-H. ZHAO[2]
[1]*Center for Astrophysics,*
University of Science and Technology of China,
[2]*Beijing Astronomical Observatory,*
Chinese Academy of Science

1. Introduction

The Large Sky Area Multi-Object Fiber Spectroscopic Telescope (LAMOST) project is a Chinese National Big Scientific Project. It has been recently been approved by The National Committee of Science and National Committee of Planning. LAMOST will get funds from our government and will start at the end of 1996. We expect to finish this project within 7 years. Here we describe the LAMOST project briefly.

The Large Sky Area Multi-Object Fibre Spectroscopy Telescope (LAMOST) is a meridian reflecting Schmidt telescope with a clear aperture of 4-meter, a focal length of 20-meter and a 5° field of view. Using active optics to control its reflecting corrector makes LAMOST a unique astronomical instrument in combining a wide field of view with large aperture. The available large linear size focal plane of 1.75-meter in diameter corresponding to its wide angular field of view of 5° could accommodate up to 4000 fibers, by which the collected light of celestial objects is fed into the spectrographs, promising a very high spectroscopic acquisition rate, say, of order of ten thousands of spectra per night, or ten million spectra every three years. LAMOST will bring Chinese astronomy into the 21st century with a leading role in the fields of large scale and large sample astronomy and astrophysics.

2. Design and Performances

2.1. GENERAL

LAMOST is a meridian reflecting Schmidt telescope with the reflecting corrector alt-azimuthally mounted. It consists of seven subsystem, that is, optical, mechanical, control, optical fibre positioning, spectrographs and CCD detectors, computer network and telescope enclosure.

Optical system: The main spherical mirror of the Schmidt telescope has a curvature of 40 meters and an area of 6.7 × 6.02 meters, consisting of 37 hexagonal segmented mirrors. Its optical axis is fixed in the meridian plane with an angle of 25 degrees to the horizontal plane. The reflecting corrector of the clear aperture of 4 meters is located at the center of the spherical mirror. It consists of 24 hexagonal segmented thin plane mirrors, which are in turn controlled by the active optics system to keep the required aspherical to remove the spherical aberration. The focal length of the optical systeem is 20 meters and its focal ratio of 5 with the field of view of 5 degrees in diameter, thus an angular area of 20 square degrees and a linear size of 1.75 meters in diameter. For more details see the references.

Mechanical structure: A quasi-meridian mounting has been adopted, with a fixed spherical main mirror and an alt-azimuthly mounted reflecting corrector. The focal plane is fixed other than field rotation.

Control System: It is necessary to control the motion of the reflecting corrector in order to track the observed sky area for the 1.5 hours around the meridian and to keep the figure of the corrector as part of an active optics system. Also it controls the precision rotation of the focal plane in order to compensate for the rotation of the field of view.

Optical fibre positioning: There are 4000 optical fibres that require precision placement at the pre-determined positions on the focal plane in order to feed into the slits of the bench spectrographs.

Spectrographs and CCD detectors: They consist of about 20 bench spectrographs with the state of the art CCD detecting systems, including low, intermediate and high spectral resolutions, located in the building underneath the focal plane.

Computer networking: The data acquisition, on-line data handling, data compression, the off-line data procession and so on, are to be connected via a local network.

Telescope enclosure: The design is to take into account the requirement to minimise the degradation of the natural atmospheric seeing by the enclosure and the mirrors, as well as assembly and maintenance.

2.2. TECHNICAL INNOVATIONS

The active optics control applied to the reflecting corrector, provides the required real time correction of the spherical aberration during tracking, leading to a combination of large field of view and large clear aperture. In this design, the optical image quality is better than 2 arcseconds (rms~ 0.7 arcseconds) at the edge of the field of view (the worst case).

TABLE 1. Summary of main technical parameters and capability

Optical:	
Clear aperture	4 meters
Curvature radius	40 meters
Focal length	20 meters
Focal ratio	5
Field of view	5 degrees
Field of view(linear)	1.75 meters
Image quality	80% encircled energy within 1.5–2.0″
Spectrograph:	
Spectral resolution	10Å at 5250Å
Spectral range	3900–9000Å
CCD QE	50%
CCD dark current	0.04 electrons sec^{-1} pixel^{-1}
CCD readout noise	5 electrons pixel^{-1}
Fibre Diameter	3.4 arcseconds (330 micrometers)
Fibre Cross section	8.55 square arcseconds
Throughput:	
Loss of central obstruction	0.2
Loss of refelection	0.3
Loss of optical fibre	0.2
Loss of spectrograph	0.75
Atmospheric transmission	0.8 at 5250Å
Total throughput for object observation	0.16
Total throughput for sky observation	0.20
Observing:	
Seeing	2 arcseconds
Additional ground seeing	1–1.5 arcseconds
Sky brightness	21 magnitude arcsec^{-2} at 5250Å
Spectroscopic hours per year	2000 hours
Integration time per observation	5400 seconds
Sky coverage	24,000 square degrees $-10° < \delta < +90°$
Limiting magnitudes	20.5 magnitude at visible
Spectra acquisition rate	10^7 per 3 years (low and int. res.)
Predicted Galaxy observations	2×10^7
Predicted QSO observations	2×10^6
Predicted Stellar observations	$> 5 \times 10^7$

The large linear size of the focal plane obtained can easily accommodate 4000 optical fibres, which is unprecedented.

The segmented mirror concept has been applied to the spherical mirror and the corrector to ease the difficulties in manufacturing large mirrors and lower the cost whilst preventing gravity deformation.

The meridian mounting has been adopted to solve the problem of the layout of long tube (long focal length) corresponding to a large Schmidt telescope. Thus a long focal length is obtained with a large clear aperture, giving a higher spectroscopic performance index $f\Omega$, where f is the focal length and Ω is the solid angle of the field of view.

The alt-azimuth mounting has the other advantages, including the fixed focal plane, the fixed optical fibre positioning system and the bench spectrographs, as well as easy accessibility to the optical fibre positioning assembly to change the observed configuration of the objects.

References

S.-G. Wang, D.-Q. Su, Y.-Q. Chu, Y.-N. Wang and X.-Q. Cui, (1995) "A Special Purpose Schmidt Telescope for Multi-Fibre Astronomical Spetroscopy," in "Wide Field Spectroscopy and the Distant Universe," ed. by S.J. Maddox and A. Aragon-Sabamanca, p. 40, World Scientific Co. Pte. Ltd.

S.-G. Wang, D.-Q. Su, Y.-Q. Chu,X.-Q. Cui and Y.-N. Wang, Applied Optics, 1996, vol.35, No.25, p.5155.

THE AAT'S 2-DEGREE FIELD PROJECT

K. TAYLOR, R.D. CANNON AND Q.A. PARKER
Anglo-Australian Observatory, Australia.

1. Introduction

The "Two-degree Field" (2dF) project at the Anglo-Australian Observatory (AAO) gives the 3.9m Anglo-Australian Telescope (AAT) a field of view two degrees in diameter at the prime focus, equipped with 400 optical fibres for multi-object spectroscopy. The basic components of 2dF are the corrector lens optics, the robot which positions the fibres and a pair of spectrographs. All these are mounted on a new 'top end ring' so that the whole assembly can be easily put on and off the telescope. Here we will give an update on the status of 2dF, highlighting features which have changed or been developed since earlier reports.

2. Components of 2dF

2.1. THE CORRECTOR LENS ASSEMBLY

The design of the corrector lens assembly has been described by Jones (1994) and its performance by Taylor and Gray (1994). It is a four-element design incorporating an atmospheric dispersion compensator (ADC), which is essential for spectroscopy over wide wavelength ranges with fibres which are only two arcseconds in diameter. The 2dF optics were fabricated by Contraves (USA) from glass made by Ohara (Japan) and delivered in June 1993. Direct imaging tests, using a temporary photographic plate holder, showed that the lens assembly met its specification of delivering sub-arcsecond images over the full field, and that the ADC gave full correction at all angles up to 65° from the zenith.

2.2. THE TWO POSITIONER GANTRIES AND THE GRIPPER

Two large X,Y gantries comprise the heart of 2dF. The upper one carries the gripper head which can *pick and place* the magnetic buttons at the ends of the optical fibres. The lower gantry carries a Focal Plane Imager (FPI), based on a Photometrics TV system, and another small TV for checking the location of fibres and the fibre plate. The two gantries, together with the tumbler mechanism, were designed and built as an in-house project in the AAO's Epping Laboratories; the gripper was manufactured to AAO design at the University of Durham. An important innovation, another design change implemented after 1991, was to use linear electric motors and encoders, rather than ballscrews for the X,Y motions of the gantries; this was the only way to get the required speed and accuracy.

The lower gantry was tested on the telescope in June 1994; the upper gantry and gripper assembly were first installed about a year later. It soon became clear that both met their specification, in terms of accuracy of positioning and freedom from flexure. Attaining the maximum speed of operation, especially for the positioner, proved more difficult. By mid-1996 the positioner was running at about a fifth of its top design speed, partly because many diagnostic tests were being run and partly out of caution. Accuracy and reliability were more important than speed at this stage.

2.3. THE SPECTROGRAPHS

A general arrangement drawing for the spectrographs is given by Gray it et al. (1993) and by Taylor (1994). The optical system consists of an off-axis Maksutov collimator, a grating and an f/1 cryogenic Schmidt camera (Jones 1994). The detectors are thinned Tektronix 1024^2 CCDs; this means that at most 200 fibres can be fed to one spectrograph and still be readily separated, hence the need for two spectrographs. There is an interchange mechanism at the entrance to each spectrograph, to enable switching between the separate sets of 200 fibres from the two field plates; this operates synchronously with the tumbler mechanism. Banks of LEDs in the spectrographs provide back-illumination during the fibre positioning phase, so that the precise location of each fibre can be determined. These must be very well shielded so that no light escapes to contaminate the neighbouring sets of fibres being used to collect data. Cooling of the cameras is by closed-cycle liquid helium to avoid having to top-up dewars.

The design and fabrication of the spectrographs was done mostly by AAO staff in Coonabarabran. The crucial aspheric plates for the cameras were completed by Gabe Bloxham at Mount Stromlo Observatory, following the untimely death of Bill James (James Optics, Melbourne). The first laboratory spectra were obtained with one spectrograph in May 1996. Good focus was obtained over the field, spectra from the individual fibres could be easily separated and scattered light was at a satisfactorily low level.

2.4. INPUT POSITION REQUIREMENTS

Given that the 2dF fibres are $2''$ in diameter, the overall positioning has to be accurate to better than $0.5''$ to avoid significant loss of light. This is a *total* error, including systematic errors in the Schmidt plates or other original source, distortions in the 2dF optics, the link between faint targets and bright guide stars, and random errors. The implication is that the input positions must have relative errors $\leq 0.15''$ (rms), although the absolute zero-point is not so critical. This is significantly better than has been achieved routinely in most wide-field astrometry; *internal* measuring errors are often at the $0.1''$ level, but final output positions are more often good to only $0.5''$.

2.5. DATA REDUCTION

Finally, it is essential to have an efficient way of reducing the data from 2dF, which will be capable of delivering several thousand spectra per night. The novel way in which this is being done, making maximum use of prior knowledge about the system, is described by Taylor it et al. (1996). The intention is that astronomers will leave the AAO with data already fully reduced.

3. Science with 2dF

The largest project planned for 2dF, which provided the primary motivation for building the instrument, is to determine the large-scale 3-dimensional structure of the local universe (to a mean redshift of about 0.1) by measuring the redshifts of a quarter of a million galaxies. There will be many other galaxy redshift projects, including deeper surveys in selected fields and studies of clusters of galaxies. A second major survey, to be carried out in parallel with the galaxy redshift survey, is to obtain spectra for 30,000 quasar candidates to explore the universe at high z. Another obvious application is to the structure and evolution of the Galaxy, and of the Magellanic Clouds, where the kinematics and chemical compositions of large samples of stars will be obtained. Many smaller "one-off" projects have been proposed, involving almost all branches of astronomy.

Most of these projects exploit the huge quantitative gain that 2dF provides but 2dF will also permit *qualitatively* new astronomy to be done. For example, to obtain good enough spectra to determine abundances for faint main sequence stars in globular clusters requires several nights of 4m telescope time. Doing such a project one star at a time was never feasible, but devoting a week of time to observing 400 such stars simultaneously may be much more reasonable.

4. Current Status of 2dF

The formal opening ceremony for 2dF took place at the AAT on 20 November 1995. At that stage all of the major components had been installed but 2dF had not yet been run as a complete system.

The first useful astronomical data were obtained in June 1996 (see the July 1996 *AAO Newsletter*). They demonstrated conclusively that 2dF can obtain spectra of the expected high quality, that the system throughput is about double that of the older Cassegrain fibre instruments, and that the on-line data reduction system works.

To commission 2dF fully will take another six months to a year, especially to get the operation up to full speed while maintaining reliability, which is essential for the highest-ranked redshift survey projects. The commissioning team will take samples of data for the major survey projects and do some smaller astronomical projects in "service observing" mode during this phase.

Overall, the project is about two years behind the original schedule and has a total external cost approaching $2.5M. The true total cost of the project, including all internal AAO staff costs and overheads, has not been calculated carefully but must be in the vicinity of $5M. The overruns are embarrassing but not too surprising, given the complexity of 2dF and the number of innovative features it contains. Had the resources needed been fully appreciated at the start we might not have embarked on the project, but what is certain is that the AAT has an excellent new instrument which gives it a unique capability to tackle some of the most important current astronomical problems.

References

Gray, P. M., Taylor, K., Parry, I. R., Lewis, I. J. and Sharples, R. M. 1993, Fibre Optics in Astronomy II, ed. P. M. Gray, Astron.Soc.Pacific Conference Series 37 145.

Jones, D. J. A. 1993, Fibre Optics in Astronomy II, ed. P. M. Gray, Astron.Soc.Pacific Conference Series 37, 355.

Jones, D. J. A. 1994, App.Optics 33 7362.

Taylor, K. and Gray, P. M. 1994, Instrumentation in Astronomy - VIII, eds. D. L. Crawford and E. R. Craine, Proc. SPIE 2198 136.

Taylor, K. 1994, Wide Field Spectroscopy of the Distant Universe: The 35th Herstmonceux Conference, eds. S. J. Maddox and A. Aragón-Salamanca, pub. World Scientific, p.15.

Taylor, K., Bailey, J. A., Wilkins, T., Shortridge, K. and Glazebrook, K. 1996, Astronomical Data Analysis Software and Systems V, eds. G. H. Jacoby and J. Barnes, Astron.Soc.Pacific Conference Series 101, 195.

SIMULTANEOUS SPECTRA OF A COMPLETE SAMPLE OF SOURCES FROM THE PMN SURVEY

Preliminary results

M.G. MINGALIEV, A.M. BOTASHEV, and V.A. STOLYAROV
Special Astrophysical Observatory, Russia

1. Introduction

The broad spectral coverage of radio sources including high frequency data are desirable for a number of reasons. Over the last decade, researchers have identified a new class of very high-redshift objects that are more closely related to normal galaxies than to quasars. Surveys with radio telescopes have been instrumental in finding these extremely distant galaxies, which are barely visible at optical wavelengths. The selection of candidates with steep radio spectra has proven a particularly effective means for finding galaxies with redshifts greater than z = 2. Secondly, they uncover flat spectrum sources which are likely to be compact and useful as radio astrometric standards and as well as candidates for further interferometric observations. Third, they provide information about physics and, in particular, about the high-energy particles responsible for radio emission. And last but not least, simultaneous observations at several wavelengths will exclude variability of sources in interpretation of their spectra. Here we present the preliminary results of an ongoing investigation of a complete sample of radio sources from the PMN Survey.

2. Selection of Sample and Observations

The adopted criteria for selection of the sample from the PMN Survey (Griffith *et al.* 1994) were:

- flux density $\geq$ 200 mJy at PMN Survey frequency (4.8 GHz)
- $00^h \leq$ RA $\leq 24^h$; $-21° \leq$ Decl. $\leq -17°$
- galactic latitude $\mid b \mid > 10°$ to exclude the galactic plane.

In all, we have selected 262 objects.

The observations were made at 2.7 cm, 3.9 cm, 7.6 cm, 13 cm, and 31 cm during three sets in July, August and December 1995 by transit way mode at the North sector of the RATAN-600 radio telescope. Each source was observed 5–8 times. As flux and coordinate standard we have used the strong radio source 1245-197. The assumed flux densities, Right Ascension and Declination are following: 1.24 Jy, 1.75 Jy, 3.0 Jy, 4.0 Jy, and 6.3 Jy at 2.7 cm, 3.9 cm, 7.6 cm, 13 cm, and 31 cm respectively; $RA_{1950.0} = 12^h 45^m 45\overset{s}{.}22$, $Dec_{1950.0} = -19°42'57\overset{''}{.}5$.

The accuracy of flux density measurements at 2.7 cm was mostly concern with signal-to-noise ratio. In general, the accuracy at the best sensitivity wavelength (7.6 cm) is 2–3% or better. The accuracy at 31 cm was reduced by strong interference from traffic, satellites and military radar systems. The accuracy for each source is given in Table.

In this paper we present the first result of radio observations of 174 sources from the above mentioned sample. To complete the radio observation the last set is scheduled on October 1996.

3. Results

Due to the good sensitivity and resolving power in RA at 7.6 cm we have improved the RA of these sources too. For most sources (~70%) the accuracy in RA equal or better than $0\overset{s}{.}1$, and $0\overset{s}{.}2$ for the remainder. This result will help us for further optical identification of these objects. The RA determined at this program were compared with positions from the PMN and Texas (Douglus 1987) Surveys. On Figures 1(a, b) one can see histograms of $\Delta\alpha/s$ normalized by the quadratic sums of the quoted rms uncertainties s (the RATAN-600 minus the PMN and the Texas positions respectively).

The main results of our program of observations are available by ftp

- site big1.sao.ru
- figure /data/ratan/mingaliev_spectra_fig
- table /data/ratan/mingaliev_spectra_table.

For determining spectra of this sources in addition to our data we have used the flux densities from the Texas 0.365 GHz Sky Survey (Douglus 1987). A match between sources in our list (after improving the RA during our observations) and Texas Survey found counterparts for ~80% of them.

4. Conclusion

For 174 radio sources from the PMN Tropical Survey: the flux densities at 2.7 cm, 3.9 cm, 7.6 cm, 13 cm, and 31 cm were measured; simultaneous

SPECTRA OF SOURCES FROM THE PMN SURVEY

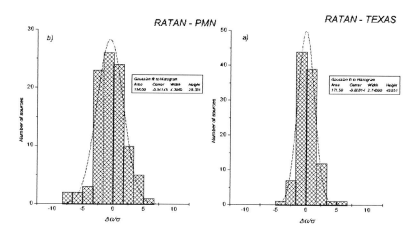

Figure 1.

spectra in wavelength region 2.7–31 cm were determined; and the accuracy in RA was improved by 5–7 times.

Further steps are follows: to complete observations of selected sample optical identification with The Digitized Sky Survey (Lasker *et al.* 1990); identification with other wavebands; to select variable sources comprising the simultaneous spectra obtained by this program with other data from other telescopes observed on different epochs.

Acknowledgements

This research is partially supported by the Russian Foundation for Basic Research Project No 95-02-04972 and INTAS Project No 94-4010.

References

Douglus J.N. 1987, Bull.Am.Astron.Soc., 19, 1048
Griffith, M.R., Wright, A.E., Burke, B.F., and Ekers, R.D. 1994, Astrophys.J.Suppl., 90, 179
Lasker, B.M., Sturch, C.R., McLean, B.J., Russell, J.L., Jenkner, H., and Shara M. 1990 Astron.J., 99, 2019

AROPS : THE ASIAGO RED OBJECTIVE-PRISM SURVEY

U. MUNARI AND R. PASSUELLO
Astronomical Observatories of Padova-Asiago, Italy

1. Background

AROPS is an objective-prism survey conducted at high galactic latitudes with the 67/92 cm (f=205 cm) Schmidt telescope of the Asiago Astrophysical Observatory (Italy). The basic technical data are:

- TP4415 high resolution plates, 20×20 cm, treated in forming gas (H_2+N_2) and covering 24 deg^2 on sky
- 4° UBK7 objective prism, for a dispersion of 1490 Åmm^{-1} at NaI D
- blue side cut-off at 4980 Å using a 2mm GG14 filter, for a recorded λ range 5000–7000 Å
- untrailed spectra to record $V \sim 16.5$ mag stars with 40 min exposures

Two main criteria have guided the selection of the AROPS fields:

- at least 15 direct imaging plates of a given field (in any photometric band, at any exposure date since 1956) must exist in the Asiago Schmidt Plate Archive. They will help in positional measurements and will also provide information about variability and colors
- the fields must be at $|b| \geq 20°$ and widely distributed in galactic latitude and longitude

151 fields were originally selected (for a total area of 3650 deg^2). An additional $\sim$100 fields (2400 deg^2) have been later added on with relaxed adherence to the above selection criteria.

2. Status Report

High quality plates have been obtained for 153 fields (as of Aug 15, 1996), with a plate rejection rate of 1 every 4 exposed. All the plates have gone through a quality assessment which has provided:

- the accurate plate center

- the magnitude of the faintest measurable stars
- star counts over five regions of $10' \times 10'$ each (0.139 deg^2 total area) on fixed positions on the plate
- the evaluation of contrast, focus, PSF, background fog level and uniformity over the whole plate field

The measured *mean* limiting magnitude for detection of stellar continuum is $V = 16$, for an *average* of 20,000 stars recorded on each plate.

3. Goals

AROPS is intended to search for, classify, catalogate and study in term of galactic distribution:

- cool stars with molecular spectra (M, S, C types)
- emission line objects of any kind
- optical counter-parts of sources from various satellite survey catalogues

4. World Wide Web

More information on AROPS, an updated status report and the list of fields/plates can be found on the *Web* at the address:
  http://www.fiz.uni-lj.si/astro/alarm/arops00/arops00.html

EUV SOURCES AND TRANSIENTS DETECTED BY THE ALEXIS SATELLITE

D. ROUSSEL-DUPRÉ[1], T. PFAFMAN[2],
J. BLOCH[1] AND J. THEILER[1]
[1]*Los Alamos National Laboratory*
[2]*Stanford University*

1. Introduction

Because ALEXIS is a spinning satellite, it is an ideal platform with which to study the time variability of the EUV cosmos. The main thrusts of this effort are to 1) detect EUV sources at known and unknown locations, 2) provide notification of transients in near real time to enable immediate follow-up from other observatories, and 3) create a time history of observed sources for comparison with previously published catalogs to aid in determining long duration variability from EUV sources.

2. Real Time Point Source Determination

Currently, the 12, 24 and 48 hour accumulation sky maps are searched as a routine part of the automated post-pass data processing. Eight on-line databases (Einstein Slew, EUVE, ROSAT-2RE, Yale Bright Star, Downes CV, TOAD CV, Gliese, and White Dwarf catalogs) are then automatically cross correlated for likely counterparts. Point source locations, source fit parameters and potential counterpart information are then sent via e-mail to the science team. Summary tables with this pertinent information and maps for these position are accessible at
    http://nis-www.lanl.gov/nis-projects/alexis/
to the satellite team for evaluation. A pager alerts the on-duty scientist to significant transient detections found in the most recent data. The web page also contains a list of locations for 1) 40 sources with good potential for detection by ALEXIS spanning a range of system types (*i.e.*, CV's, RS CVn's, ALGOL systems, flare stars, *etc.*), 2) 12 EUV bright sources and

3) associated near-by blank fields as control locations, the count rate if it exceeds a certain threshold in a one degree box and other information about the detection are kept in a database. This can then be searched on-line to provide information about possible long term variability from these systems as well as use the blank field locations to provide a quick-and-dirty idea about the detection statistics. In addition to the on-line catalogs, we also use SIMBAD to search for likely candidates and the WEB SKYVIEW page to compare with other catalogs and other observations from various experiments. If a source passes all of the credibility tests, we then send out e-mail alerts to variable star observers around the world requesting ground based observations to help identify the optical counterpart.

3. Types of Point Sources

ALEXIS observes both steady sources (mostly bright white dwarfs) and transients. The transient or variable sources observed to date are 1) cataclysmic variables: VW Hyi, U Gem, AR UMa, 2) flare stars: Alpha Cen, RE J2353-702 and 3) unknown short duration transients. The cataclysmic variables show significant variability between outbursts both in overall brightness and spectral character. In addition, because of the continuous monitoring capability of ALEXIS, 2 EUV outbursts or "flashes" have been observed four days in advance of the optical turn-on for VW Hyi. Forty high probability unidentified short duration transients have been observed in an 11 month period. These unknown transients have durations ranging from 4–48 hours, and have no clear optical counterpart. Four of these systems have been observed by EUVE for as long as 24 hours without detection, although by the time EUVE was on target, the source was no longer bright in the ALEXIS instruments.

4. Summary

Because the ALEXIS attitude reconstruction algorithm has been evolving and changing through December 1995, the sky maps that are currently available and that have been searched for point sources have not been produced systematically with the same software. Therefore, it is difficult to ascertain the ALEXIS transient event rate, although the best estimate is about one per week. In the future, we will be reprocessing all archived data which will allow us to systematically determine the noise threshold level. Then the detailed source catalog including variability information can be generated and a better determination of the transient rate observed in each energy channel.

Acknowledgements

These results represent the dedicated efforts of C. Little, M. Kennison, K. Ramsey, S. Ryan, S. Fletcher, B. Dunne, A. McNeil, P. Patel, P. Ratzlaff, S. Stem, and J. Wren. This work was supported by the Department of Energy.

THE UIT SURVEY OF THE ULTRAVIOLET SKY BACKGROUND

W.H. WALLER[1,2] AND T.P. Stecher[2]
[1]*Hughes STX*
[2]*NASA Goddard Space Flight Center*

AND

The Ultraviolet Imaging Telescope Science Team
http://fondue.gsfc.nasa.gov/UIT/UIT_HomePage.html

1. Introduction

When viewed from above the Earth's atmosphere, the nighttime ultraviolet sky background is profoundly dark. Recent measurements indicate that the diffuse UV sky background is up to 100 times (5 magnitudes) fainter than the equivalent visible background as measured from the ground. Much of this difference can be attributed to the Sun's lower emissivity at UV wavelengths, leading to reduced irradiation of and scattering by the interplanetary dust. Because the resulting Zodiacal light is so much weaker in the UV, a comprehensive characterization of the UV sky can yield important information on the more distant Galactic and extragalactic backgrounds and, ultimately, on their material origins (see Brosch, this volume, p. 57).

Despite recent concerted efforts, the strength and spatial distribution of the UV background remain controversial topics. Estimates range from a few hundred photons s^{-1} cm^{-2} sr^{-1} $Å^{-1}$ ($\sim$27 mag $arcsec^{-2}$) with no obvious spatial distribution to several thousand "photon units" with a strong gradient toward the Galactic midplane (cf. Bowyer 1991; Henry 1991). Models of the stellar UV radiation field depend critically on the distribution and scattering properties of the interstellar dust, yielding significantly different predictions as a function of grain albedo and phase function (cf. Murthy & Henry 1995). Herein, we summarize recent results from an analysis of UV images obtained by the Ultraviolet Imaging Telescope (Waller *et al.* 1995).

2. The UIT Experiment

As part of the December 1990 *Astro-1* Spacelab mission on the Space Shuttle *Columbia*, the Ultraviolet Imaging Telescope (UIT) obtained 361 NUV (~2500 Å) images and 460 (~1500 Å) images of the sky in 66 separate pointings—each with a 40 arcmin field of view and a resolution of ~2 arcsec. The resulting images enable discrimination between the diffuse sky and discrete objects (stars, nebulae, and galaxies with $m_{UV} < 20$ mags) for true background measurements. Another 78 target fields were imaged in the FUV during the March 1995 *Astro-2* mission. Processing and calibration of the *Astro-2*/UIT images is nearing completion.

The UIT shared the Spacelab's Instrument Pointing System with the Hopkins Ultraviolet Telescope (HUT) and the Wisconsin Ultraviolet Photo-Polarimeter Experiment (WUPPE). Simultaneous spectral observations by the HUT could be compared with UIT's FUV backgrounds, thereby constraining the effects of atmospheric nightglow on UIT's FUV backgrounds.

2.1. EFFECTS OF AIRGLOW

Photometric analysis of the UIT images has yielded positive detections of FUV and NUV backgrounds in both the daytime and nighttime skies. The FUV backgrounds are dominated by the effects of airglow—even at night—correlating with the OI ($\lambda\lambda 1304, 1356$) line emission measured by HUT. An excess background of roughly 700 photon units (~26 mag arcsec^{-2}) indicates Galactic and extragalactic contributions. Total background intensities similar to this excess are found in the deep FUV-B1 images that were obtained during the March 1995 *Astro-2* mission—a time when the Sun was at minimum activity, producing airglow levels ~3 times lower than experienced during the *Astro-1* mission.

2.2. EFFECTS OF STRAY LIGHT

Stray light from the Sun dominates the daytime NUV backgrounds, while stray light from UV-bright stars (just beyond the field of view) can occasionally produce an appreciable effect at night. The resulting backgrounds can be fit by a single "Baffle Function." The Moon's deviation from this fit indicates that stray Moonlight is a negligible contributor to the measured backgrounds.

2.3. ZODIACAL BACKGROUNDS

Away from the Galactic plane, the nighttime NUV backgrounds are correlated at the 98% confidence level with the Zodiacal UV light predicted from visible-light measurements—both backgrounds decreasing with ecliptic latitude. These relations are best fit with a NUV/VIS "color" of 0.5 ± 0.2 (where the solar emissivity spectrum gives a color of unity) and with an extrasolar component at high galactic latitude of 300 ± 300 photon units.

2.4. GALACTIC BACKGROUNDS

Both the FUV and NUV intensities show strong dependences on Galactic longitude and latitude, reaching the highest levels in diffuse regions next to structured nebulosity (*e.g.*, Cygnus, Vela, and Gum fields). The blue ($FUV-NUV$) colors at these high levels ($\sim 10^4$ photon units) are consistent with scattering of ambient OB starlight by galactic dust. The location of the dust is uncertain, but is probably associated with the adjoining nebulosity.

2.5. EXTRA-GALACTIC BACKGROUNDS

The nighttime NUV intensities—after subtraction of a Zodiacal component with a NUV/Vis color of 0.5 ± 0.2—yield residual intensities that correlate with FIR measurements of the corresponding fields. Extrapolation of this NUV-FIR relation to negligible FIR intensities indicates an extragalactic NUV emission component of 200 ± 100 photon units. Such an estimate for the "cosmic" UV background supports the low intensities that have been proposed in the debate over the strength and structure of the UV background (cf. Henry 1991; Bowyer 1991; Brosch, this volume, p. 57).

3. Conclusions

By imaging in selected UV "windows" that do not include the OI airglow emission or significant Zodiacal light, one can reduce the sky background to 300 photon units (27 mag arcsec^{-2}) or fainter. Such a dark sky is ideal for pursuing studies of the dim outer regions of nearby galaxies, low-surface-brightness galaxies within the local supercluster, as well as faint primeval galaxies much farther away (O'Connell 1987). Ultraviolet imaging experiments such as the wide-field UIT and narrow-field HST/WFPC2 and HST/FOC cameras can benefit from the darker skies and subsequently enhanced contrasts that are available in these UV "windows." Future surveys with more sensitive UV detectors than are currently available should be able to more fully characterize and exploit the dark-sky advantage that UV imaging affords (cf. Brosch, this volume, p. 57).

Acknowledgements

UIT research is funded through the Spacelab Office at NASA Headquarters under Project number 440-51.

References

Bowyer, S. 1991, Ann.Rev.Astron.Astrophys., 29, 59
Brosch, N., this volume, 57
Henry, R. C. 1991, Ann.Rev.Astron.Astrophys., 29, 89
Murthy, J., & Henry, R. C. 1995, Astrophys.J., 448, 848
O'Connell, R. W. 1987, Astron.J., 94, 876
Waller, W. H., *et al.* 1995, Astron.J., 110, 1255

Part 3. The Interstellar Medium

PHYSICAL PROCESSES IN THE LARGE SCALE ISM FROM DUST OBSERVATIONS

F. BOULANGER
Institut d'Astrophysique Spatiale, Université Paris, France

Abstract. Over the last two decades observations of dust emission in the infrared have played an important role in the development of research on the interstellar medium. The study of the spectral energy distribution has led to the discovery of small dust particles including the large aromatic molecules (PAHs). Infrared sky images have been used to study the structure of interstellar matter, the evolution of dust within the interstellar medium and the star formation efficiency of interstellar clouds.

1. Introduction

In this conference we have been given a panorama of sky-surveys available over the whole electromagnetic spectrum. Most of these have contributed to our understanding of the interstellar medium (ISM) but biased by my own research experience, I review physical processes in the ISM from the perspective opened by the analysis of dust emission in infrared sky images. Since the interpretation of these observations has made use of data at other wavelengths, this topic illustrates well the benefits of data intercomparison across the electromagnetic spectrum. I will mostly refer to observations obtained with the Infrared Astronomy Satellite (IRAS), the Cosmic Background Explorer (COBE) and the Infrared Space Observatory (ISO). But data obtained with other instruments or space projects like the japanese satellite IRTS and the Air-Force project MSX (Price, this volume, p. 115) are also relevant. A general description of surveys in the infrared is given by Beichman in this volume (p. 27).

In the $100\mu m$ IRAS images, diffuse emission with a filamentary structure reminiscent of cirrus clouds is present over most of the sky. In this review I will show that cirrus observations have provided new insights about many key processes in the evolution of interstellar matter. In particular, the composition of dust (Section 2) and its evolution within the interstellar medium, the small-scale structure of clouds and the chemical transition between atomic and molecular gas (Section 3). A key motivation of research on interstellar matter is to understand physical parameters controlling the onset and the efficiency of star formation. Infrared Surveys in the mid and far infrared provide an important statistical information on star formation which I also discuss in the paper (Section 4).

2. Dust Composition

2.1. EMISSION SPECTRUM

Infrared observations of ordinary interstellar matter, located far away from any peculiar heating source, have been mostly obtained from space with cooled telescopes because the emission is faint and extended. These observations, in particular those of cirrus clouds optically thin to Galactic star light, constitute a reference for the study of dust emission properties. In Figure 1, I have gathered data providing the complete spectral energy distribution of dust in the Solar Neighborhood heated by the local mean interstellar radiation field. The emission power which is distributed over a large range of wavelengths from the near-infrared to millimeter wavelengths, originates from particles emitting over a wide range of temperatures. To account for the spread in temperatures it is necessary to extend the size distribution of dust grains down to molecules with a few tens of atoms (Désert et al. 1990, Dwek et al. 1997).

At long wavelengths the emission spectrum from large grains is well fit by a single Planck curve with an emissivity proportional to ν^2 and $T_d = 17.5K$ (Boulanger et al. 1996a). The dust emissivity,

$\tau_\lambda/N_H = 1.0\,10^{-25}\,(\lambda/250\mu m)^{-2} cm^2$ for $\lambda > 250\mu m$, is remarkably close to the value obtained by Draine and Lee (1984) for a mixture of compact graphite and silicate grains and is much smaller than values predicted for porous or fractal grains (Wright 1987, Mathis and Whiffen 1989). At wavelengths shorter than $100\mu m$ the observed emission is well in excess of the single temperature fit. This excess emission is attributed to particles small enough to be heated to temperatures much higher than the equilibrium temperature of the large grains by the stochastic absorption of photons. The presence of such small particles was first advocated on the basis of near and mid-IR observations of the Galactic plane and reflection nebulae (Sellgren et al. 1983). By providing complete sky maps of the near and mid-IR emission from interstellar matter, IRAS and DIRBE have demonstrated the ubiquity and importance of these small particles. Léger and

Puget (1984) associated the emission from small particles with the presence of a well defined set of emission bands at 3.3, 6.2, 7.7, 8.6 and 11.3 μm in the spectra of a wide range of celestial objects. These emission bands being characteristic of C-C and C-H bonds in aromatic molecules, they proposed that the smallest particles are large polycyclic aromatic hydrocarbons (PAHs). It is now experimentally demonstrated that the 12μm diffuse emission measured by IRAS is associated with PAHs. The presence of the 3.3 and 6.2 μm emission features in the spectrum of the diffuse emission from the Galaxy was first evidenced by the Arome balloon experiment (Giard *et al.* 1994, Ristorcelli *et al.* 1994). Spectra from the Galactic plane emission obtained recently with the ISO and IRTS satellites show the presence of the full set of PAHs features (Mattila *et al.* 1996, Onaka *et al.* 1996, Tanaka *et al.* 1996). The instruments on board of ISO are also sufficiently sensitive to detect the emission bands in the emission from high latitude cirrus clouds (Boulanger *et al.* 1996b and 1997). The presence of all the known emission features from aromatic hydrocarbon in the spectrum of the Galaxy and nearby clouds strongly supports the existence of large aromatic molecules in the most general physical conditions in interstellar clouds. The abundance of Carbon in PAHS with less than a few 100 atoms is estimated to represent 15% of the total cosmic abundance of carbon.

The temperature fluctuations associated with the quantization of the heating radiation significantly affects the emission spectrum of grains up to sizes of 50 Å (Draine and Andersson 1985). Particles in the intermediate size range between PAHs and large dust grains are often referred to as very small grains (VSGs). For the local interstellar radiation field, VSGs emission dominates the emission from other dust components from 15 to 60μm. Since no emission feature has yet been observed in this wavelength range the chemical nature of VSGs is unknown. However, intermediate size grains must be related to the PAHs and the large grain. It is thus speculated that the VSGs could be small coal like particles (carbon clusters with variable degrees of hydrogenation and mixed hybridation, Papoular *et al.* 1996) which can be seen as aggregates of PAH molecules. Another possibility is that the size distribution of silicate grains extends down to small sizes. Infrared spectroscopy is a powerful mean to identify dust composition. The analysis of the spectroscopic information obtained with ISO should thus lead to significant advances in our knowledge of the chemical composition of dust.

2.2. EVOLUTION OF DUST WITHIN THE ISM

Data on scattering, absorption and emission from dust grains provide independent evidence for the evolution of interstellar dust from place to place in

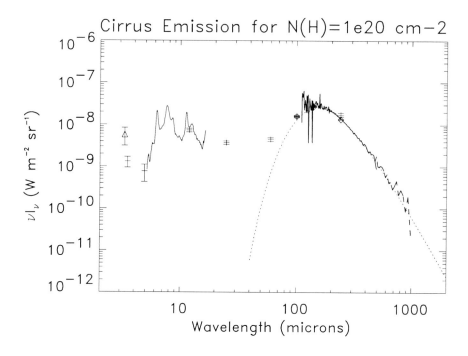

Figure 1. The dust emission spectrum from the near-IR to millimeter wavelengths. This figure combines (1) the 3.3μm emission of cirrus associated with the molecular ring (Giard *et al.* 1994), (2) an ISOCAM spectrum between 5 and 16 μm measured at the northern edge of the dense cloud in Ophiuchus (Boulanger *et al.* 1996b), (3) measurements of the high latitude cirrus emission from the COBE satellite with the Diffuse Infrared Background Experiment (DIRBE) (Bernard *et al.* 1996) and the Far Infrared Absolute Spectrometer (FIRAS) (Boulanger *et al.* 1996a).

the interstellar medium. The interstellar extinction curve is known to exhibit considerable variations in the UV and optical from one line of sight to another. An important conclusion of the analysis of the IRAS and COBE all-sky images is that there is substantial evidence for changes in the dust size distribution and emission properties within clouds of moderate overall opacity ($A_V < 2$mag) to stellar light (Weiland *et al.* 1986, Boulanger *et al.* 1990, Laureijs *et al.* 1991). The emission from PAHs and very small grains in the 12, 25 and 60μm bands of IRAS, relative to that from large dust grains at 100μm, varies by one order of magnitude from cloud to cloud and within clouds. Models of the IR emission from clouds show that these color variations trace changes in the abundance of PAHs and very small particles (Bernard *et al.* 1993). The spectral studies made with the IRAS data have been extended to the whole dust emission spectrum with the COBE data. Lagache *et al.* (1997) have shown that, where the abundance of small dust particles is low, the equilibrium temperature of large grains is lower than the mean cirrus value. This is an interesting result showing that the change in abundance of very small grains is correlated with a change in the temperature of large grains. With an angular resolution of a few arcminutes observations made with the balloon-borne instrument Pronaos have led to the detection of dense condensations in nearby clouds where the dust temperatures is as low as 12 K (Ristorcelli *et al.* 1997).

The infrared color variations have been correlated with changes in the UV part of the extinction curve (Boulanger *et al.* 1994). However, the amplitude of none of the spectral features in the extinction curve is linearly proportional to the abundance of the mid-IR emitters. This means that none of these spectral features is specific to the mid-IR emitters which is not unexpected since the size range of particles susceptible to contribute to spectral features in the UV extinction, up to sizes of a few 100 Å, is much larger than that of mid-IR emitters (size < 30 Å).

Two processes are assumed to play an important role in the evolution of dust within the interstellar medium. First, model calculations suggest that dust in the low density components of the ISM is efficiently destroyed by supernovae shock waves. The efficiency of this process is demonstrated by the large variations of depletions between the different phases of the ISM (*e.g.*, Sofia and Cardelli 1994). Recent theoretical work (Jones *et al.* 1996) and observations (Zagury *et al.* 1997) suggest that the shattering of grains in grain-grain collisions is an effective process to produce small particles. The effect of this process on the grain size distribution has only been investigated in fast shocks (100 km/s and higher) traveling in the low density intercloud phase of the ISM Jones *et al.* estimate the velocity threshold for grain shattering in grain-grain collisions to be as low as 1 km/s. Since this velocity is typical of the relative motions between dust particles induced by turbulent motions (Falgarone and Puget 1995) grain shattering could also be an effective process to form small particles within interstellar clouds. Second, investigations based on dust absorption features seen in the infrared indicate that dust grains accrete gas molecules in dense clouds and that icy mantles are able to evolve chemically towards refractory compounds through photolysis. It is most likely that this evolution process also affects the small particles which can grow in size by mutual coagulation or can stick on large grains. Serra *et al.* (1992) and Marty *et al.* (1994) have suggested that in dense clouds PAH molecules could be bound together by metals and thus be in the form of organometallic complexes. This is a possible explanation of the observed correlation between the cold component of the infrared emission (the component with a low abundance of small particles) with the ^{13}CO emission from dense molecular gas (Laureijs *et al.* 1991, Abergel *et al.* 1994).

3. Infrared Cirrus

3.1. INFRARED EMISSION TRACER OF INTERSTELLAR GAS

Among the various observational means of imaging local interstellar matter: optical images (Guhathakurta and Cutri 1994), 21 cm emission from H I (see Kulkarni and Heiles 1987), and CO surveys at high latitude (Magnani *et al.* 1996), the IRAS images remain unique in providing a complete view at the intricate structure of this matter over a wide range of angular scales and brightness. In this section we describe the observational basis which supports the use of the far-IR brightness as a tracer of both atomic and molecular gas within cirrus.

Several studies based on the IRAS data showed that the 100 μm emission away from the Galactic plane and away from the principal nearby molecular complexes is generally tightly correlated with the distribution of the neutral atomic gas (*e.g.*, Boulanger and Pérault 1988; Deul and Burton 1993). This correlation analysis has been extended to longer wavelengths using the DIRBE and FIRAS data (Boulanger *et al.* 1996). The far-infrared spectrum and data points in Figure 1 are the result of this correlation in regions of HI column density smaller than $5\,10^{20}\,cm^{-2}$. In these low column density regions, the $I_\nu(100\mu m)$ emission for $N(HI) = 10^{20} H\,cm^{-2}$ is 0.85 MJy/sr and 0.54 MJy/sr for the IRAS and DIRBE data, respectively. The two numbers differ due to a known calibration difference between the two experiments. The far IR-HI correlation shows that dust and gas are well mixed and that, at high latitude, the intensity of dust heating is quite uniform. The emission ratio also provides an empirical conversion factor between far-infrared brightness and gas column density.

The validity of the far-IR emission as a tracer of gas distribution has been extended to regions of higher column density by comparing the 100 μm brightness with the visible extinction. Most published studies apply to clouds at intermediate latitudes with moderate extinctions, Av up to 2-3 mag (*e.g.*, Boulanger *et al.* 1997). The $I_\nu(100\mu m)/A_v$ values found in these translucent molecular clouds are 2 to 3 times lower than the mean value of 15.9 MJy/sr per mag for atomic clouds at high latitude (IRAS value). The reduction factor fits with estimates of the effect of the radiation field attenuation on the IR emission from non-transparent clouds (Bernard *et al.* 1993).

3.2. ATOMIC AND MOLECULAR GAS IN CIRRUS

Gas in Cirrus is dominantly atomic but CO observations (*e.g.*, Magnani *et al.* 1996) and the comparison between IRAS and H I data suggest that many cirrus features contain some molecular gas. The presence of molecular gas is often inferred from the existence of an infrared excess emission with respect to what is expected from the atomic gas seen in emission in the H I 21cm line (Désert *et al.* 1988, Boulanger *et al.* 1996). In the all-sky studies with 30 to 40' resolution many high latitude molecular clouds are heavily diluted. The comparison between IRAS, HI and CO observations is thus the most conclusive where H I observations with resolution comparable or better than IRAS have been obtained (*e.g.*, Herter *et al.* 1990, Joncas *et al.* 1992). With higher angular resolution one may clearly identify clouds for which the IR-H I correlation extends down to the IRAS resolution and others for which there is no correlation (*e.g.*, de Vries *et al.* 1987). Many of the latter cirrus coincide with high latitude molecular clouds seen in emission in CO (*e.g.*, Heithausen *et al.* 1993).

The CO emission is far from being simply related to the column density of molecular hydrogen in cirrus. Based on a large and deep CO survey of cirrus clouds around the North Celestial Pole, Heithausen *et al.* (1993) showed that there is not a one to one correlation between the two intensities and that there is no threshold of $I_\nu(100\mu m)$ for the detection of CO. Several studies have concluded to the existence of a gas component seen in the infrared but not in emission in H I nor in CO (Reach *et al.* 1994, Meyerdierks and Heithausen 1996). For example, the ratio between gas mass as derived from the far-infrared emission and the CO emission varies by one order of magnitude among clouds in the Chameleon complex (Boulanger *et al.* 1997). We want to stress that the reported variations between molecular mass and CO emission are local in nature and that they do not question large-scale estimates of molecular masses derived from CO data. In particular, the mass of molecular gas in the Solar Neighborhood derived from CO observations is known to be in good agreement with that inferred from Copernicus measurements of H_2 absorption lines in the UV (Savage *et al.* 1977).

Measurements of the CO column density, together with investigations of physical conditions indicate that cirrus clouds span the critical regime where the abundance of CO starts to become significant but in which most of the carbon is still in atomic form (van Dishoeck *et al.* 1991, Gredel *et al.* 1992). Theoretical models (van Dishoeck and Black 1988 and Viala *et al.* 1988) show that in this regime, small fluctuations in involved physical parameters can lead to large variations in observable CO line intensity. The abundance of CO depends not only on the total column density but also on the gas density, the far-UV radiation field and the dust extinction in the far-UV. Various observations suggest that the presence of dense gas is correlated with an excess in the far-UV extinction curve (Cardelli and Clayton 1991, Jenniskens *et al.* 1992, Boulanger *et al.* 1994). Dust may thus facilitate the onset of formation of self-shielded CO by attenuating radiation coinciding with photo-dissociative transitions. It is most likely that this is only one of the many processes which connect the dynamical, thermal and chemical evolutions of interstellar matter.

3.3. DENSITY STRUCTURE

Most cirrus studies concentrate on localized features and little is known about the large scale organization of cirrus. The lack of systematic distance estimates has prevented so far the construction of a 3-D representation of the nearby interstellar medium which will locate cirrus clouds with respect to the local cavities of coronal gas and the energy sources, stellar winds and supernovae explosions, which have created them. This knowledge may be important to understand the morphology of cirrus clouds and the origin of their turbulent motions. The X-ray all-sky maps from ROSAT have been used to identify several instances of shadows due to the absorption of X-ray emission from the coronal gas by intervening cirrus clouds (Mebold et al. 1994). The Hipparcos catalog (Mignard, this volume, p. 399) contains 30,000 stars with measured parallax in between 100 and 200 pc for the Sun. This distance information combined with the stellar reddenings opens the prospect of better understanding the large scale organization of cirrus. The three dimension structure of cirrus could also be mapped by combining the Hipparcos data with IRAS searches for heated dust around stars (Gaustad and van Buren 1993).

The infrared sky images from both IRAS and DIRBE have been used to characterize statistically the self-similar structure of interstellar matter. Gautier et al. (1992) have used IRAS images to measure the power spectrum of high latitude cirrus clouds. The shapes of the power spectra were found to be well represented by a power law. Two dimensional analyses yield indices near -3 from several tens of degrees to the $5'$ angular resolution of IRAS (Gautier et al. 1992, Kogut et al. 1996). Based on these results, Gautier et al. deduced that the brightness contrast (i.e, the column density contrast for the matter traced by the far-infrared emission) increases with angular size, θ, as $\theta^{0.5}$. But Gautier et al. also realized that the most striking features of the IRAS images correspond to a non-Gaussian tail in the distribution of brightness fluctuations. Thus, the sky structure is not fully characterized by power spectra which only measure the standard deviation of the brightness fluctuations. Abergel et al. (1996) went further in this analysis by using wavelet transforms which allow to study the scaling law between brightness fluctuations and angular size as a function of position in the image. This work shows that the power law breaks at the position of filaments which have an excess of power on small angular sizes. In the image analyzed by Abergel et al., these singular structures correspond to regions of cold IRAS emission which are known from the correlation with ^{13}CO emission to trace dense gas. Observations of molecular lines at radio wavelengths provide a complementary perspective on the structure of cirrus clouds. The CO emitting gas has been shown to be structured on scales much smaller than the IRAS resolution. Small scale clumping is also necessary to explain the observed excitation of the CO rotational transitions. Gas densities in CO emitting pieces of cirrus are estimated to be at least 10^3 H$_2$cm^{-3} (Falgarone et al. 1991 and van Dishoeck et al. 1991). The excitation of other molecules such as CS provides even higher density estimates in some cirrus cores (Reach et al. 1995). The minimal density value of 10^3 H$_2$cm^{-3} is one to two orders of magnitude larger than the mean densities one gets from the mean column densities and structure sizes derived from the IRAS images, especially where CO is detected and the mean column density of gas as estimated from $I_\nu(100\mu m)$ is low. The velocity information provided by radio observation has also been used to show that the turbulent energy within cirrus clouds greatly exceeds the binding energy of gravity (e.g., Heithausen 1966). The existence of a self similar structure over many decades of sizes with large density contrast down to unresolved scales is a fascinating aspect of cirrus clouds which remains to explained.

4. Star Formation

In the Galactic Plane the infrared emission is made of a diffuse component and localized sources respectively accounting for 3/4 and 1/4 of the IR luminosity of the Galaxy.

The various groups which have analyzed the IRAS data (Pérault 1987, Cox and Mezger 1988, Sodroski et al. 1989, Bloemen et al. 1990) and more recently the DIRBE data (Sodroski et al. 1997) agree on the following conclusions. The heating of Galactic dust is dominated by O, B and A stars. Older stars contribute only 30% of the overall luminosity of the Galaxy. A large fraction of the luminosity of dust is provided by stars of intermediate mass, late B and A stars with ages in the range 10^7 to a few 10^8 yrs. This is an important conclusion which implies that the IR emission provides an information on star formation distinct from that of radio observations measuring the thermal emission from H II regions. From the Solar Neighborhood to the molecular ring, the radial distribution of the summed luminosity of O, B and A derived from the infrared data follows that of ionizing stars derived from radio observations. The enhanced rate of star formation currently observed in the molecular ring has thus been lasting for at least a few 10^7 years: it is not a recent burst.

The brightest sources were known from balloon observations prior to the IRAS survey (Hauser et al. 1984 and references therein). They are all associated with known luminous star forming regions (*e.g.*, Myers *et al.* 1987). IRAS allowed the discovery of a large number of ultra-compact H II regions which represent the youngest high mass star in the Galaxy (Wood and Churchwell 1989). These stars have been found to concentrate in an annulus of massive star formation coincident but about 30 % narrower in radius than the Molecular Ring (Bronfman *et al.* 1996). But, since only a minor fraction of molecular clouds in the Galaxy is associated with high mass stars, little is known about star formation efficiency for the bulk of the molecular gas in the Galaxy. The ISO satellite has opened the possibility to extend this statistical study to lower mass stars.

In the wavelength range 5–17μm ISOCAM, the camera of the Infrared Space Observatory (ISO) is able to detect sources more than 100 times fainter than IRAS. One of the major research goals being pursued with this new instrument is the study of low mass star formation. With the available sensitivity it is possible to detect pre-main sequence stars of Solar mass within the Molecular ring, the region of highest concentration of gas and star formation, in the Galaxy. The preliminary analysis of ISO-CAM images of the Galactic plane (Pérault *et al.* 1996) shows that half of the detected sources have no near-infrared counterpart in the K images of the DENIS survey, the near-IR southern sky survey being carried out at La Silla. The only plausible candidates for such a large number of cold sources ($1300/\deg^2$) are dusty young stars. It is foreseen that ISO observations in the Galactic plane will allow to detect 50,000 new young stars and thus provide statistical information about star formation efficiency and luminosity function of young stars for a large set of molecular ring clouds. Another interesting result of the ISOCAM observations is the detection of dark features which are likely to be due to absorption of the diffuse Galactic emission by opaque gas condensations. The most contrasted features would correspond to very dense filaments with visible extinction Av of at least 25 mag. This provides another interesting perspective on star formation since many of this absorbing structures could contain proto-stellar condensations.

5. Conclusion

Infrared surveys have opened new perspectives on interstellar matter. The emission from dust has been observed and mapped from the near-IR to millimeter wavelengths. The spectral energy distribution and spectroscopy have been used to investigate the size distribution and composition of interstellar dust and its evolution within the interstellar medium. The IRAS sky images have revealed the intricate structure of interstellar matter with a yet unsurpassed horizon in terms of sky coverage and sensitivity. These maps have prompted numerous studies of the structure of interstellar clouds at other wavelengths. In particular, CO, H I and optical observations have broadened the IRAS perspective by providing images on smaller scales and velocity information. Soon these observations will be compared with images obtained with ISO. The infrared surveys also provide a census of young embedded stars allowing to statistically study the star formation efficiency and the stellar luminosity function for a large set of clouds.

Since a significant fraction of the conference was devoted to the analysis of large data-bases, I want to stress that the problems raised by the analysis and intercomparison of extended emission from Galactic interstellar matter are different in nature from those raised by the study of catalogs. Research is also badly needed to develop tools to statistically characterize and compare images.

References

Abergel, A., Boulanger, F., Mizuno, A. and Fukui, Y. 1994, Astrophys.J., 423, L59
Abergel, A., Boulanger, F., Delouis, J.M., Dudziak, G., Steindling, S. 1996, Astron.Astrophys., 309, 245
Bernard, J.P. et al. 1996, in Unveiling the Cosmic Infrared Background, ed. E. Dwek, AIP Conf. Proceedings 348, p. 105
Bernard, J.P., Boulanger, F., and Puget, J. L. 1993, Astron.Astrophys., in press
Bloemen, J.B.G.M., Deul, E. R., and Thaddeus, P. 1990, Astron.Astrophys., 233, 437
Boulanger, F. and Pérault , M. 1988, Astrophys.J., 330, 964
Boulanger, F., Falgarone, E., Puget, J.L., and Helou, G. 1990, Astrophys.J., 364, 136
Boulanger, F., Prévot, M. L., and Gry, C. 1994, Astron.Astrophys., 285, 956
Boulanger, F. et al. 1996a, Astron.Astrophys.312, 256
Boulanger, F. et al. 1996b, Astron.Astrophys.315, L325
Boulanger, F., Bronfman, L., Dame, T.M. and Thaddeus, P. 1997, Astron.Astrophys., in press
Bronfman, L., Nyman, L.A., amd May, J. 1996, Astron.Astrophys.Suppl., 115, 1
Cardelli, J. A. and Clayton, G. C. 1991, Astron.J., 101, 1021
Cox, P. and Mezger, P. G. 1989, Astron.Astrophys.Rev. 1, 49
Désert , F. X., Bazell, D. and Boulanger, F. 1988, Astrophys.J., 334, 815
Désert , F. X., Bazell, D. and Blitz, L. 1990, Astrophys.J., 355, L51
Deul, E. R. and Burton, W. B. 1990, Astron.Astrophys., 230, 153
de Vries, H. W., Heithausen, A. and Thaddeus, P. 1987, Astrophys.J., 319, 723
Draine, B. T., and Lee, H. M. 1984, Astrophys.J., 285, 89
Draine, B.T. and Anderson, N. 1985, Astrophys.J.292, 494
Dwek, E. et al. 1997, Astrophys.J.475, 565
Falgarone, E., Phillips, T. G. and Walker, C. K. 1991, Astrophys.J., 378, 186
Falgarone, E., Puget, J.L. 1995, Astron.Astrophys.293, 840
Gaustad, J. E., van Buren, D. 1993, Publ.Astron.Soc.Pacific105, 1127
Gautier, T.N., Boulanger, F., Pérault, M., Puget, J.L. 1992, Astron.J., 103, 1313
Giard, M.,Lamarre, J.M., Pajot, F., Serra, G. 1994, Astron.Astrophys., 286, 203
Gredel, R., van Dishoeck, E. F., de Vries, C. P. and Black, J. H. 1992, Astron.Astrophys., 257, 245
Hauser, M.G. et al. 1984, Astrophys.J.285, 74
Heithausen, A., Stacy, J. G., de Vries, H. W., Mebold, U. and Thaddeus, P. 1993, Astron.Astrophys., 268, 265.

Heithausen, A. 1996, Astron.Astrophys.314, 251
Herter, T., Shupe, D. L. and Chernoff, D. F. 1990, Astrophys.J., 352, 149
Jenniskens, P., Ehrenfreund, P. and Désert, F. X. 1992, Astron.Astrophys., 265, L1
Joncas, G., Boulanger, F., Dewdney, P. E. 1992, Astrophys.J., 397, 165
Jones, A.P., Tielens, A.G.G.M. and Hollenbach, D.J. 1996, Astrophys.J.469, 740
Kogut, A. et al. 1996, Astrophys.J.460, 1
Lagache, G., Abergel, A., Boulanger, F., Puget, J.L. 1997, Astron.Astrophys.in press
Laureijs, R. J., Clark, F. O., and Prusti, T. 1991, Astrophys.J., 372, 185
Léger, A., Puget, J.L. 1984, Astron.Astrophys.137, L5
Magnani, L. Hartmann, D., Speck, B.G. 1996, Astrophys.J. Supp.106, 447
Marty, P., Serra, G., Chaudret, B. and Ristorcelli, I. 1994, Astron.Astrophys.282, 916
Mathis, J.S. and Whiffen, G. 1989, Astrophys.J.341, 808
Mattila, K. et al. 1996, Astron.Astrophys.315, L353
Mebold, U., Kerp, J., Moritz, P. Engelmann, J. and Herbstmeier, U. 1994 in The First Symposium on the Infrared Cirrus and Diffuse Interstellar Clouds, R.M. Cutri and W.B. Latter (eds), ASP Conf. Ser. 58, 45.
Meyerdierks, H., Heithausen, A. 1996, Astron.Astrophys.313, 929
Myers, P.C., et al. 1986, Astrophys.J.301, 398
Onaka, T., Yamamura, I., Tanabe, T., Roellig, T., Yuen, L. 1996, Publ.Astron.Soc.Japan 48, L41
Papoular, R., et al. 1996, Astron.Astrophys.315, 222
Pérault , M. 1987, Thèse d'Etat, Université Paris VII
Pérault , M. et al. 1996, Astron.Astrophys.315, L165
Puget, J.L. and Léger, A. 1989, Ann.Rev.Astron.Astrophys. 27, 161
Reach, W. T., Koo, B. C., Heiles, C. 1994, Astrophys.J., 429, 672
Reach, W.T., Pound, M.W., Wilner, D.J., and Lee, Y. 1995, Astrophys.J.441, 244
Ristorcelli, I. et al. 1994, Astron.Astrophys.286, L23
Ristorcelli, I. et al. 1997, Diffuse Infrared Radiation and the IRTS, ed. T. Matsumoto
Savage, B.D., Bohlin, R.C., Drake, J.F., Budich, W. 1977, Astrophys.J.216, 291
Sellgren, K., Werner, M. W. and Dinerstein, H. L. 1983, Astrophys.J., 271, L13
Serra, G. et al. 1992, Astron.Astrophys.260, 489
Sofia, U.J., Cardelli, J.A. and Savage, B.D. 1994, Astrophys.J.430, 650
Sodroski, T.J. et al. 1997, Astrophys.J.480, 173
Tanaka, M., Matusmoto, T., Murakami, H., Kawada, M., Noda, M. Matsuura, S. 1996, Publ.Astron.Soc.Japan 48, L53
van Dishoeck, E. F., and Black, J. H. 1988, Astrophys.J., 334, 771
van Dishoeck, E. F., Black, J. H., Phillips, T. G., and Gredel, R. 1991, Astrophys.J., 366, 141
Viala, Y. P., Letzelter, C., Eidelsberg, M. and Rostas, F. 1988, Astron.Astrophys., 193, 265
Weiland, J.L., Blitz, L., Dwek, E., Hauser, M. G., Magnani, L., and Rickard, L. J. 1986,Astrophys.J., 306, L101
Wood D.O.S., Churchwell, E. 1989 Astrophys.J.340, 265
Wright, E.L. 1987, Astrophys.J.320, 818
Zagury, F., Boulanger, F., and Jones, A.P. 1997, Astron.Astrophys.in press

RELATIONS BETWEEN STAR FORMATION AND THE INTERSTELLAR MEDIUM

Y. FUKUI AND Y. YONEKURA
Department of Astrophysics, Nagoya University

Abstract. We review observational results concerning star formation and dense molecular clouds, the interstellar medium most relevant to star-formation process, as well as future prospects.

1. Introduction

Stars form in interstellar molecular clouds. Recent ideas on star-formation are reviewed in the literature (see *e.g.*, Shu *et al.* 1987 and references therein). Our knowledge on star formation is obtained mainly by observations of molecular and infrared emission from cold interstellar medium, while the optical data on young stellar objects (YSOs) place important constraints on star-formation theories (*e.g.*, Herbig & Bell 1988).

The $J = 1\text{-}0$ ^{12}CO emission at 2.6 mm was most extensively used to trace molecular clouds in the Galaxy in 1970's and 1980's (see Combes 1991). The most important knowledge obtained through ^{12}CO surveys is that there are 3,000–4,000 giant molecular clouds (GMCs) of $\sim 10^6\ M_\odot$ in the Galactic disk, being confined into a spiral arm pattern (Scoville *et al.* 1987; Dame *et al.* 1987). These ^{12}CO observations are however either of low resolution and large coverage or of high resolution and small coverage. The goal of next-generation CO surveys is obviously to cover a larger sky area at resolution high enough to resolve individual sites of star formation in an optically thin probe.

Far-infrared observations made with the *IRAS* (*Infrared Astronomical Satellite*) provided a unique tool to probe candidate protostars that emit most of the radiation at far-infrared wavelengths over the whole sky in 1980's. The relatively high angular resolution, a few arc minutes, was good enough to resolve these objects. A protostar (or a protostar candidate, to be exact) is generally invisible, deeply embedded in dense molecular gas, and is observable only at far-infrared and/or sub-millimeter wavelengths. The *IRAS* catalog provides a comprehensive list of such protostellar objects with a sensitivity limit of ~ 1 $L_\odot$ in bolometric luminosity in typical nearby star-forming regions within a few hundred parsecs (Beichman *et al.* 1986). These objects are perhaps mature protostars close to the end of the mass accretion phase as suggested by their luminosity of $\gtrsim 1\text{--}10$ $L_\odot$, high for a low-mass protostar. Next-generation infrared satellites will reveal even earlier stages of a protostar that should appear as a very faint infrared source of $\lesssim 0.1$ $L_\odot$.

Most of the recent observational studies are focused on less than a few tens of well-known regions of star formation. Accordingly, a small fraction of the protostellar *IRAS* sources have been studied yet, making it uncertain how individual star formation is related to more global properties of star formation. One of such important properties is the stellar initial mass function (IMF), which indicates that most of the stars in the Galaxy are of solar mass, and that the number of stars rapidly decreases with stellar mass (Scalo 1986). We still do not understand the origin of the IMF.

We shall review a recent ^{13}CO survey covering ~ 17 % of the sky with the Nagoya 4-meter millimeter-wave telescopes. This survey has provided a rich sample of star-forming dense clouds within 1–2 kpc of the Sun. The dataset combined with the *IRAS* point source catalog indicates that the ^{13}CO clouds better represent star-forming regions on the Galactic scale than ^{12}CO. Physical properties of the ^{13}CO clouds are derived by using the sample of nearly 500 clouds, and average characteristics of star formation in the ^{13}CO clouds are summarized.

2. Star Formation in Giant Molecular Clouds

We begin with reviewing star formation in GMCs. It is suggested that most of the stars form in GMCs mainly from the fossil records of star formation like OB associations. The number of GMCs we can study in detail is not large; only several GMCs are located within 1 kpc of the Sun. The best studied GMCs are Orion B and Orion A clouds located at ~ 500 pc, for which high-resolution molecular images, and extensive datasets for near and/or far-infrared sources are available.

In the Orion B cloud five dense CS cores (density $\gtrsim 10^4$ cm^{-3}) and three near-infrared clusters are distributed as shown by unbiased surveys made within the spatial extent of the $J = 1\text{--}0$ ^{12}CO emission (Lada *et al.* 1991). Three of the CS cores well coincide with the infrared clusters, which are interpreted as that star formation occurs almost exclusively as a rich cluster in a dense cloud core.

On the other hand, the dense molecular gas (CS and $C^{18}O$) in the Orion A cloud is distributed throughout the cloud more or less continuously, and so is star formation (Fukui et al. 1994; Nagahama et al. 1996). The $J = 1$–0 $C^{18}O$ distribution illustrates that protostellar $IRAS$ sources, molecular outflow sources, and candidate T Tauri stars are all distributed over the cloud with no strong clustering toward individual peaks of dense gas except for the Orion KL region. Infrared imagings indicate that there are several infrared clusters toward luminous protostellar sources (Ori KL, L1641-N, L1641-S, etc., e.g., Fukui et al. 1986; Hodapp 1994), but that they are not so rich in member as in Orion B.

3. Star Formation in Small Clouds

The cloud complex in Taurus is located at ~ 150 pc, and is active in low-mass star formation as shown by studies of T Tauri stars. This complex having a total mass of $\sim 7,000$ $M_\odot$ is not associated with young OB associations nor SNRs, and YSOs are distributed with much lower number density than in Orion clouds, making it one of the best places to test theories on low-mass star formation with no external triggering.

The ^{13}CO distribution of the complex is compared with ~ 100 YSOs including T Tauri stars and protostellar $IRAS$ sources (Mizuno et al. 1995). Relatively loose clustering of YSOs toward ^{13}CO peaks indicates that the ^{13}CO clouds are formation sites of these YSOs whose ages are $\lesssim 10^6$ yrs. A comparison with the $C^{18}O$ distribution on the other hand indicates that protostellar $IRAS$ sources, invisible objects younger than T Tauri stars, are almost always within the $C^{18}O$ cores, indicating that the $C^{18}O$ cores are the sites of *ongoing* star formation (Onishi et al. 1996a, 1996b).

The ^{12}CO emission, used to study the Galactic structure or the giant molecular clouds, is no longer a good tracer of individual star formation on a scale of ~ 1 pc. Instead, the ^{13}CO emission that represents density of $\sim 10^3$ cm^{-3} better traces the distribution of YSOs. This is also true for other star-forming clouds of various masses like Orion A, L1251, L1340, IC5146, etc. (e.g., Sato et al. 1994).

4. A ^{13}CO $J = 1$–0 Survey of Star-Forming Clouds

The preceding discussion suggests that ^{13}CO emission is a good probe of star-forming molecular gas. A ^{13}CO $J = 1$–0 survey along the Galactic plane has been carried out by using the two 4-meter millimeter-wave telescopes at Nagoya University since 1990 (e.g., Fukui & Mizuno 1991; Dobashi et al. 1994, 1996; Yonekura et al. 1996). The observations are made at a grid spacing of $8'$ in l and b with a $2.7'$ beam at 0.1 km s^{-1} velocity resolution. The coverage extends to $\pm\ 20°$ in b over $0°$–$230°$ in l corresponding to ~ 17 % of the sky (Figure 1). 340,000 points have been observed at a 3 σ noise level equivalent to 1×10^{21} cm^{-2} in molecular column density. One of the telescopes, named NANTEN telescope, was moved to the Las Campanas Observatory in Chile in 1996 spring in order to extend the survey to the southern sky (Figure 2). Detailed analyses of these ^{13}CO data have been made for regions of Cygnus, Cassiopeia, Perseus, Gemini, and Auriga. The ^{13}CO distribution is highly clumpy and is unambiguously divided into respective ^{13}CO clouds at a 3 σ noise level. Thus defined ^{13}CO clouds are used to derive physical parameters of the high-density interstellar molecular gas, and to study the mass spectrum, morphology, and dynamical properties of star-forming clouds.

One of the important outcomes of the ^{13}CO survey is that the ^{13}CO clouds are used to identify associated $IRAS$ point sources whose luminosities cannot be otherwise estimated. The association is reliable since most of the ^{13}CO clouds within 1–2 kpc are located at relatively high galactic latitude where confusion with distant galactic sources is negligibly small. Even close to the galactic plane, $i.e.$, at $|b| < 1°$, more than 70% of the association is considered to be real according to a statistical test by Yonekura et al. (1996).

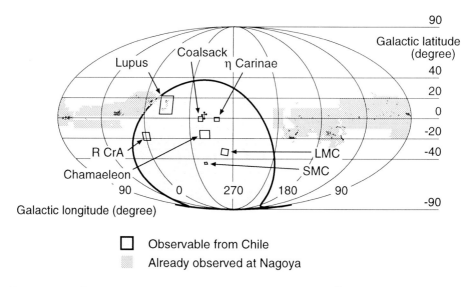

Figure 1. A figure that represents the progress of the Nagoya ^{13}CO survey. The regions where observations have been already finished are represented by hatches. Regions which cannot be observed at Nagoya ($\delta < -25°$) are represented by brighter hatches. Some well-known star-forming regions in the southern sky are also shown.

Figure 2. The 4-meter millimeter-wave telescope named NANTEN (=southern sky in Japanese) installed at Las Campanas Observatory in Chile.

5. ^{13}CO Cloud Properties

In the following we shall present a summary on the cloud properties and star formation in them. We shall confine ourselves to ^{13}CO clouds of $\gtrsim$ 10 $M_\odot$, and protostellar $IRAS$ sources of $\gtrsim$ 10 $L_\odot$, limitations from the detection limits in the ^{13}CO and $IRAS$ surveys, respectively. The number of the cloud analyzed here is 478 and that of the associated $IRAS$ sources is 471.

(1) *Cloud mass spectrum*: The mass spectrum of the ^{13}CO clouds is well represented by a power law of an index value of -1.7; $dN_{\text{cloud}}(M_{\text{cloud}})/dM_{\text{cloud}} \propto (M_{\text{cloud}})^{-1.7}$. A list of the index value for various star-forming regions in the literature indicates that the value is remarkably uniform in a range from -1.3 to -1.8.

(2) ^{13}CO *line width ΔV –size R relation*: For the ^{13}CO clouds this relation is very poorly seen as $\Delta V(^{13}\text{CO}) \propto R^{0.18}$ with a correlation coefficient of < 0.4. The power-law index value is smaller than 0.5, a value often cited in previous works (*e.g.*, Larson 1981).

(3) *Virial mass–LTE mass relation*: Virial mass of the ^{13}CO clouds is different from LTE mass calculated from the integrated intensity for an assumed excitation temperature of the CO $J = 1$–0 transition, 10 K. The relation between the two masses is represented as follows; $M_{\text{virial}} \propto (M_{\text{LTE}})^{0.64}$. This implies that virial mass tends to become larger than LTE mass for small clouds. This deviation is significant particularly for clouds of $\lesssim 10^3$ $M_\odot$, suggesting that small clouds are not gravitationally

bound; external pressure may confine them, or alternatively, small clouds are short-lived, with a timescale of $R/\Delta V \sim 10^6$ yrs.

(4) *Luminosity function of protostellar IRAS sources*: We identified 471 *IRAS* sources having FIR color of log $[F(25\ \mu\text{m})/F(60\ \mu\text{m})] \leq 0$ to be associated with the ^{13}CO clouds. All the *IRAS* sources thus identified show spectral energy distribution typical to protostars or T Tauri stars, and are definitely detected at three *IRAS* bands centered at 12, 25, and 60 μm. The flux densities were used to calculate the bolometric luminosity, L_{IR}, for the cloud's distance. The luminosity function of the *IRAS* sources is represented as follows; $dn_{\text{IR}}(L_{\text{IR}})/dL_{\text{IR}} \propto (L_{\text{IR}})^{-1.45}$. The power-law index seems fairly universal among the regions analyzed here. If we assume a mass to luminosity relation $L_{\text{IR}} \propto m^{3.45}$ for main-sequence stars, the luminosity function can be converted as follows; $dn_{\text{IR}}(m)/dm \propto (m)^{-2.55}$. This relation is consistent with the Scalo's IMF, suggesting that the IMF is universal.

(5) *Number of IRAS sources vs. cloud mass*: The number of associated *IRAS* sources in a ^{13}CO cloud is related to the LTE mass as follows; $n_{\text{IR}}(M_{\text{LTE}}) \propto (M_{\text{LTE}})^{0.62}$. This relation can be further used to calculate star formation efficiency (SFE) as follows; SFE $= \Sigma m/(M_{\text{LTE}} + \Sigma m) \propto (M_{\text{LTE}})^{-0.25}$. The dependence of SFE on the cloud mass is very weak and may be described with a small negative power index, suggesting that small clouds can be equally important in star formation on the galactic scale as well as massive clouds. We should note that the statistics is still to be improved by including other regions, and that the mass-to-luminosity relation may be uncertain for protostars whose luminosity is not yet well connected to mass.

6. Future Prospects

Our observational knowledge on young stellar objects is still quite limited even for the solar neighborhood of $\lesssim$ 1–2 kpc. This warrants further efforts to make surveys for YSOs at various wavelengths. We expect the following surveys will provide important steps forward in this coming several years:

(1) The existing optical surveys for T Tauri stars are limited to some well-known regions of star formation. X-ray and near-infrared surveys will provide more comprehensive and complete knowledge on how and where the young stellar objects, mainly T Tauri stars, are distributed. These surveys include 2MASS, DENIS, and next-generation X-ray satellites such as XMM, ABRIXAS, and AXAF in addition to ROSAT and ASCA.

(2) It is becoming more and more convincing that we are beginning to see protostars which are still growing in mass by accumulating material from the interstellar space. Far-infrared satellites are now observing

relatively mature protostars close to the end of the mass accretion stage as indicated by their high luminosity, $\gtrsim$ 1–10 $L_\odot$, as a low-mass protostar. Next-generation IR satellites, including IRIS and SIRTF in addition to ISO, will be sensitive enough to detect even fainter sources at 100 μm–1 mm having $\lesssim$ 0.1 $L_\odot$ within 1 kpc. These results will provide a strong observational basis to understand the whole evolutionary scenario of a protostar from the early stages of a gaseous condensation and of formation of the initial hydrostatic protostellar core.

(3) Ongoing ^{13}CO and mm/sub-millimeter surveys both for the northern and southern sky will provide a complete sample of dense cloud cores and nearby star-forming regions. The NANTEN telescope at Las Campanas in Chile will be one of the powerful instruments toward this direction.

We are grateful to the organizing committee for the financial support to participate this fruitful symposium. This research was in part financially supported by the Grants-in-Aid for Scientific Research by the Ministry of Education, Science, and Culture (Nos. 05402003, 06452020, and 07640357). Collaborators of the Nagoya ^{13}CO survey are Y. F., Y. Y., H. Ogawa, A. Mizuno, T. Nagahama, T. Onishi, A. Kawamura, A. Obayashi, K. Tachihara, K. C. Xiao, N. Yamaguchi, A. Hara, T. Hayakawa, S. Kato, N. Mizuno, B. G. Kim, J. P. Bernard, and K. Dobashi.

References

Beichman, C. A., et al. (1986), Astrophys.J., 307, 337
Combes, F. (1991), Ann.Rev.Astron.Astrophys., 29, 195
Dame, T. M., et al. (1987), Astrophys.J., 322, 706
Dobashi, K., Bernard, J. P. and Fukui, Y. (1996), Astrophys.J., 466, 282
Dobashi, K., et al. (1994), Astrophys.J.Suppl., 95, 419
Fukui, Y. and Mizuno, A. (1991), in E. Falgarone, et al. (eds.), "Fragmentation of Molecular Clouds and Star Formation," Dordrecht, Reidel, 275
Fukui, Y., et al. (1994), in T. Montmerle, et al. (eds.), "Cold Universe," Edition Frontieres, Gif sur Yvett, 157
Fukui, Y., et al. (1986), Astrophys.J.Lett., 311, L85
Herbig, G. H. and Bell, K. R. (1988), Lick Obs. Bull., No.1111
Hodapp, K.-W. (1994), Astrophys.J.Suppl., 94, 615
IRAS Point Source Catalog (1988), Joint IRAS Science Working Group
Lada, E. A., Bally, J. and Stark, A. A. (1991), Astrophys.J., 368, 432
Larson, R. B. (1981), Mon.Not.R.astron.Soc., 194, 809
Mizuno, A., et al. (1995), Astrophys.J.Lett., 445, L161
Nagahama, T., et al. (1996), to be submitted to Astrophys.J.
Onishi, T., et al. (1996a), Astrophys.J., 465, 815
Onishi, T., et al. (1996b), Astrophys.J., submitted
Sato, F., et al. (1994), Astrophys.J., 435, 279
Scalo, J. M. (1986), Fundam. Cosmic Phys., 11, 1
Scoville, N. Z., et al. (1987), Astrophys.J.Suppl., 63, 821
Shu, F. H., Adams, F. C. and Lizano, S. (1987), Ann.Rev.Astron.Astrophys., 25, 23
Yonekura, Y., et al. (1996), Astrophys.J.Suppl., submitted

A DENIS SURVEY OF STAR FORMING REGIONS

E. COPET
Observatoire de Paris

1. Introduction

A complete census of embedded stellar population can be made by exploring in the infrared large areas of the sky in which giant molecular clouds extend. Very recently, thanks to the of large format IR array detectors, studies of young stellar population in GMCs, have been undertaken by different authors (*i.e.*, Lada *et al.* 1991) but all these observations were limited to relatively small regions of the whole GMCs, the DENIS project (Epchtein, this volume, p. 106) surveys the south hemisphere at I, J and K_s bands, including most of these clouds.

2. Orion A

The Orion molecular cloud complex is the best studied GMC of our Galaxy, given its short distance (D=450 pc; *e.g.*, Genzel & Stutzki 1989). During the proto-survey period, the DENIS instrument has observed this well-known cloud at J and K_s band.

2.1. OBSERVATIONS AND RESULTS

The observations of the region were made in february 1995 using the DENIS standard procedure (Epchtein *et al.* 1994). For the purpose of this study, 6 "DENIS strips" were used to cover the 3 square degrees around the Trapezium. The data reduction and the source extraction method is presented in more details in Copet *et al.* (in preparation).

We estimate the completeness limit of our sample at 15.6 and 13.8 at J and K_s band, respectively. The number of sources extracted above the completeness limit are 7032 at K_s and 7621 at J. The spatial distribution of the stars detected at the K_s band is presented in Figure 1. The J and K_s

A DENIS SURVEY OF STAR FORMING REGIONS

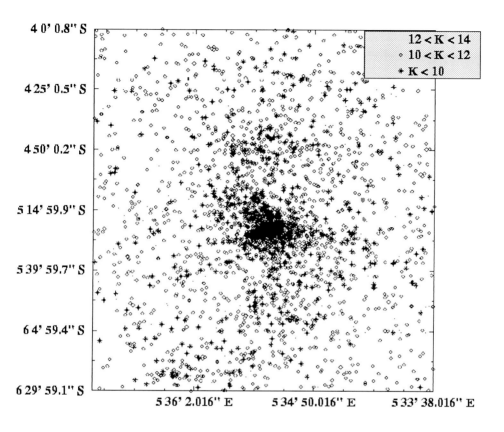

Figure 1. Spatial distribution of the sources detected at K_s band according 3 ranges of magnitude. The equinox of the coordinates is 2000.0

luminosity functions (not shown) present a clear excess of high luminosity stars in the vicinity of the Trapezium. The (J−K) color of the stars follows the CO density distribution, but some very red objects (J−K > 3) could be found far away the CO density peaks.

Using a molecular CS survey of this region (Tatematsu *et al.* 1993) and correlating our catalog with the CS peaks location, we have detected inside the CS cores 114 sources at J and 906 sources, in the K_s band. Probably as only 10 % of these sources are associated with the cores, then, usually rich clusters of embedded objects are not observed.

3. Chamaeleon

The Chamaeleon region has been mapped at several wavelengths: IR (Gauvin *et al.* 1992), X (Huenemoerder *et al.* 1994). The DENIS project has

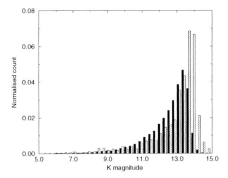 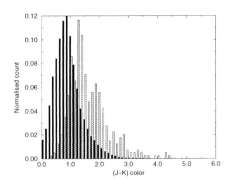

Figure 2. Histograms obtained with the DENIS data in the Chamaeleon region. The left side presents the K luminosity function in the Cha I region (grey bars) while the right side represents the distribution of the (J−K) color in the same area. The black bars are the distribution derived from a 30 degree long strip for comparison. The counts are normalised to unity.

already surveyed a large area on the Cha I and Cha II clouds in the J and K_s bands. We present in Figure 2 the luminosity function and the color histogram for the Cha I region. The number of K_s sources detected in this region is around 3000.

We note a very faint stellar population in the Cha I cloud (K>14) close to our detection limit and an excess of reddenning, but with only one color, we cannot disentangle between an intrinsic reddenning of the star or an extinction due to the cloud material.

4. Conclusion

The preliminary results on these two star-forming regions show that the nominal performances of the instrument are reached. The comparison with ISOCAM data (Kaas A.A., *private communication*; Nordh *et al.* 1996) shows a good correlation between the stellar population detected in the Near and Far infrared bands and provide an excellent base to study these kind of regions at higher sensitivity.

References

Epchtein N., *et al.*, 1994 Astrophys.Space.Sc. 217, 3
Genzel R., Stutzki J., 1989 Ann.Rev.Astron.Astrophys. 27, 41
Lada C.J., DePoy D.L., Merrill K.M., Gatley I., 1991 Astrophys.J. 374, 533
Huenemoerder D.P., Lawson W.A., Feigelson E.D., 1994 Mon.Not.R.astron.Soc. 271, 967
Nordh L., Olofsson G. *et al.*, 1996 Astron.Astrophys. in press
Tatematsu *et al.*, 1993 Astrophys.J. 404, 643

TRACING THE INTERSTELLAR MEDIUM IN OPHIUCHUS ACROSS 14 ORDERS OF MAGNITUDE IN FREQUENCY

S.W. DIGEL[1,2], S.D. HUNTER[2] AND S.L. SNOWDEN[2,3]
[1] *Hughes STX*
[2] *NASA Goddard Space Flight Center*
[3] *Universities Space Research Association*

1. Introduction

The spectral range of ground and space-based observations has expanded rapidly in recent years. Essentially complete sky coverage is available in bands spanning more than 14 orders of magnitude in frequency, with varying sensitivities and resolutions. Remarkably, extended emission (and absorption) associated with the interstellar medium can be seen across this range. This poster illustrated the range of information now at hand for just a single interstellar cloud complex, that in Ophiuchus. This complex, well known for its abundant on-going star formation, is relatively nearby ($\sim$125 pc), and well removed from the plane, offering the advantages of large angular size and little confusion in most bands from background Galactic emission.

TABLE 1. Data Sources

Image	Bands	Reference
H I	21 cm	Burton 1985; Digel et al. 1992, unpub.
CO	2.6 mm	de Geus et al. 1990
IRAS	12, 60, 100 μm	Wheelock et al. 1994
DIRBE	1.2, 2.2, 3.5 μm	Hauser et al. 1995
DSS	0.6 μm	Postman et al., in preparation
ROSAT	0.75, 1.5 keV	Snowden et al. 1994
EGRET	>100 MeV	Hunter et al. 1994; Digel & Hunter 1994

Owing to the limited space available, only a sample of the images from various bands for the central part of the Ophiuchus complex are presented here (Figure 1); for descriptions of the observations and the analyses, the reader is referred to the references cited in Table 1. Analyses involving more than one band have been used to determine the spatial distributions, column densities, and temperatures of the interstellar gas and dust in Ophiuchus, as well as the line-of-sight distribution of X-ray emission in the Galaxy and the density of cosmic rays.

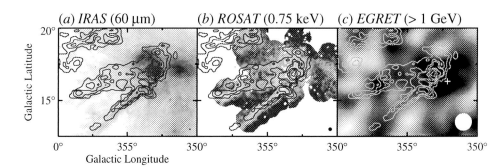

Figure 1. Overlay of contours of integrated CO intensity (de Geus *et al.* 1990) in the central part of Ophiuchus on images of (*a*) infrared, (*b*) X-ray, and (*c*) high-energy γ-ray intensity. The infrared emission traces dust heated by stars and star-forming regions, and is well correlated with the CO, which traces dense interstellar gas. The X-ray emission is strongly anticorrelated with the CO, as the interstellar gas and dust absorbs the soft X-ray background. The γ-ray emission is primarily from π^0 decay after cosmic-ray interactions in the interstellar gas. The white cross marks the position of quasar PKS 1622-253, a γ-ray source that complicates the analysis. The contour levels are 5, 10, 15, 25, 35, and 45 K km s^{-1}. Intensity ranges (white-black) are 5–250 M Jy sr^{-1} (infrared, logarithmically scaled), (0.4–3) $\times$ 10^{-5} s^{-1} arcmin^{-2} (X-ray), and (0.2–1.8) $\times$ 10^{-5} cm^{-2} s^{-1} sr^{-1} (γ-ray). The X-ray image is a mosaic of pointed observations and point sources have been removed.

References

Burton, W. B. 1985, Astron.Astrophys.Suppl., 62, 36
de Geus, Bronfman, & Thaddeus 1990, Astron.Astrophys., 231, 137
Digel, S. W., & Hunter, S. D. 1994, Proc. 2nd Compton Symp., ed. C. E. Fichtel *et al.* (New York: AIP Press), pp. 484–488
Hauser, M. G. *et al.* 1995, COBE DIRBE Expl. Supp., Vers. 2.0, COBE Ref. Pub. No. 95-A (Greenbelt, MD: NASA/GSFC)
Hunter, S. D. *et al.* 1994, ApJ, 436, 216
Snowden, S. L., Digel, S. W. & Freyberg, M. J. 1994, AAS HEAD meeting, Napa Valley
Wheelock, S. L. *et al.* 1994, IRAS Sky Survey Atlas Expl. Supp., JPL Publ. 94-11 (Pasadena: JPL)

A NEW CO SURVEY OF THE MONOCEROS OB1 REGION

R.J. OLIVER[1], M.R.W. MASHEDER[1] AND P. THADDDEUS[2]
[1] *Dept. of Physics, University of Bristol*
[2] *Center for Astrophysics*

1. Survey

This survey in the $(J = 1 \to 0)$ line of CO at 115 GHz was conducted in 1993 and 1994 using the 1.2m millimeter-wave telescope at the CfA (see Oliver et al. 1996 for more complete description). The survey area, $\ell = 196.0°$ to $\ell = 206.5°$, $b = -1.5°$ to $b = +3.5°$ was covered on a square grid in ℓ and b with spacings of $3'.75$ (0.4 FWHM) to give uniform sensitivity. All the spectra were position switched against positions measured be free of CO to 0.05 K. Each spectrum has 256 channels 0.25 MHz wide, giving a resolution of 0.65 km s^{-1} over 166 km s^{-1}. The channel to channel noise temperature in the original spectra was 0.24 K (RMS) which was reduced to 0.115 K per beam by smoothing to $10'$.

2. Results

This survey shows much more molecular gas in the Mon OB1 region and more complicated spatial and kinematic structures than previously known. Most of the emission is concentrated in two distinct velocity ranges (see Figure 1). The strongest emission is from gas in the Local arm at velocities between -5 and $+10$ km s^{-1}; the weaker is at velocities from $+15$ km s^{-1} at $\ell = 196°$ to $+35$ km s^{-1} at $\ell = 206.5°$ and is from the Perseus arm.

The large-scale velocity structure is not, however, completely described by two spiral arms. Notable departures are: (a) weak inter-arm emission at $\ell > 203°$, (b) two clouds with velocities that put them at kinematic distances beyond the Perseus arm (clouds 33 and 34 in the Figure 1) and (c) emission between $\ell = 197.5°$ and $\ell = 201°$ at 'forbidden' negative velocities ranging from -5 to -15 km s^{-1} (clouds 4 and 9 in the Figure 1), which cannot result from differential Galactic rotation in the third quadrant.

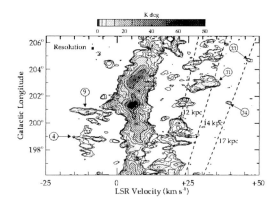

Figure 1. Longitude-velocity plot for CO detected toward the Mon OB1 region. The grey-scale ranges from 0.05 K deg (white) to 74.0 K deg (black). Contour levels are 0.05, 0.1, 0.2, 0.4, 0.8, 1.6, 3.2, 6.4, 12.8, 25.6, and 51.2 K deg.

The strongest CO emission ($\int T_{\mathrm{mb}} dv > 16$ K km s^{-1}) is well defined in discrete cores which are coincident with the young open clusters NGC 2245, NGC 2247, NGC 2264 and other regions which have been independently identified as star forming. A striking result of this new survey is the large area in which weak CO emission ($\int T_{\mathrm{mb}} dv < 5$ K km s^{-1}) is now detected; 80% of the area surveyed has emission $> +3\sigma$, 60% of which is below the detection limit of Blitz's pioneering survey. This weak emission is spatially diffuse and fragmented, extending as halos around the strong emission previously reported towards the Cone Nebula and Mon R1

The strong H I emission of the Perseus arm has a weak molecular counterpart. The molecular clouds identified in this distant arm are comparable in mass and physical dimensions to an unbiased sample of local clouds. Moreover, the luminosity-linewidth relation for this unbiased sample of outer Galaxy clouds shows no significant deviation from that determined for the Brand and Wouterloot (1995) inner Galaxy cloud sample.

The association of Perseus arm clouds with colour-selected *IRAS* point sources provides evidence that they are forming massive stars and that the efficiency of such star formation is approximately independent of cloud mass. The discovery of these outer Galaxy clouds involved in massive star formation reaffirms the importance of studying the distribution of CO in the outer Galaxy.

References

Brand J., and Wouterloot J.G.A., 1995, Astron.Astrophys., 303, 851
Oliver R.J., Masheder M.R.W., Thaddeus P., 1996, Astron.Astrophys.315, 578.

AN Hα SURVEY OF THE GALACTIC PLANE

Q.A. PARKER[1] AND S. PHILLIPPS[2]
[1] Anglo-Australian Observatory, Australia,
[2] University of Bristol, England

1. Introduction

We describe a major new Anglo-Australian proposal for a U.K. Schmidt Telescope (UKST) Hα survey of the Southern Galactic plane, Magellanic clouds and selected regions. The agreed survey will use a new 12×12 inch monolithic Hα interference filter of very high specification in combination with Tech Pan film. Tech Pan offers significant advantages for this work due to an inherent sensitivity at Hα and its extremely fine grain, high resolution, exceptional DQE, excellent imaging and low noise (*e.g.*, Parker *et al.* 1994). The combination of Tech Pan and a narrow band Hα filter will provide a survey of unprecendented area coverage, depth and resolution, superior to any previous optical survey of ionized gas in the galaxy. It should to lead to exciting new discoveries and avenues of research.

2. The Need for the Survey

Considering the great importance of variable star formation within and between galaxies it is surprising how little survey work has been done. Gunn made a systematic Hα survey in the 1950's using a 6inch telescope and coarse grained emulsion whilst other work has mainly concentrated on relatively small areas of interest for specific study (*e.g.*, Russeil, this volume, p. 186) or else is of low resolution (*e.g.*, Dennison *et al.*, this volume, p. 182). The only existing UKST wide area Hα survey work dates from the late 1970's (Davis, Elliot & Meaburn 1976). It was mainly carried out by coarse grained (though fast) 098 emulsion and a far from optimum Hα filter. Many parts of the plane remain to be covered, particularly the outer extensions beyond a few degrees from the Galactic equator. The northern Milky Way above Dec $= -20°$ has not been surveyed at all at good resolution. Progress in other wavebands highlights the paucity of the optical counterpart for the study of Galactic gas. A clear need for a high resolution optical survey to complement the studies at other wavebands is seen.

3. The Scientific Aims of the Survey

Hα emission lines from HII regions are one of the most direct optical tracers of current star formation. These lines also trace ionized gas in the ISM revealing, for example, stellar outflows in regions masked by strong reflection nebulae, shocks from high velocity galactic HI clouds, optical couterparts of supernova remnants, emission nebulosity close to young stellar sources and stellar wind-blown bubbles, sheets, filaments, *etc.* The spatial extent and detailed morphology of HII regions, OB associations and the wide variety of structures (shells, holes, bubbles, filaments and arcs) over a range of scales from a few arcseconds to tens of degrees can be particularly well studied by Hα imaging.

The nearest star forming complexes may lie as close as 100pc with physical sizes of tens of parsecs. Such structures often present large angular sizes (a degree or more) yet can exhibit fine detail at the arc-second level. If we wish to study the interaction of ionized structures with their large scale environment we clearly need surveys of considerable extent and at good resolution. CCDs cannot yet match the wide-area coverage, uniformity and resolution of the UKST/Tech Pan combination to which a wide angle, yet deep Hα survey of the Galactic plane is well suited. The high resolution of Tech Pan Hα imaging should enhance our ability to resolve out point sources from more extended emission (*e.g.*, detection of more distant planetary nebulae in the Magellanic clouds). Furthermore we should provide better definition of the sharp shock fronts seen around ionized gas clouds and be able to investigate the morphology and environment of Herbig-Haro objects and find more distant or less extended examples. We expect to determine accurate surface brightness and its variations in extended regions across the entire survey via independent CCD calibration.

Of particular interest on the large scale will be comparisons between Hα emission and other indicators of interstellar gas and/or star formation activity. These include Giant molecular clouds and the general molecular ISM traced by CO observations, radio continuum emission, γ-rays, HI, dust clouds or IRAS far infra-red flux. This survey should complement the radio maps being obtained by the ATNF, MOST, the new Parkes HI multibeam survey as well as those from mm wave telescopes.

4. Filter Specification

Barr associates was commissioned to supply a very high specification large monolithic Hα filter. Stringent optical requirements were necessary as the filter is to be used in a converging f/2.48 beam and the excellent imaging of Tech Pan must not to be seriously compromised. A 3-cavity design was adopted with ion-assisted deposition of refractory oxide on both sides of a SCHOTT RG630 R-band filter. The filter will have a clear aperture of 280mm minimum with thickness: 5.5 ± 0.5mm so as to fit inside existing UKST plateholders. The bandpass FWHM should be 70 ± 10Å with CWL of ~ 6590Å and 0.01% of peak out of band transmission. The transmitted wavefront should be $\lambda/4$ per 25mm or better with peak transmission $\geq 75\%$ across CA with 5% max variations.

5. Survey Size, Timescale and Availability

The survey is timely in respect of telescope loading as most competing photographic surveys are drawing to an end and there are few UKST non-survey projects when the Galactic plane is well placed. The narrow-band nature of the Hα filter means that the survey can proceed in grey/bright time when the sky is too bright for normal observations. The survey should commence towards the end of 1996 and will initially include 160 standard UKST fields. This will then be extended to the Galactic plane's outer regions and to declinations from +0 to +15 degrees. Exposures will be of the order of 3 hours. The survey will take 3–4 years to complete and represents the largest and most important new UKST photographic survey.

The survey will be made available to the astronomical community as quickly as possible though the consortium which involves groups in the U.K. (Bristol, Cardiff and ROE) and Australia (Sydney, Wollongong and AAO) will have some initial scientific exploitation rights. Original films will be scanned on the SuperCOSMOS machine at 10μm resolution to produce pixel maps of each field. Data will be disseminated on CD-ROM (1.8GB per scanned film). A small number of survey film copies may be made according to demand. The survey's photometric integrity will be assessed via independent narrow band photometry with CCDs on other telescopes and with reference to previously studied objects over a range of UKST fields.

References

Davies ,R. S., Elliot, K. H., & Meaburn,J. 1976, Mem.R.astron.Soc.,81,89.
Parker, Q. A., Phillipps, S., Morgan, D. H., 1995, Proceedings of "The Future Utilisation of Schmidt Telescopes," IAU Colloq. 148, p. 96, Chapman *et al.* (eds)

AN IMAGING SURVEY OF THE GALACTIC H-ALPHA EMISSION WITH ARCMINUTE RESOLUTION

B. DENNISON, J.H. SIMONETTI,
G.A. TOPASNA AND C. KELLEHER
Institute for Particle Physics and Astrophysics
Virginia Polytechnic Institute and State University

1. Introduction

We are presently carrying out a northern hemisphere survey of the Galactic Hα emission. Our instrument, the Virginia Tech Spectral Line Imaging Camera (SLIC) utilizes a fast objective lens ($f/1.2$) with a cryogenically-cooled TK 512×512 CCD. A filter wheel in front of the lens allows us to select interference filters, including a narrowband Hα filter and a broader bandpass continuum filter in a line free part of the spectrum. The fast optics in combination with the low noise CCD result in sub-Rayleigh sensitivity at confusion limited levels. (1 Rayleigh = $10^6/4\pi$ photons cm^{-2} s^{-1} sr^{-1}.) This corresponds to an emission measure sensitivity of ≈ 1 pc cm^{-6}. Parameters of our system are given in Table 1.

Our survey with its $1\rlap{.}'6$ resolution is complementary to Reynold's and collaborators' WHAM (Wisconsin H-Alpha Mapper) survey which collects detailed spectral information with approximately 50$'$ resolution. Other efforts in the southern hemisphere are also complementary to ours. These include a similar survey by J. Gausted, P. McCullough and D. Van Buren, a galactic plane Schmidt survey by Q. Parker and collaborators, and Fabry-Perot observations by D. Russeil and collaborators at Marseille. See papers by McCullough (p. 184), Parker (p. 179) and Russeil (p. 186), this volume.

2. Survey Availability

Our survey has covered 1.3 sr (as of 9/96). The survey already reveals a wealth of structure including very faint filaments away from the galactic plane. Because the survey is CCD based, it contains the full range of

TABLE 1. SLIC Parameters

Filter Bandpass	1.75 nm
CCD Quantum Efficiency	80% @ 650 nm
CCD Dark Current	10^{-3} e^{-1} s^{-1}
Focal Length	58 mm
Pixel Size	$1'.6$
Circular Field Diameter	$10°$
Tracking Precision	$2''$

surface brightness from sub-Rayleigh structures to 10^3 Rayleigh features (and brighter) near the plane, without saturation of the brighter features. The resolution of the survey makes it ideal for comparison with IRAS maps and X-ray observations, as well as HI observations in order to study the relationship between the various phases of the ISM and the warm ionized medium.

We are presently calibrating the existing observations and shall soon make them available as FITS images at our web site:

http://www.phys.vt.edu/~astrophy/halpha.html

We have also constructed mosaics of several extended regions near the galactic plane. These are available as GIF and JPEG images at our web site. The mosaics are presented in a nonlinear display to show qualitatively the detailed structure.

3. Early Scientific Results

We have carried out deep Hα observations of fields in which other groups have observed apparent anisotropies in the cosmic microwave background. Our results rule out any significant contribution from foreground galactic free-free emission (Simonetti *et al.* 1996).

We have also discovered a supershell inflated by stellar winds from young stars associated with the HII region W4. This structure was previously thought to be an open galactic chimney.

This research is supported by NSF grant AST-9319670 and a grant from the Horton Foundation to Virginia Tech.

References

Simonetti, J.H., Dennison, B. and Topasna, G.A., 1996. Astrophys. J. Letters, 458, L1

INTERSTELLAR CIRRUS OBSERVED IN BALMER Hα

P.R. MCCULLOUGH
Astronomy Dept., University of Illinois

1. Introduction

We present Hα images with 0.1° resolution and fields of view larger than 10°. In some regions of the sky, the Balmer Hα emission is correlated positively with IRAS 100 μm emission. Observations of such sensitivity and angular scale as these provide a new view of the interstellar medium of our Galaxy (see also Dennison *et al.*, this volume, p. 182) and may allow us to distinguish between Galactic foreground and cosmic background for both the free-free emission and the thermal dust emission associated with the warm ionized medium of the Milky Way.

2. Results

In one field at galactic latitude −65° (McCullough 1997), the sensitivity is limited in part by confusion: the anisotropy of the Hα surface brightness is ∼0.2 Rayleighs peak-to-valley, typically, which corresponds to an emission measure of ∼0.5 cm^{-6}-pc or an R magnitude of 32.7 per square arc second.

In this paper we present a mosaic centered on M31, which is the overexposed ellipse in the center. By blinking the Hα image with IRAS images made with SkyView (skyview.gsfc.nasa.gov), we notice that some objects are visible both in Hα and in the infrared, at 60 and 100 μm. (There are also objects visible in Hα but not in the infrared, and vice versa.) In the particular case below, the objects common to the infrared and Hα images are a few arcuate filaments, all with the intriguing property of being concave (not convex) as viewed from the Galactic plane, which is 15° above the top of the images below. The arcs have radii of curvature of 5° to 10°. They are unremarkable in the confusion of the IRAS images viewed alone but are seen clearly by blinking with the Hα image. Presumably they are shells of gas and dust expanding away from the Galactic plane.

INTERSTELLAR CIRRUS OBSERVED IN BALMER Hα

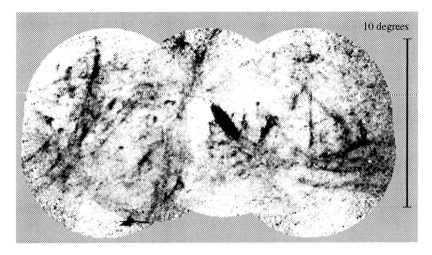

Figure 1. This Hα image is centered on 0h45m +41° [2000], (l,b) = (121°.6,−21°.8). It may be compared with the infrared image below. The filaments have surface brightnesses of ∼1 Rayleigh.

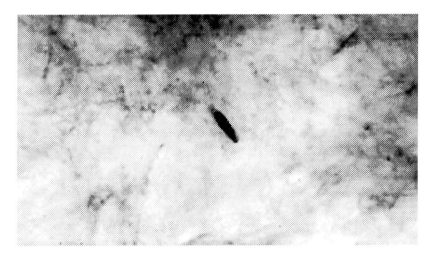

Figure 2. This IRAS 100 μm image is of the same region as the Hα image above.

Acknowledgements

We are pleased to acknowledge the creators of SkyView, a service that reduces the tedium of overlaying images.

References

McCullough 1997, Astron.J. submitted.

THE MARSEILLE OBSERVATORY Hα SURVEY: COMPARISONS WITH CO, 6 CM AND IRAS DATA

D. RUSSEIL[1], P. AMRAM[1], Y.P GEORGELIN[1],
Y.M GEORGELIN[1], M. MARCELIN[1], A. VIALE[1],
E. LE COARER[2] AND A. CASTETS[2]
[1] *Observatoire de Marseille*
[2] *Observatoire de Grenoble*

1. Introduction

The Marseille Observatory Hα survey supplies Hα velocities of the ionized hydrogen over large zones of the sky towards the galactic plane. This survey, led at the ESO La Silla, uses a 36 cm telescope equiped with a scanning Fabry-Perot interferometer and a photon counting camera (Le Coarer *et al.* 1992). About 250 fields ($39' \times 39'$) toward the galactic plane have already been covered (see Figure 1) with a spatial resolution of $9'' \times 9''$ and a spectral resolution of $5 \, \text{km} \, \text{s}^{-1}$. This allows us to observe the discrete HII regions and the diffuse ionized gas widely distributed between them and to separate the distinct layers found along the line of sight. HII regions are often grouped on the molecular cloud surface, then CO, radio continuum and recombination lines surveys of the galactic plane are also essential to distinguish the HII region-molecular cloud complexes met on the line of sight, and in order to take dynamical effects into account, such as the champagne effect, for the kinematic distance determination. Indeed, the spiral structure pattern determination requires avoiding any artificial spread by clearly identifying the giant complexes composed of molecular clouds, HII regions, diffuse ionized hydrogen widely surrounding them, and exciting stars. On the other hand the ionized gas data (Hα and recombination lines) associated with IRAS data help us to study the nature of the young objects constituent of these complexes and to assess their detectability. We present two fields from the Hα survey and parallel large scale investigations.

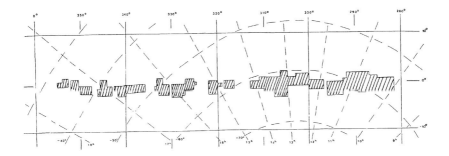

Figure 1. The coverage of the Hα sky survey between l=280° and 0°

2. The l=234°, b=0° Direction

The discrete HII regions met in this direction seem isolated, but the Hα data compared with CO and stellar ones, allow to identify two distinct HII regions-molecular complexes respectively at 2.1 kpc and 4.2 kpc. In S305 and S307, a "champagne" effect is observed (Russeil *et al.* 1995). If such an effect is neglected it will give an erroneous derived kinematic distance.

3. The l=298°, b=0° Direction

We enumerate the different kinematic components detected at the Hα wavelength along this line of sight.

- The first diffuse emission ($V_{lsr} \sim -3 \, km \, s^{-1}$) can be associated with the interstellar medium linked to the Sco-Cen association which is situated at about 130 pc (Degeus *et al.* 1989).
- The second diffuse emission ($V_{lsr} \sim -25 \, km \, s^{-1}$) at about 2.8 kpc accounts for the crossing of the near part of the Carina arm.
- The third diffuse emission ($V_{lsr} \sim -40 \, km \, s^{-1}$) and the HII region RCW64 at about 5.4 kpc (Brand 1986) present a strong rotation curve departure.
- The far complex: 8 radio sources (Caswell & Haynes 1987) are detected in Hα ($V_{lsr} \sim +25 \, km \, s^{-1}$). The 5 GHz radio continuum, IRAS and CO emission morphologies associated with kinematic distances allow to group the sources into one single complex at 10 kpc (located at the far part of the Carina arm). The IRAS and the 5 GHz radio continuum data suggest that the 3 radio sources without Hα counterpart can be explained by some absorbing cloud on the line of sight rather than by being burried inside the clouds.

4. Parallel Surveys

- The IRAS map investigation of the second field, has revealed 4 far-infrared extended sources without Hα nor radio counterpart: their nature remains ambiguous from color criteria only. In order to find potential common feature, we have itemized them through the fourth galactic quadrant: 177 sources have been selected. The study of the color ratios of these sources is in progress.
- A CO survey of the southern Milky Way has already been made with a 8.8' resolution (Bronfman *et al.* 1989), very different from the 9" resolution of the Hα survey. It allows to get only the large scale structure of molecular cloud. But, the hydrogen ionized by a newly formed star can exhibit a particular motion with respect to the molecular cloud within which was born. Then, to identify these particular regions we have begun a CO survey of the galactic plane, between l= 282° and l= 353°, with the SEST radiotelescope taking advantage of its 45"resolution to make a CO velocity probing of the Hα field for some particular areas and to measure the CO profile towards localized regions. About 130 regions has already been observed. At a first examination, certain highly emitting regions present no ^{13}CO counterpart and evident velocity departures.

5. Conclusion

The Hα survey of the Milky Way supplies optical observations of the galactic HII regions and allows to observe the large scale distribution of the ionized diffuse components. Comparing it with surveys at other wavelengths, we can identify and determine the distance of giant complexes and estimate the general absorption along the line of sight useful for the spiral structure study. This requires also the knowledge of the early type stars distribution. On the other hand, the use of multi-wavelength surveys would allow to identify the deeply embedded stars, to quantify the absorption, to derive the physical conditions of the HII regions, to establish the energy budget and eventually to clarify the nature of unclassified objects.

References

Brand J. 1986 Ph. D. Thesis, University of Leiden
Bronfman L., Alvarez H., Cohen R.S., Thaddeus P. 1989 Astrophys.J.Suppl., 71, 481
Caswell J.L., Haynes R.F. 1987 Astron.Astrophys., 171, 261
Degeus E.J., de Zeeuw P.T., Lub J. 1989 Astron.Astrophys., 216, 44
Le Coarer E., Amram P., Boulesteix J., *et al.* 1992 Astron.Astrophys., 257, 389
Russeil D., Georgelin Y.M., Georgelin Y.P., Marcelin M. 1995 Astron.Astrophys.Suppl., 114, 557

A LARGE-SCALE CO IMAGING OF THE GALACTIC CENTER

T. OKA[1], T. HASEGAWA[2], F. SATO[3], H. YAMASAKI[3],
M. TSUBOI[4] AND A. MIYAZAKI[4]
[1] *The Institute of Physical and Chemical Research (RIKEN)*
[2] *Institute of Astronomy, The University of Tokyo*
[3] *Department of Astronomy and Earth Sciences,*
Tokyo Gakugei University
[4] *Institute of Astrophysics and Planetary Science,*
Ibaraki University

1. Introduction and Observations

Molecular gas in the Galactic center region is spatially and kinematically complex, and its physical conditions are distinctively different from those of molecular gas in the Galactic disk (*e.g.*, Morris 1996). Relative paucity of current star formation activity, despite the abundance of dense molecular gas in this region, is one of the problem at issue.

Using the 2×2 multi-beam SIS receiver at the NRO 45m telescope (beamwidth $16''$), we have made CO high resolution mapping observations of the Galactic center region. We have collected about 44,000 ^{12}CO (J=1–0) spectra and over 13,000 ^{13}CO (J=1–0) spectra with $34''$ grid spacing. The ^{12}CO data cover almost the full extent of the Galactic center molecular cloud complex.

2. Morphology and Kinematics

Our CO images with extremely wide spatial dynamic range provide innovative view of the molecular gas in the Galactic center region (Oka *et al.* 1996, Hasegawa *et al.* 1996).

- Enormous number of molecular arcs and/or shells.
- Sharp emission edges and filamentary structures.
- An high velocity expanding molecular ring ($d \sim 50$pc) near the center.
- A molecular "smoke" originated from the central 10 parsecs.
- Large molecular flare at $l \simeq 1.3°$ consisting of many filaments.

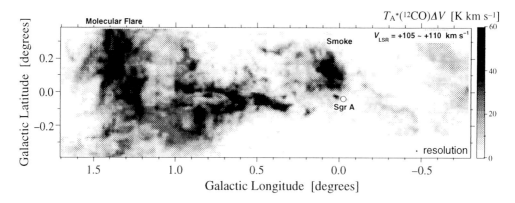

Figure 1. A gray-scale image of ^{12}CO ($J=1$–0) emission in the velocity range $V_{\rm LSR}=+105$ to $+110$ km s^{-1}. Molecular flare at $l \simeq 1.3°$ is associated with a number of molecular shells/arcs.

The morphology and kinematics of molecular gas strongly suggest that there are enormous number of supernova remnants in the Galactic center region, and that the region has experienced an era of active star formation in the recent past. A burst of star formation with a short duration time will have peak in the mechanical energy released as superwind about 5×10^7 years after a starburst (Heckman *et al.* 1993). The Galactic center may be currently in a "wind dominated" phase with quiescent star formation after a starburst.

The central region of the Galaxy may have been experienced recurrent bursts of star formation. A concentration of AGB stars in the central 100 pc (Lindqvist *et al.* 1991) could be remnants of ancient starbursts.

References

Hasegawa, T. *et al.* 1996, Nature, submitted
Heckman, T. M., Lehnert, M. D., and Armus, L. 1993, in "The Environment and Evolution of Galaxies," eds. J. M. Shull and H. A. Thronson, Jr. (Dordrecht: Kluwer), 455
Lindqvist, M., Winnberg, A., Habing, H.J., and Matthews, H.E. 1991, Astron.Astrophys.Suppl., 92, 43
Morris, M. 1996, in IAU Symposium 170, "CO:Twenty-five Years of Millimeter-wave Spectroscopy," in press
Oka, T., Hasegawa, T., Sato, F., Tsuboi, M., and Handa, T. 1996, in IAU Symposium 170, "CO:Twenty-five Years of Millimeter-wave Spectroscopy," in press

SURVEY OF CORRELATED FIR, HI, CO, AND RADIO-CONTINUUM EMISSION FEATURES IN THE MULTI-PHASE MILKY WAY

W.F. WALL[1,2,3] AND W.H. WALLER[3,4]
[1]*INAOE, México*
[2]*NASA/GSFC, USA*
[3]*StarStuff Incorporated*
[4]*Hughes STX, NASA/GSFC USA*

1. Introduction

The interstellar medium (ISM) is rich with structure on varying size scales, reflecting its diverse energetics and dynamics. A step toward understanding this structure is to enhance the visibility of the structure on finer size scales. We present maps of a section of the Galactic plane filtered with the Median Normalized Spatial Filter (MNSF, a median smoothed map is *divided* into the original map), which emphasizes higher latitude emission relative to that of the plane itself (see also Waller *et al.*, this volume, p. 194). The maps also illustrate the spatial correlations between the interstellar dust and the various gas phases: ionized, atomic, and molecular. The dust is represented by the IR emission maps of the IRAS survey, the ionized gas by the radio continuum maps at 1.4 GHz (Reich 1978, northern sky) and 2.3 GHz (Jonas & Baart 1995, southern sky), the atomic gas by surveys of the HI 21 cm line emission (Heiles & Habing 1974, Weaver & Williams 1973, Cleary *et al.*1979, Kerr *et al.* 1986), and the molecular gas by surveys of the CO $J = 1 \rightarrow 0$ line emission (Dame *et al.* 1987).

2. Processing

A rectangular window $15°$ wide in longitude and $0°\!.5$ high in latitude was used in the spatial filtering. The MNSF was used for all the maps, except for the lower signal-to-noise CO maps, in which unsharp masking (*i.e.*, smoothed map *subtracted* from the original map) was used. Maps of the dust-gas correlations were made by running a 2°5-square box over the filtered IRAS and gas emission maps and, using the Spearman test, zeroing all positions in the IRAS map with probabilities of correlation (between gas and dust emission intensities) less than 0.9. The maps of correlated dust emission features were color-coded: red for correlation with the radio continuum, green for the HI line, blue for the CO line. The three color-coded maps were then superposed to form a true-color image.

3. Results

The resultant fine-scale maps are shown in Figure 1. Comparison of the top and bottom panels of Figure 1 show that the fine-scale structure in the dust emission at 100 μm is in overall agreement with that of the gas. This is expected since the dust pervades most gas phases. The disagreements that do exist are sometimes attributable to the different way in which the CO map was filtered; the unsharp masking does *not* emphasize the CO emission away from the plane, as does the MNSF processing.

Clearly visible in the middle panel is a large number of FIR features that are correlated with the atomic phase. Analysis of positions within 11° of the Galactic plane for all longitudes shows that $\sim 70\%$ of the FIR structures are correlated with structured HI emission. The preponderance of atomic counterparts to the FIR structures evident away from the mid-plane indicates energetics that are sufficient to re-organize the ISM without ionizing it. Also, it should be mentioned that many of the correlated FIR-HI features away from the mid-plane may also correlate with the molecular gas, which can be tested when a higher signal-to-noise ^{12}CO J $= 1 \to 0$ map becomes available.

Acknowledgements

This research was supported in part by a NASA ADP contract to StarStuff Incorporated and by a CONACyT Grant to INAOE.

References

Cleary, M. N., Heiles, C., & Haslam, C. G. T. 1979, Astron.Astrophys.Suppl., 36, 95
Dame, T. M., Ungerechts, H., Cohen, R. S., de Geus, E. J., Grenier, I. A., May, J., Murphy, D. C., Nyman, L.-Å., & Thaddeus, P. 1987, Astrophys.J., 322, 706
Heiles, C. & Habing, H. J. 1974, Astron.Astrophys.Suppl., 14, 1
Jonas, J.L. & Baart, E. E. 1995, Astrophys.Sp.Sc., 230, 351
Kerr, F. J., Bowers, P. F., Jackson, P. D., & Kerr, M. 1986, Astron.Astrophys.Suppl., 66, 373
Reich, W. 1978, Astron.Astrophys., 64, 407
Weaver, H. & Williams, W. 1973, Astron.Astrophys.Suppl., 8, 1

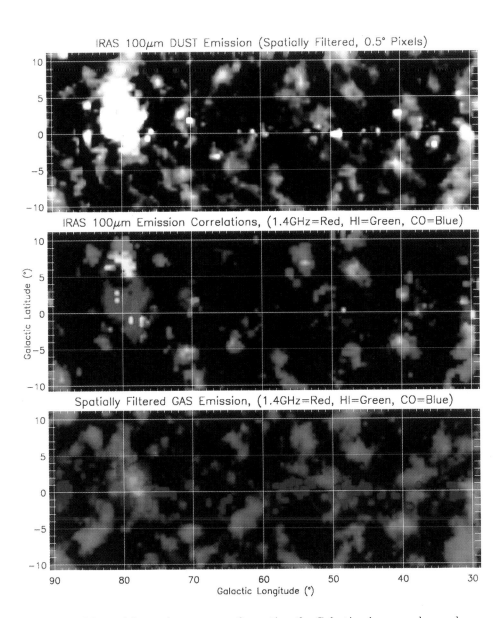

Figure 1. Maps of fine-scale structure of a section the Galactic plane are shown above. The top panel shows the dust emission at 100 μm and the bottom panel shows the corresponding map of gas emission. The middle panels represent correlations of the 100 μm emission with those of the gas components.

SURVEY OF FINE-SCALE STRUCTURE IN THE FAR-INFRARED MILKY WAY

W.H. WALLER[1,2,3], F. VAROSI[1,2,3],
F. BOULANGER[4] AND S.W. DIGEL[1,3]
[1] *Hughes STX*
[2] *StarStuff Inc.*
[3] *NASA/GSFC*
[4] *Institut d'Astrophysique Spatiale, Université Paris XI*

1. Mapping the Galactic ISM

What is the general morphology of the diffuse interstellar medium? Is it mostly uniform or clumpy? Are the clumps mostly in the form of spheroidal clouds, sinuous filaments, extended sheets, or discrete shells? And do the *clumps* or the *voids* better define the overall structure? By addressing these morphological questions, one can better constrain the dynamical processes that are most responsible for shaping and energizing the ISM.

By mapping the FIR emission from dust that has been warmed by the interstellar radiation field (ISRF), one can trace both the cool and warm phases of the diffuse ISM. These two phases represent most of the mass in the diffuse ISM. Recent data products produced by IPAC from the IRAS mission database provide the best resolved and most complete mapping of the Galactic FIR emission. And through spatial filtering, the strong gradient in brightness towards the Galactic midplane can be eliminated, thereby revealing the fine-scale FIR structure throughout the Galaxy.

2. Revealing Structure in the FIR Milky Way

We have produced a $360° \times 60°$ mural of $100\mu m$ emission in the Milky Way from $60° \times 60°$ mosaics. These mosaics were made from the IRAS Infrared Sky Atlas "plates" using the *SkyView* Virtual Observatory (found at http://skview.gsfc.nasa.gov/skyview.html). By applying a median normalizing spatial filter, we were able to eliminate the strong gradient in

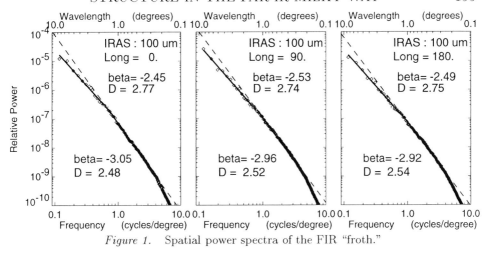

Figure 1. Spatial power spectra of the FIR "froth."

brightness towards the Galactic midplane. The resulting images reveal a "froth" of superposed filaments, voids, and shells (see Figure 2).

This fine-scale structure extends all the way down to the Galactic midplane. Moreover, it scales in intensity with the smoothly varying background, independent of latitude, thus suggesting that the fine-scale residual emission is co-extensive with the smooth background. *We conclude that the fine-scale structure is not merely of local origin, but consists of both nearby and more distant features in the disk.*

Although we had expected to find morphological evidence for supernova-driven "worms" or "chimneys" rooted in the Galactic plane, our processing shows the FIR fine-scale structure to be more complex (*e.g.*, less coherent and less rooted) as viewed in projection. The observed FIR "froth" is just beginning to be identified with other tracers of the interstellar medium (*e.g.*, CO, HI, and radio-continuum—cf. Wall and Waller, this volume, p. 191).

Analysis of the spatial statistics shows that the FIR fine-scale structure is self-similar with a angular power-law exponent of $\beta \approx -3$ and and a fractal dimension of D$\approx$ 2.5—similar to that found in isolated cirrus and molecular clouds. On scales larger than 1.5°, the power-law exponent flattens to $\beta \approx -2.5$, perhaps indicating a change in the characteristic structure (see Figure 1). This could be due to different dynamical inputs organizing the small and large-scale structures (*e.g.*, turbulence and diffusion on small scales vs. macroscopic winds and shock fronts on larger scales).

Acknowledgements

This research was supported in part by a NASA Astrophysics Data Program contract (#NAS5-32591) to StarStuff Incorporated.

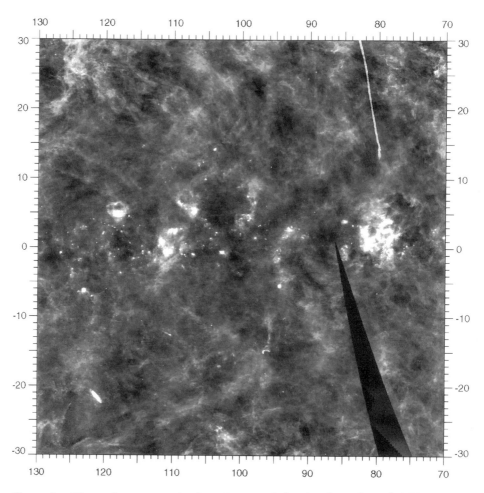

Figure 2. Fine-scale structure in the 100 μm emission for the 60° × 60° field centered at $(\ell, b) = (100°, 0°)$. Prominent emitting structures include the Cygnus star-forming region at $(G78 + 2)$, IC 1396 at $(G99 + 4)$ which seems to form part of a giant shell, NGC 7822 at $(G118 + 6)$ which also shows a shell-like morphology, and the galaxy M31 at $(G121 - 21)$. The remaining fine-scale features probably represent a superposed mix of nearby and more distant filaments, voids, and shells (see also Wall & Waller, this volume, p. 191).

Part 4. Galactic Structure

FORMATION AND EVOLUTION OF THE MILKY WAY

S.R. MAJEWSKI
Department of Astronomy, University of Virginia

1. Introduction

To paint with broad brush strokes, the spectrum of Galactic formation models has as extrema what may be termed the "fast and smooth" and the "slow and lumpy" scenarios. Appropriate or not, to ascribe as champions of these views the works of, respectively, Eggen *et al.* (1962, ELS hereafter) and Searle & Zinn (1978, SZ hereafter) has become *de rigueur*—though it is prudent to bear in mind the soft focus of historical perspective, and the maxim that "no one remembers what one has said, only what others say one has said" (on this point, see Sandage 1990 and Majewski 1993, section 1). If recently published textbooks (whose task is to relay pedantically what the experts have said) mirror our present understanding of the formation of the Milky Way, then it would seem that the experts have done little to negotiate the different formation pictures to offer a coherent compromise:

> [Figure 18.24 shows the] schematic sequence of the collapse and condensation of a large cloud of gas and dust to make globular clusters and the Galaxy's disk. Because of its original spin, the matter eventually makes a disk with a central bulge. *The stars in the globular clusters form in the cloud before the disk had developed. The entire process took less than one billion years.* [Zeilik 1996].

> Astronomers reason that, early on, our Galaxy was rather irregularly shaped, with gas distributed throughout its volume. *Possibly it formed via the merger of several systems, as depicted...* When the stars formed during these stages, there was no preferred direction in which they moved and no preferred location in which they were found. *In time*, rotation caused the gas and dust to fall to the Galactic plane and form a spinning disk...*The older stars were left behind, forming the halo...* New stars forming in the disk inherit its overall rotation and so orbit the Galactic center on ordered, circular orbits. [Chaisson & McMillan 1996].

These elementary descriptions (emphases mine) touch on some of the concepts typically allied to the "fast and smooth" scenario—*e.g.*, fast collapse and Galactic "spin-up" during collapse, abundance gradients between and within Galactic populations, and relatively smooth (*i.e.*, "organized" or sequenced) spatial, kinematical and chemical transitions between populations—as well as notions often coupled to the "slow and lumpy" model—namely, an extended (halo) configuration, unorganized kinematically and chemically, born of the merger of initially isolated systems. The theme of this brief is that our understanding of the Galaxy may not be nearly as confused as the above synopses would imply; progress is being made toward a description which, not surprisingly, is a compromise of the extreme models often proffered. This is not to say that a universally accepted formation scenario exists; however, recent observational successes have yielded a number of important, previously missing, pieces to the puzzle. At the simplest level, the surfacing compromise involves a "slow and lumpy" scenario for the extended parts of the Galaxy, referred to here as the "halo," with aspects of the "fast and smooth" paradigm reserved for the interior regions of the Galaxy, the thin and Intermediate Population II (IPII), "thick" disks.

I conclude with an appeal to large, systematic surveys as the promise toward resolving the perplexities of Galactic structure and formation.

2. A Slow Formation for the Halo

Accurate relative age dating of star clusters is perhaps one of the most important products of the CCD photometry age. The result of this work is clear evidence for a rather protracted formation epoch for globular clusters (even excluding the so-called "disk" globulars). Whatever the absolute value for the *mean* age of the globular clusters (a topic still of some considerable debate), it is clear that the globular cluster age dispersion is substantial—at least several Gyr, but in any case much larger than a "rapid" collapse scenario would imply. The bevy of "rogue" clusters of youthful age (2–4 Gyr less than the mean cluster age) is now large enough (Rup 106, Pal 12, Ter 7, N6366, Arp 2, IC4499; see Gratton & Ortolani 1988, Stetson *et al.* 1989, Buonanno *et al.* 1993, 1994, Da Costa *et al.* 1992, Da Costa & Armandroff 1995) that they can no longer be considered exceptional cases, but rather they represent an important aspect of Milky Way construction.

Even more startling a threat to old paradigms is that while the age spread of halo globular clusters widens, so too does the age spread of open clusters (Janes & Phelps 1994) and other "disk" populations. The result is a growing disk/halo age overlap that breaks down the traditional view of a distinct "age of halo formation" followed by an "age of disk formation." The age of the open cluster Be 17 is now given as 12^{+1}_{-2} (Kałużny 1994, Phelps 1997). Though subject to vagaries of absolute age scales, through relative age dating schemes (Janes & Phelps 1994) it is clear that this object is *older* than the youngest halo globular clusters. The latter are older than 47 Tuc, the prototypical "disk" globular cluster. Coupled with this blurring in distinctive ages is a blurring of the distinction between "globular" and "open" clusters; several objects, *e.g.*, Lyngå 7 (Ortolani *et al.* 1993) and BH176 (Kałużny 1995) may be "transitional" between the two cluster types.

An upper age for the disk similar to that provided by the open clusters is also yielded by Strömgren photometry studies of disk field stars (Edvardsson *et al.* 1993, Nordström *et al.* 1996), as well as the results of combining the most recent white dwarf luminosity functions with the latest white dwarf cooling theory (Hernanz *et al.* 1994, Isern *et al.* 1995, 1997).

It is now evident that halo globular clusters were forming even after the Galactic disk had initiated star formation. This provokes the vexing question of the location of the formation sites for the young halo globular clusters (Section 3). Relevant to this question is the fact that, when divided into "young" and "old" halo globular groups (on the basis of their horizontal branch morphology, *i.e.*, those exhibiting the most significant second parameter effect), distinctively different chemical, kinematical and spatial distributions obtain. The "young" halo clusters have more extreme kinematics, are distributed in an extended, spherical distribution, and exhibit no metallicity gradients, while the "old" clusters as a group exhibit less extreme kinematics and are distributed in a flattened configuration (van den Bergh 1993, Zinn 1993a, Da Costa & Armandroff 1995; see also the earlier work of Rodgers & Paltoglou 1984). The "old halo" globular clusters and their apparently natural extension into the disk globulars, when arranged according to [Fe/H], *do* demonstrate kinematical and spatial gradients reminiscent of the ELS "spin-down" concept (Zinn 1993). It has been suggested (Zinn 1993) that the disk+old halo globulars may trace the spin-down of a single, dissipational the Galactic disk. Similarities between properties of this latter set of globulars and those of the Galactic IPII suggest they represent the same Galactic population (Majewski 1993, 1995; Section 5).

On the other hand, the null or retrograde mean velocity of the "young halo" globulars, their lack of an [Fe/H] gradient, and their spatial distribution are inconsistent with their formation as part of a grand collapse that may have formed the inner Galaxy. Coupled with their apprently younger ages, an origin tied to accretion is implied, along the tenets of the SZ scheme. If the "old halo+disk" component represents some form of grand collapse of a primordial (but eventually self-enriching), Galaxy-sized gas cloud, the *later* formation of some "young halo" globulars requires a reservoir of gas that did not participate in the collapse, but either left behind in, or introduced into, the outermost parts of the early Galactic system.

3. The Role of the Galactic Satellites and High Velocity Clouds

Maps (*e.g.*, Wakker 1991a) of the H I high velocity clouds (HVCs) make evident two points, *assuming that the HVCs are at large distances and not associated with the Galactic plane*: (1) There is H I in the halo now. Though presently meager on galactic mass scales, the presence of *any* such gas at the present stage of Galactic evolution encourages contemplation of previously larger, available reservoirs. (2) If represented by the distribution of this gas, the halo is "lumpy" now. Murphy *et al.* (1995) find 37% of lines of sight to have high velocity H I with $N_H = 7 \times 10^{17}$ cm^{-2}. Much of this is concentrated into large, often elongated complexes, like the Magellanic Stream and Complexes A, M and C. Unfortunately, there is still only minimal data bearing on the distances to the HVCs, but the weight of evidence seems to support cloud distances (especially when coupled to their velocities) commensurate with halo membership for HVCs (Danly 1989, Danly =it et al. 1993, de Boer *et al.* 1994, Keenan *et al.* 1995, Wakker *et al.* 1996).

Are the HVCs primordial clouds left *in situ*? The gas metallicity is low but highly variable from complex to complex—*e.g.*, 0.002 to > 0.07 the solar value for Ca$^+$/H I (but these measurements are subject to many uncertainties, such as net exchange with dust; Schwarz *et al.* 1995). The general conclusion is that this gas cannot be *entirely* primordial (Schwarz *et al.*; Wakker 1991b gives a more complete treatment of HVC origin scenarios).

Was this gas stripped out of Galactic satellites? The stringy appearance of the various complexes is reminiscent of the Magellanic Stream, which is generally accepted to represent gas pulled out of the Magellanic system. There are some intriguing correlations between the orientations of these HVC chains and aligned families of Galactic satellites, such as the proposed "Magellanic" and "Fornax-Leo-Sculptor" planes (Kunkel 1979, Lynden-Bell 1982). The additional correlation of second parameter globular clusters to these planes (Lynden-Bell 1982, Majewski 1994, Lynden-Bell & Lynden-Bell 1995, Fusi Pecci *et al.* 1995) as well as the increasing evidence for tidal disruption of satellites (see below), suggests closer cosmogonical ties between satellite galaxies, young halo clusters, halo field stars and HVCs should these spatial correlations prove more than chance. Kinematical data will be essential to testing this hypothesis. For example, a simple accounting of the presently restricted distribution of the statistical phase space {spatial position, radial velocity, proper motion estimated from orientation} by Kunkel *et al.* (1997) suggests an original population of 10^3–10^4 dwarf galaxies; since observational evidence for even short look-back times does not support such large overpopulations, Kunkel *et al.* propose that tidal disruption of formerly larger satellites may have spawned an abundance of débris particles populating a small portion of their statistical phase space. Analysis of the growing number of absolute proper motions for Galactic satellites shows a remarkable coincidence between the orbits of the Magellanic Clouds, Ursa Minor and Draco, and consistent with a trailing Magellanic Stream in the common orbital plane (see Majewski *et al.* 1997); however, recent evidence is less supportive of the Fornax-Leo-Sculptor alignment of satellites (Majewski *et al.* 1997) and clusters (Dauphole *et al.* 1996). The discovery (Mirabel *et al.* 1992) of dwarf galaxies forming in the tidal débris of interacting galaxies provides an especially relevant paradigm to consider in the context of these possible alignments.

Another obvious location for post-collapse halo gas reservoirs is in the dwarf satellites themselves. As a group, they demonstrate a remarkable age spread, even excluding the presently gas-rich Magellanic system. Apart from Sagittarius (Sgr), all of the satellite dSph's contain old stellar populations, but half of them also have evidence for intermediate-aged populations, as young as 8–10 Gyr old in Leo II (Mighell & Rich 1996) and ≈ 6 Gyr old in Carina (Smecker-Hane *et al.* 1994). Star formation histories are clearly varied from satellite to satellite. For example, Carina shows evidence for distinct star formation bursts whereas Leo II had an extended star formation phase wherein most of the stars formed throughout a 6–8 Gyr period. The individual nature of this star formation leads to interesting complexities, such as the fact that Carina has two stellar populations with different ages but about the same [Fe/H], whereas Sgr has two populations with the same age but different [Fe/H] (Sarajedini & Layden 1995).

The existence of multiple star generations in individual dSphs raises the question of how such small galactic systems sustain a major burst of star formation, yet retain enough gas to instigate a succeeding burst? The answer may be related to the apparently large mass-to-light ratios suggested by internal velocity dispersions, though it is clear that a high M/L explanation for (at least some of) the high velocity dispersions is still disputed by tidal model enthusiasts (Kuhn *et al.* 1996, Burkert 1997; see Pryor 1996 and Irwin & Hatzidimitriou 1995 for recent advocations to the dissenting viewpoint). The possibility of some kind of pressure confinement is raised by recent results of Weiner & Williams (1996), whose detection of Hα emission on the leading edges of three major H I clouds in the Magellanic Stream suggests the presence of a large density, $n_{\rm H} \approx 10^{-4}$ cm^3, of hot, ionized gas at large distances from the Galaxy. (Note that Moore & Davis 1994 argue for such a hot gas phase based on the present configuration and dynamics of the Magellanic Stream.) These results also support a ram-pressure stripping origin, rather than a tidal origin, for the Magellanic Stream.

4. More Lumpy Structure in the Halo

The above comments on the origin of the Magellanic Stream notwithstanding, it is now clear that tidal intrreactions do play a role in the formation of the Galactic halo. The example of the tidally disrupted Sgr dwarf galaxy (Ibata *et al.* 1995) confirms earlier suspicions that the destruction of Galactic satellites contributes both clusters and stars to the halo milieu. Satellite mergers should leave behind fossil evidence in the form of phase space substructure for field stars. Tidally disrupted stellar systems produce long-lived, coherent streams of stars strung out along the orbit of the decaying parent object (see models in McGlynn 1990, Moore & Davis 1994, Johnston *et al.* 1996), analogous to the streams of meteoroid débris left along the paths of comets orbiting (and slowly destroyed by) the sun. To the extent that mergers contribute to the formation of the halo dictates whether tidal streams are an extra signature overlaying a dynamically relaxed stellar population (formed by other processes), or whether the halo has a more complex structure, like "a can of worms" (Majewski *et al.* 1996b).

Eggen (cf. 1996a,b and references therein) has long championed the idea of the existence of moving groups of metal poor, high velocity stars in the solar neighborhood. Other evidence for halo phase space substructure has been hinted at by various tentative findings of possibly more distant halo moving groups (see references in Majewski *et al.* 1996a), typically manifested as unexpected clumpings in position and radial velocity in *in situ* surveys of halo stars. That the Sgr dwarf was discovered in a similar way, albeit with much greater statistical significance, suggests a logical connection between these moving groups and disrupted dwarf satellites; presumably the less significant "detections" of radial velocity groups reported by others correspond to older, now more tenuous, tidal streams.

We (Majewski *et al.* 1994, 1996b) have been investigating evidence of apparent phase space substructure in a deep survey of stellar proper motions, photometry and spectroscopy towards the North Galactic Pole (SA57). Proper motion data alone (Majewski 1992) gave rise to the unexpected result that stars beyond about 5 kpc from the Galactic plane (the distance at which the IPII stellar density drops off sufficiently that the halo dominates) showed a significant mean *retrograde* rotational velocity. This alone is problematical for grand collapse scenarios for the halo, and implies a significant contribution of halo stars formed in some other way. Among the retrograde stars, Majewski (1992) identified a candidate halo "moving group," which subsequent spectroscopic analysis has supported by way of independent coherence in the group radial velocities. Full velocity information for stars in this magnitude-limited survey reveals, moreover, that the halo appears to contain a high degree of dynamical clumpiness: very few halo stars in the SA57 sample do *not* appear to belong to one of several, relatively distinct velocity clumps. This evidence suggests a dynamically young, *lumpy* structure for the halo, which may be *dominated* by the débris of tidally disrupted stellar agglomerations, either globular clusters or dwarf satellite galaxies, which *slowly* (over a Hubble time) have been dissolving into the "melting pot" that is the stellar halo.

5. The Division of Old Stellar Populations

Distant halo stars, many kpc from the Galactic mid-plane, exhibit extreme kinematics with significant phase space substructure that suggests an origin by way of, and perhaps dominated by, accretion events (Majewski *et al.* 1996b). At the same time, the IPII or "extended"/"thick" disk shows evidence of being a ubiquitous, vertically extensive, yet still predominantly flat, structure, and it contains both extremely metal-poor stars and stars that rotate more slowly with increasingly larger distances from the Galactic plane (in the interest of brevity, the reader is referred to Majewski 1995 for references supporting these statements). Therefore, it is useful to consider a new description of Galactic field star populations that parallels Zinn's division of the Galactic globular cluster system into: 1) a "younger," spheroidally distributed population that is non- or retrograde rotating and with an origin possibly tied to accretion processes; and 2) an "older," flattened distribution showing ELS "spin-up" and with a origin possibly related to dissipational collapse. Such a paradigm for both globular clusters and stars provides not only economy of hypothesis, but simultaneously resolves a number of various Galactic survey results previously considered inconsistent with one another (see Majewski 1993).

This physical description of stellar chemical, spatial and kinematical distributions points to a model of stellar origins that essentially represents a marriage of the SZ and ELS models. There appear to be (Sandage 1990, Majewski 1993) two populations of older, metal poor stars in the Galaxy: one closely aligned with traditional notions of a more or less spherically distributed, kinematically extreme Galactic "halo," but now associated with an *accreted*, SZ-like formation; and a second population with spatial and kinematical properties usually reserved for the Galactic IPII, or "extended"/"thick" disk population, but having an origin more along the lines of the global collapse envisioned by ELS (albeit on a slower, dissipational timescale). The connection or independence of the latter population to the Galactic "thin" disk is still a subject of great controversy; however, a number of lines of evidence support a similarly elderly age for both the thin disk and IPII (Section 2; references in Majewski 1995). Subsequent discussions of these new "dual halo" Galaxy scenarios refer to the two old, metal-poor populations, respectively, as the "accreted" and "contracted" halo (Norris 1994), or the "high" and "low" halo (Carney et al. 1996).

6. What's Old is New Again: Systematic Surveys of the Galaxy

At the turn of this century, Kapteyn (1906) devised the *Plan of Selected Areas* ("SAs") as a means to "attack" systematically the problem of understanding "the sidereal world," *i.e.*, the Milky Way system. There ensued a period of great activity, whereby substantial amounts of effort the world over was devoted to contributing photometry, astrometry, and spectroscopy of stars in Kapteyn's 206 SAs. The grand scope and initial perceived importance of the *Plan* was such that coordination was essential, and this prompted the eventual creation of *IAU Commission 32: Selected Areas* as well as the *Subcommittee on Selected Areas of IAU Commission 33: Structure and Dynamics of the Galactic System*. Coordination of the *Plan* was also the subject of two of the earliest IAU Symposia (Nos. 1 and 7).

Since the mid-part of this century—when it was discovered that the spiral nebulae were extragalactic systems, and also coincident with the rise in emphasis on star *clusters* as a tool for Galactic astronomy (Paul 1981)—activity on the SAs has unfortunately strongly declined (IAU Comm. 32 no longer exists). However, the wisdom and value of a systematic and coordinated astrometric, photometric and spectroscopic approach to studying the Milky Way is now more obvious. There is growing evidence that various subsystems of the Galaxy (*e.g.*, the bulge with its bar, the disk with its warp, and the apparently dynamically unrelaxed halo with its gaseous and tidal, stellar streams) are highly asymmetric, and therefore not described adequately by global models derived from only a few lines of sight (a common practice). Kapteyn's original vision of a fully integrated photometric, astrometric and spectroscopic survey has never been fully realized, though the decline in SA activity has overlapped with the development of modern instrumentation that might be brought to bear on the program with far more efficiency, precision and depth than he could have imagined.

To be sure, forays along this path are being made (for example, those by the Basel group, *e.g.*, Fenkart 1989; Sandage 1983; the Besancon group, *e.g.*, Ojha *et al.* 1996; and by the author and collaborators, *e.g.*, Siegel *et al.* 1997), but a satisfactory Galactic stucture "solution" may not be possible without a grand, all-out, "full-sky" attack as Kapteyn envisioned. Such systematic optical surveys of field stars will need to be integrated both with complementary work on Galactic clusters and satellites, as well as information at other wavelengths on the gaseous phases of the Milky Way: *e.g.*, X-ray surveys will provide necessary constraints on/checks for diffuse, ionized gas, while much more work is needed to understand the relationship of the H I radio data, especially the HVCs, to the stellar populations of the Milky Way and its satellite system. Ultimately, all of these Galactic studies will need to be combined with similar studies of external galaxies before a satsifactorily complete picture of the formation and evolution of normal, Milky Way-like galaxies can be obtained.

References

Buonanno, R., *et al.* 1993, Astron.J., 105, 184
Buonanno, R., *et al.* 1994, Astrophys.J., 430, 121
Burkert, A. 1997, Astrophys.J.Lett., 474, L99
Carney, B. W., Laird, J. B., Latham, D. W. & Aguilar, L. 1996, Astron.J., 112, 668
Chaisson, E. & McMillan, S. 1996, "Astronomy Today, 2nd Ed.," (Prentice Hall: Englewood Cliffs, NJ), p. 499
Da Costa, G. S. & Armandroff, T. E. 1995, Astron.J., 109, 2533
Da Costa, G. S, Armandroff, T. E. & Norris, J. E. 1992, Astron.J., 104, 154
Danly, L. 1989, Astrophys.J., 342, 785
Danly, L, Albert, C. E. & Kuntz, K. D. 1993, Astrophys.J.Lett., 416, L29
Dauphole, B., *et al.* 1996, Astron.Astrophys., 313, 119
de Boer, K., *et al.* 1994, Astron.Astrophys., 286, 925
Edvardsson, B., *et al.* 1993, Astron.Astrophys., 275, 101
Eggen, O. J. 1996a, Astron.J., 112, 1595
Eggen, O. J. 1996b, Astron.J., 112, 2661
Eggen, O. J., Lynden-Bell, D. & Sandage, A. R. 1962, Astrophys.J., 136, 748
Fenkart, R. 1989, Astron.Astrophys.S, 81, 187
Fusi Pecci, F., Bellazzini, M., Cacciari, C. & Ferraro, F. R. 1995, Astron.J., 110, 1664
Gratton, R. G. & Ortolani, S. 1988, Astron.Astrophys.Suppl., 73, 137
Hernanz, M., *et al.* 1994, Astrophys.J., 434, 652
Ibata, R. A., Gilmore, G. & Irwin, M. J. 1995, Mon.Not.R.astron.Soc., 277, 781
Irwin, M. & Hatzidimitriou, D. 1995, Mon.Not.R.astron.Soc., 277, 1354
Isern, J., *et al.* 1995, in "The Formation of the Milky Way," eds. E. J. Alfaro & A. J. Delgado, (Cambridge University Press, Cambridge), p. 179
Isern, J., *et al.* 1997 ASP Conf. Ser. 112, eds. A. Burkert, D. H. Hartmann & S. R. Majewski, (ASP: San Francisco), p. 181
Janes, K. A. & Phelps, R. L. 1994, Astron.J., 108, 1773
Johnston, K.V., Hernquist, L. & Bolte, M. 1996, Astrophys.J., 465, 278
Kałużny, J. 1994, Acta Astron., 44, 247
Kałużny, J. 1995, Astron.Astrophys., 300, 726
Kapteyn, J. C. 1906, "Plan of Selected Areas," (Hoitsema Brothers: Groningen)
Keenan, F. P., *et al.* 1995, Mon.Not.R.astron.Soc., 272, 599
Kuhn, J. R., Smith, H. A. & Hawley, S. L. 1996, Astrophys.J.Lett., 469, L93
Kunkel, W. E. 1979, Astrophys.J., 228, 718
Kunkel, W. E., Demers, S. & Irwin, M. 1997, in preparation
Lynden-Bell, D. 1982, Observatory, 102, 202
Lynden-Bell, D. & Lynden-Bell, R. M. 1995, Mon.Not.R.astron.Soc., 275, 429
Majewski, S. R. 1992, Astrophys.J.Suppl., 78, 87
Majewski, S. R. 1993, Ann.Rev.Astron.Astrophys., 31, 575
Majewski, S. R. 1994, Astrophys.J.Lett., 431, L17
Majewski, S. R., Munn, J. A. & Hawley, S. L. 1994, Astrophys.J.Lett., 427, L37

Majewski, S. R. 1995, in "The Formation of the Milky Way," eds. E. J. Alfaro & A. J. Delgado, (Cambridge: Cambridge University Press), p. 199
Majewski, S. R., Hawley, S. L. & Munn, J. A. 1996a, in ASP Conf. Ser. 92, eds. H. Morrison & A. Sarajedini, (ASP: San Francisco), p. 119
Majewski, S. R., Munn, J. A. & Hawley, S. L. 1996b, Astrophys.J.Lett, 459, L73
Majewski, S. R., Phelps, R. L. & Rich, R. M. 1997, in ASP Conf. Ser. 112, eds. A. Burkert, D. H. Hartmann & S. R. Majewski, (ASP: San Francisco), p. 1
McGlynn, T. A. 1990, Astrophys.J., 348, 515
Mighell, K. & Rich, R. M. 1994, in ASP Conf. Ser. 92, eds. H. Morrison & A. Sarajedini, (ASP: San Francisco), p. 528
Mirabel, I. F., Dottori, H. & Lutz, D. 1992, Astron.Astrophys., 256, L19
Moore, B. & Davis, M. 1994, Mon.Not.R.astron.Soc., 270, 209
Murphy, E. M., Lockman, F. J. & Savage, B. D. 1995, Astrophys.J., 447, 642
Nordström, B., et al. 1997, in ASP Conf. Ser. 112, eds. A. Burkert, D. H. Hartmann & S. R. Majewski, (ASP: San Francisco), p. 145
Norris, J. 1994, Astrophys.J., 431, 645
Ojha, D. K., et al. 1996, Astron.Astrophys., 311, 456
Ortolani, S., Bica, E. & Barbuy, B. 1993, Astron.Astrophys., 273, 415
Paul, E. R. 1981, Journal for the History of Astronomy, 12, 77
Phelps, R. L. 1997, Astrophys.J., in press
Pryor, C. 1996, in ASP Conf. Ser. 92, eds. H. Morrison & A. Sarajedini, (ASP: San Francisco), p. 424
Rodgers, A. W. & Paltoglou, G. 1984, Astrophys.J., 283, L5
Sandage, A. 1983, in "Kinematics, Dynamics, and the Structure of the Milky Way," ed. W.L.H. Shuter, (Dordrecht: Reidel), p. 315
Sandage, A. 1990, J.R.Astron.Soc.Canada, 84, 70
Sarajedini, A. & Layden, A. C. 1995, Astron.J., 109, 1086
Schwarz, U. J., Wakker, B. P. & van Woerden, H. 1995, Astron.Astrophys., 302, 364
Searle, L. & Zinn, R. 1978, Astrophys.J., 225, 357
Siegel, M. H., Majewski, S. R., Reid, I. N. & Thompson, I. 1997, in ASP Conf. Ser. Vol., in press, ed. R. Humphries, (ASP: San Francisco)
Smecker-Hane, T. A., Stetson, P. B., Hesser, J. E. & Lehnert, M. D. 1994, Astron.J., 108, 507
Stetson, P. B., et al. 1989, Astron.J., 97, 1360.
van den Bergh, S. 1993, Astron.J., 105, 971
Wakker, B. P. 1991a, Astron.Astrophys.Suppl., 90, 495
Wakker, B. P. 1991b, in IAU Symposium 144, ed. H. Bloemen, (Kluwer: Dordrecht), p. 27
Wakker, B., et al. 1996, Astrophys.J., 473, 834
Weiner, B. J. & Williams, T. B. 1996, Astron.J., 111, 1156
Zeilik, M. 1996, "Astronomy: The Evolving Universe, 7th Ed.," (Wiley: New York), p. 429
Zinn, R. 1993, in ASP Conf. Ser. Vol. 48, eds. Smith, G. & Brodie, J., (ASP: San Francisco), p. 38

by combining counts and proper motions, since the quasi null-rotation of the halo is easily identifiable (Soubiran 1993).

Specific tracers, such as RR Lyrae or BHB (Layden 1995; Kinman 1994) allow one to identify the density and kinematics of different halo sub-components. Carney et al. (1994) obtained an important local kinematic halo sample, and Beers et al. (1996) build a non-kinematically biased sample of 1936 nearly local halo stars with low metallicity. Such samples allow one to determine the halo and thick disk local kinematic properties.

For distant stars, radial velocities are known in a dozen Galactic directions, with a few tens of stars in each direction. Modeling the observed radial velocity distribution of RR Lyrae or BHB (Arnold, 1990; Flynn et al. 1996; Sommer-Larsen et al. 1994) with different anisotropic velocity distributions allows one to deduce the possible distribution functions (DF) and to deduce limits on realistic potentials. Most of these stars are distant and their tangential velocity is unknown, so models with different tangential velocities and very different potentials are found to be equally probable. Without accurate proper motions, just a lower limit is obtained on the total Galactic mass distribution.

Globular clusters (GC) are important halo tracers at large distances giving information on the potential up to 40 kpc (Dauphole & Colin 1995; Kochanek 1996). GC samples are probably not complete (Da Costa 1995), detection biases are unknown, and there is a lack of reliable proper-motions for the more distant GC. These two elements prevent an accurate definition of the potential up to 40 kpc, but are compatible with a flat or rising rotation curve up to that radius.

A similar analysis based on satellites of the Galaxy allows one to explore the potential further away. The analysis is based on a small number (25) of objects with only 3 measured and usable proper-motions (Kochanek 1996). The assumption of stationarity is certainly doubtful due to typical crossing time of orbits.

Metal-poor Blue Main sequence Stars near the solar circle have been discovered by Preston et al. (1994) as a new kinematic population. These stars have an isotropic velocity dispersion and relatively large mean rotational velocities of about $128\,\mathrm{km\,s^{-1}}$. Preston et al. suggest they are probably accreted from dwarf spheroidal satellites. It will be essential to extend the detection of this population out the galactic plane, since it will be the best potential tracer at intermediate distances around 3–6 kpc out of the Galactic plane.

3.2. LOCAL TRACER : ESCAPE VELOCITY

A classic method (for example : Carney et al. 1988; Kochanek 1996) for probing the Galactic potential at very large radius consists of determining the local escape velocity from the velocity distribution of high velocity stars. In the solar neighbourhood, it is estimated to be between 450 and 650 km s^{-1}. If stars with larger velocities evaporate and leave the Galaxy, we get a lower limit to the total mass of the Galaxy. The observed escape velocity implies a galactic total radius $R_{lim} \geq 34$ kpc for a flat rotation curve up to R_{lim} and no mass beyond that radius (Cudworth 1990). This approach suffers from various difficulties. Firstly, large velocities stars are selected from large proper motion surveys and are biased towards objects with the largest proper motion errors, although in principle this can be corrected. Secondly, it is not excluded that these large velocity stars are visiting evaporated stars from a satellite of the Galaxy, like the Magellanic Clouds or a neighbouring dwarfs galaxy.

3.3. LOCAL TRACER : POTENTIAL FLATTENING

The velocity distribution of local stars is not only a mixture of the history of various stellar populations, but also reflects the potential. This may be shown, for example, at $z = 0$ for a spherical potential and a flat rotation curve ($=v_c$). In this case, the DF may be written as:

$$f(r, z = 0; v_r, v_\theta, v_z) = e^{-\frac{v_r^2}{2\sigma_r^2}} * e^{-\frac{v_z^2}{2\sigma_z^2}} * e^{-\frac{v_z^2}{2\sigma_z^2}(\frac{v_c^2}{v_\theta^2}-1)(\frac{\sigma_z^2}{\sigma_r^2}-1)} * fct(r, v_\theta)$$

The correlation between vertical the v_z and tangential v_θ velocities is non-existent for a plane-parallel potential. This correlation term is important for disk populations with large velocity dispersions. Analysing the Gliese catalogue of nearby stars, we estimate that the flattening of the potential is $\sim 0.6 \pm 0.4$ (Pichon & Bienaymé 1996). Analysing a larger local sample will allow one to evaluate the potential flattening with a better accuracy.

3.4. VELOCITY ELLIPSOID INCLINATION : POTENTIAL FLATTENING

Exact dynamical modeling of halo tracers requires complete 3D models. However, correlations between radial and vertical motions may be just obtained from the orientation of the velocity ellipsoid, and probe the shape of the potential without need for accurate measurements of the density distribution. In the case of a Stäckel potential, this orientation does not depend on the velocity dispersion. In a spherical potential, the ellipsoid points towards the Galactic centre while in a plane-parallel potential it stays parallel to the Galactic plane.

The Majewski *et al.* (1996) sample at large distances above the Galactic plane can be examined with this purpose, although they remark that it shows substructures and is probably not kinematically mixed. Features in their (u,w) velocity plot (Figure 2) may reflect initial conditions. However, if the star groups came from an initial cluster with very small internal dispersion, the resulting structure observed today, and partially mixed in the halo, will plot such a (u,w) diagram and will just trace an isopotential pointing towards ($R = -5$ kpc, z=0) beyond the galactic center (if $<z> = 5\,kpc$ is the mean distance of these stars). This shows that the potential is oblate and can be reproduced with a nearly spherical dark matter distribution. Thick disc stars should also be good potential tracers at heights 2 to 5 kpc above the galactic plane. However, their ellipsoid inclination will be smaller.

References

Arnold R., 1990, Mon.Not.R.astron.Soc. 244, 465
Beers T.C., Sommer-Larsen J., 1996, Astrophys.J.Suppl. 96, 175
Bienaymé O., Séchaud N., 1996, Astron.Astrophys., (submitted)
Boulares A., Cox D.P., 1990, Astrophys.J. 365, 544
Carney B.W., Latham D.W., Laird J.B., 1988, Astron.J. 96, 560
Carney B.W., Latham D.W., Laird J.B., 1994, Astron.J. 107, 2240
Crézé M., 1991, IAU Symp. 144, 313
Cudworth K.M., 1990, Astron.J. 99, 590
Da Costa G.S., 1995, Publ.Astron.Soc.Pacific 107, 58
Dauphole B., Colin J., 1995, Astron.Astrophys. 300, 117
Dehnen W., Binney J., 1996, ASP Conf. Series 92, eds. H. Morrison & A. Sarajedini, 391
de Vaucouleurs G., Pence W.D., 1978, Astron.J. 83, 1163
Durand S., Dejonghe H., Acker A., 1996, Astron.Astrophys. 310, 97
Dwek E., Arendt R.G., Hauser M.G. *et al.*., 1995, Astrophys.J. 445, 716
Eggen O.J., Lynden-Bell D., Sandage A., 1962, Astrophys.J. 136, 748
Fich M., Tremaine S., 1991, Ann. Rev. Astron. Astrophys. 29, 409
Flynn C., Sommer-Larsen J., Christensen P.R., 1996, Mon.Not.R.astron.Soc.
Gilmore G., Wyse R., Jones J.B., 1996, Astron.J. 109, 1095
Gilmore G., Wyse R., Kuijken K., 1989, Ann. Rev. Astron. Astrophys. 27, 555
Kinman T.D., Suntzeff N.B., Kraft R.P., 1994, Astron.J. 108, 1722
Kistiakowsky V., Helfand D., 1995, Astron.J. 110, 2225
Kochanek C., 1996, Astrophys.J. 457, 228
Kuijken K., 1995, IAU Symp. 164, 198
Kuijken K., Tremaine S., 1994, Astron.J. 421, 178
Larsen J.A., Humpreys R.M., 1994, Astrophys.J. 436, L149
Layden A.C., 1995, Astron.J. 110, 2288
Majewski S.R., Munn J.A., Hawley S.L., 1996, Astrophys.J. 459, L73
Malhotra S., 1995, Astrophys.J. 448, 138
Ojha D. K., Bienaymé O., Robin A. C., Mohan V., 1994a, Astron.Astrophys. 284, 810
Ojha D. K., Bienaymé O., Robin A. C., Mohan V., 1994b, Astron.Astrophys. 290, 771
Ojha D. K., Bienaymé O., Robin A. C. *et al.*, 1996, Astron.Astrophys. 311, 456
Pichon C., Bienaymé O., 1996, in preparation
Preston G.W., Beers T., Shectman S.A., 1994, Astron.J. 108, 538
Reid N., Majewski S.R., 1993, Astrophys.J. 409, 635

Robin A.C., Crézé M., Mohan V., 1992, Astron.Astrophys. 265, 32
Robin A.C., Haywood M., Crézé M. *et al.*, 1996, Astron.Astrophys. 305, 125
Searle L., Zinn, 1978, Astrophys.J. 225, 357
Sevenster M.N., Dejonghe H., Habing H.J., 1995, Astron.Astrophys. 299, 689
Sommer-Larsen J., Flynn C., Christensen P.R., 1994 Mon.Not.R.astron.Soc. 271, 94
Soubiran C., 1993, Astron.Astrophys. 274, 181
te Lintel Hekkert P., 1990, Ph D Thesis "The evolution of OH/IR stars and their dynamical properties," Leiden University
Tremaine S., 1993, AIP Conf. Proc. 278, eds. S.S. Holt *et al.*, 599
Weinberg M. D., 1996, Astron.Soc.Pacific Conf. Series 91, ed. R. Buta *et al.*, 516

USING MULTI-WAVELENGTH ALL SKY INFORMATION TO UNDERSTAND LARGE SCALE GALACTIC STRUCTURE.

R.L. SMART, R. DRIMMEL AND M.G. LATTANZI
Osservatorio Astronomico di Torino

1. Introduction

Our galaxy was first seen to be warped, similarly to other galaxies, in 21cm HI surveys (Kerr 1957). Since then a large database on our and other warps has been built but they remain a puzzle for theorists trying to construct a general warp theory. Warps are observed to be common rather than rare features of spiral galaxies, which implies they are long lived. They are observed in both isolated and multiple galaxy systems in the gas but the stellar warp is often difficult to detect. The galactic warp is always below b=10° making it difficult to observe because of absorption and confusion, to add to this problem the sun appears to lie near to the line of nodes and to view the warp at it's largest extent we must observe through a large part of the disk. A theoretical model must evoke driving forces that explain the longevity, commonality, and independence of environment that we see in warps.

2. Previous Work

Observations of the HI 21cm line indicate the southern half of the galactic warp curving back towards the plane after reaching an amplitude of 1.5 kpc at radius 16kpc, while the northern half continues to rise, reaching 4kpc at 23 kpc (Burton 1991). Wouterlott *et al.* (1990) analyzed CO molecular cloud observations and showed these are consistent with the HI observations.

The stellar warp is difficult to observe for many reasons: absorption in the plane, confusion of warp stars with foreground or background stars, an apparent (related?) cutoff in the stellar density near the radius the warp begins, and a possible non-coincidence of the gas and stellar warp makes it difficult to predict where the stellar warp is. Early work by Guibert *et al.* (1978) showed that the positions of young star populations are consistent to a plane model both with and without a warp. More recent work on OB stars (Cameron-Reed 1996 and references therein) and Cepheids (Efremov *et al.* 1981) implies the young star plane is warped. This maybe confused because of their proximity to their birthplace in the HI plane.

Carney & Seitzer (1993) show an overabundance of stars in the HI warped direction compared to fields at the same longitude but opposite latitude. Djorgovski & Sosin (1989) examined AGB stars in the IRAS Point Source Catalog showing a warped plane. These both indicate that the old stellar disk is also warped. Freudenreich *et al.* (1994) looked at the results of the Diffuse Infrared Background Experiment on COBE to map out the surface brightness of the dust plane in our galaxy and found it was consistent with the HI observations but indicated different parameters. All of these three studies examine the projected warp, *e.g.*, there is very little distance information and, therefore, room for other interpretations.

Therefore, for our galaxy, on a first examination observations indicate a consistent warped plane for all the constituents, young and old stars, dust and gas with that caveat that there is room for different warp parameters of the constituents. Observations of our galaxy and other galaxies indicate these parameters are different. Freudenreich *et al.* (1994) mentioned above found different parameters for the gas and dust. Miyamoto *et al.* (1994), looking at proper motions of old K/M giants and young O/B stars, found that the plane parameters differed significantly, with the young stars following the HI plane and the old stellar plane unwarped. Florido *et al.* (1991) observed three galaxies in four colors, basically sampling different constituents, and found that the warp had a significant variation. Finally on a more fundamental note, if all the constituents follow the same warped plane it strongly implies that the driving force is gravity as other possibilities, *e.g.*, intergalactic winds or magnetic fields, would effect dust and gas quite differently than the stars. However, Battaner *et al.* (1991) found a relationship between warp orientations of nearby galaxies which is difficult (if not impossible) to explain with warps originating from gravitational fields.

3. Problems and Future Work

One of the main problems here is with the differing modes of interpretation. With HI and CO data distances are found using a rotation curve and gas velocities, for the OB and cepheids usual photometric distance techniques are employed and for the old stars and dust observations minimal distance

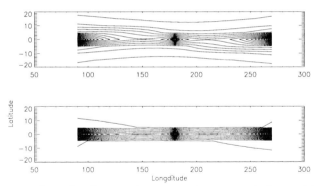

Figure 1. Contour plots of stellar number counts in a warped plane without and with absorption. When absorption is added the warp effect in number counts disappears.

information is known. An example of the many problems here is shown in Figure 1, which compares the number counts of a warped stellar plane with and without absorption from a simple galactic model. Similar examples can be made for other systematic errors introduced by, for example, non-circular rotation curves, miscalibration of photometric distance estimates, *etc.*

At Torino we have undertaken two projects to examine the warp of our galaxy. One is to interpret the integrated observations of the dust and old stars using three dimensional models. The second is to compare the HIPPARCOS observations of O/B stars to simulated galactic models of these stars. Using HIPPARCOS results has a number of advantages, they are all sky and complete to a well understood level, the kinematical information provides more constraints than just spatial data alone, they are very precise and free from systematic zonal effects that hindered the Miyamoto *et al.* work. Finally, TYCHO, the HIPPARCOS star mapper, will provide precise colors on a consistent system for finding accurate photometric parallaxes.

Acknowledgements

RLS thanks the SOC for support. to attend this meeting. We thank Stefano Casertano, James Binney and David Spergel for useful comments.

References

Battaner E., Floridao E. & Sanchez-Saavedra M.L. 1990 Astron.Astrophys. 236, 1.
Binney J. 1992, Ann.Rev.Astron.Astrophys. 30, 51
Burton W.B. 1991 SAAS-FEE 21, eds. W.B. Burton, B.G. Elmgreen & R. Genzel.
Cameron Reed B. 1996 Astron.J. 111,804.
Carney B.W. & Seitzer P. 1993 Astron.J. 105, 2127.
Djorgovski S. & Sosin C. 1989 Astrophys.J. 341, L13.

Efremov Y.N., Ivanov G.R. & Nikolov N.S., 1981 Astron.Sp.Sc 75, 407.
Freudenreich H.T., Berriman G.B., Dwek, E., Hauser, M.G., Kelsall T., Moseley S.H., Silverberg R.F., Sodroski T.J., Toller G.N., & Weiland J.L. 1994 Astrophys.J.Lett. 429, L69.
Florido, E., Battaner, E., Sanchez-saavedra, M. L.,Prieto M. & Mediavilla, E. 1991 Astron.Astrophys. 242, 301.
Guibert J., Lequeux J. & Viallefond F. 1978 Astron.Astrophys. 68, 1
Kerr F.J. 1957 Astron.J. 62, 93.
Miyamoto M., Yoshizawa M., & Suzuki S. Astron.Astrophys. 194, 107.
Wouterloot J.G.A., Brand J., Burton W.B., & Kwee K.K. 1990 Astron.Astrophys. 230, 21.

STRUCTURE AND KINEMATICAL PROPERTIES OF THE GALAXY AT INTERMEDIATE GALACTIC LATITUDES

D. K. OJHA[1], O. BIENAYMÉ[2] AND A. C. ROBIN[3]
[1] *Institut d'Astrophysique de Paris*
[2] *Observatoire de Strasbourg*
[3] *Observatoire de Besançon*

Abstract. We have carried out a sample survey in UBVR photometry and proper motions in various directions in the Galaxy. Three fields in the direction of galactic anticentre, centre, and antirotation have been surveyed. Using our new data together with wide-area surveys in other fields available to date, we discuss the radial and vertical structure of the Galaxy. Our results confirm that the thick disk population is distinct from other populations based on their kinematical and spatial distribution. The most probable value of scale height for the thick disk component is determined to be $h_z \simeq 760 \pm 50$ pc and a local density of $\simeq 7.4^{+2.5}_{-1.5}\%$ relative to the thin disk. The ratio of the number of thick disk stars in our galactic centre region to that in anticentre region yield $h_R \simeq 3 \pm 1$ kpc for the scale length of thick disk. These values are in perfect agreement with the recent determination given by Robin *et al.* (1996).

1. Fields Surveyed

The 3 fields chosen are in the direction of galactic anticentre ($l = 167°$, $b = 47°$; Ojha *et al.* 1994a, 1996b; hereafter GAC1,2), galactic centre ($l = 3°$, $b = 47°$; Ojha *et al.* 1994b; hereafter GC) and antirotation ($l = 278°$, $b = 47°$; Ojha *et al.* 1996b; hereafter GAR).

2. Structural and Kinematical Parameters of the Thick Disk

We have used the combination of 3 intermediate latitude fields (GAC1,2, GC & GAR) to derive the structural parameters of the thick disk population (Ojha *et al.* 1996a). The thick disk characteristics are, $h_z \simeq 760 \pm 50$ pc and local density $\simeq 7.4^{+2.5}_{-1.5}\%$ relative to the thin disk. We deduce that the scale length of the thick disk is $h_R \simeq 3\pm1$ kpc. Robin et al. (1996) recent determination gives $h_R \simeq 2.8\pm0.8$ kpc, $h_z \simeq 760\pm50$ pc, with a local density of 5.6±1.0 % of the disk.

To perform the kinematical separation, we have used a maximum likelihood method (SEM algorithm : Celeux & Diebolt 1986) in order to deconvolve the multivariate Gaussian distributions and estimate the corresponding parameters. SEM also gives an estimation of the proportions and densities of each population along the line of sight distance. By comparing the star count ratio between the two data sets (GAC1,2 & GC) in each distance bin, we obtain the scale length of thick disk is $h_R \simeq 3.6\pm0.5$ kpc.

From comparison of proper motion distributions in 4 fields (GAC1,2, GC, GAR and NGP (Soubiran 1993)), the kinematical results can be estimated on a base line of 5 kpc (Ojha et al. 1996a). By combining the kinematical results from 4 fields, we have derived the velocity ellipsoid of the thick disk population. The mean kinematic parameters are summarized in Table 1. The data constrain the asymmetric drift of the thick disk, which is found to be 53±10 km/s with respect to the Sun.

TABLE 1. The mean kinematic parameters of thick disk (in km/s) derived from 4 fields (GAC1,2, GC, GAR & NGP). σ_W is determined for the most probable vertical potential (see Ojha et al. 1996a). V_{lag} is with respect to the Sun.

	σ_U	σ_V	σ_W	V_{lag}
Thick disk	67±4	51±3	40	−53 ±10

References

Celeux G., Diebolt J., 1986, Rev. Statistique Appliquée 34, 35
Ojha D.K., et al., 1994a, Astron.Astrophys. 284, 810 (GAC1)
Ojha D.K., et al., 1994b, Astron.Astrophys. 290, 771 (GC)
Ojha D.K., et al. 1996a, Astron.Astrophys. 311, 456
Ojha D.K., et al., 1996b, Astron.Astrophys. (in preparation) (GAR or GAC2)
Robin A.C., et al. 1996, Astron.Astrophys. 305, 125
Soubiran C., 1993, Astron.Astrophys. 274, 181

PROPER MOTIONS IN THE BULGE: LOOKING THROUGH PLAUT'S LOW EXTINCTION WINDOW

R.A. MÉNDEZ[1], R.M. RICH[2], W.F. VAN ALTENA[3],
T.M. GIRARD[3], S. VAN DEN BERGH[4] AND S.R. MAJEWSKI[5]
[1] *European Southern Observatory*
[2] *Columbia University, Astronomy Department*
[3] *Yale University, Astronomy Department*
[4] *Dominion Astrophysical Observatory*
[5] *University of Virginia, Department of Astronomy*

1. A Survey for Proper-motions

We are conducting the deepest and largest photographic proper-motion survey ever undertaken of the Galactic bulge. Our first-epoch plate material (from 1972–3) goes deep enough ($V_{lim} \sim 22$) to reach below the bulge main-sequence turnoff. These plates cover an area of approximately $25' \times 25'$ of the bulge in the low-extinction ($A_v \sim 0.8$ mag) Plaut field at l= 0°, b= $-8°$, approximately 1 kpc south of the nucleus. This is the point at which the transition between bulge and halo populations likely occurs and is, therefore, an excellent location to study the interface between the dense metal-rich bulge and the metal-poor halo.

In order to study the formation of the bulge and its chemical evolution one would like to measure proper-motions, radial velocities, and abundances for members of that stellar population, in a way similar to that of the seminal effort by Eggen, Lynden-Bell & Sandage in 1962, which addressed the oldest stellar population in our Galaxy.

2. Survey Description and Preliminary Results

Our project is based on a unique sample of twenty photographic plates of a Galactic bulge field on the minor-axis at b= $-8°$ obtained by Sidney van den Bergh in 1972–3 using the Kitt Peak 84-inch and the Palomar 200-inch telescopes. The 100-inch telescope at Las Campanas has been used to obtain thirteen second epoch plates in 1993; deep intermediate epoch plates of this field (1979) were obtained by Jeremy Mould also at Las Campanas.

The plates were digitized using the the Yale PDS 2020G laser interferometer/microdensitometer measuring machine. Analysis of the Yale-PDS microdensitometer data routinely yields star centroids to 1/20 of a pixel (20 mas). A star measured on five plates in each color will have its position known at least 2–3 times better than this, and over the 21 yr baseline we expect errors no larger than 0.5 mas yr^{-1} in each color. This corresponds to approximately 20 km s^{-1} at the distance of the Galactic Center. This is comparable to Spaenhauer et al. (1992), and matches the accuracy with which radial velocities in our spectroscopic follow-up will be measured.

Preliminary results, based on only three first epoch and three last-epoch plates spanning 21 years (Méndez et al. 1996) indicate that it is possible to obtain proper-motions with errors less than 1 mas yr^{-1} for a about 5,000 stars down to V=18, without color restriction. For the subsample with errors less than 1 mas yr^{-1} we derive proper-motion dispersions in the direction of Galactic longitude and latitude of 3.378 ± 0.033 mas yr^{-1} and 2.778 ± 0.028 mas yr^{-1} respectively. These dispersions agree with those derived by Spaenhauer et al. (1992) in Baade's window.

We expect to measure CCD photometry in the B, V, and I passbands, and proper-motions for an unbiased sample of approximately 30,000 stars in our minor-axis field. We hope to further obtain radial velocities and low resolution abundances for about 5,000 stars. A large, unbiased sample is important because much of the outcome depends on dividing the data into subsamples as a function of abundance or kinematics. Radial velocities from our spectroscopic survey will be extremely important as a complement to the proper-motions to confirm the presence of a stellar bar in the Bulge.

References

Méndez, R.A., Rich, R.M., van Altena, W.F., Girard, T.M., van den Bergh, S., & Majewski, S.R., 1996, in "The Galactic Center," 4th ESO/CTIO Workshop, Astron.Soc.Pacific Conference Series, 102, R. Gredel, ed. (ASP, San Francisco) p. 345
Spaenhauer, A., Jones, B.F., Whitford, E., 1992, Astron.J., 103, 297

GALACTIC STRUCTURE WITH THE APS CATALOG OF THE POSS I

J.A. LARSEN
University of Minnesota

1. Introduction

The maturity of all-sky surveys hold great promise for galactic structure studies. Surveys such as the APS Catalog of POSS I allow large scale studies of stellar distributions where the data is very complete to a reasonably faint magnitude, uniformly reduced, and independently calibrated. The APS data covers a large portion of the sky, allowing studies where the primary source of uncertainty is no longer small-number statistics.

Figure 1 shows the locations and coverage of the APS Galactic Structure Fields. These fields are 16 square degree patches of the sky placed every 10 degrees in galactic latitude and 45 degrees in galactic longitude. The survey is designed to probe the Galaxy as Kapteyn suggested, but in a more statistically significant manner. This article presents APS results to date and discusses future work in which all-sky surveys will be valuable.

2. Galactic Structure from the APS Catalog

The distribution of stars outside of the Galaxy's disk has been studied. Larsen and Humphreys (1994) have used counts from seven POSS I fields, 16 square degrees each, including the NGP plus six in the $l = 90°/270°$ plane with $|b| = 40°$ and $50°$. The the ratio of blue stars ($O - E < 1.2$) between these fields has a mean value of 0.5. If one assumed only halo stars were present a value $c/a \leq 0.5$ for a deprojected $r^{1/4}$ law would be derived. When corrected for possible contamination by quasars and a standard model of a thick disk, the ratio increases to only 0.6. We concluded from this that there is a significant population of stars above the galactic plane in a flattened distribution in addition to those already ascribed to an

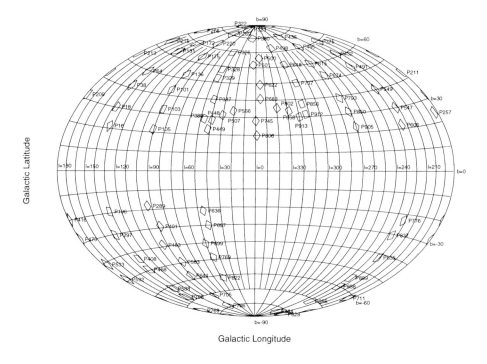

Figure 1. The APS Galactic Structure Fields.

extended or thick disk. Whether this excess represents halo or additional thick disk stars has yet to be determined.

APS counts have been used to determine the Sun's distance ($Z_\odot$) above the galactic plane (Humphreys and Larsen 1995). Star counts were taken in 12 Palomar Sky Survey fields—six each at the North and South galactic poles—192 square degrees. All stars with $15 < O < 18$ and $O - E > 1.8$ were selected to isolate a disk population. The total counts show significantly more stars in the six fields at the SGP indicating that the Sun is above the galactic plane as defined by neutral hydrogen. The observed ratio of N(SGP)/N(NGP) is 1.11 ± 0.02 implying $Z_\odot = 20.5 \pm 3.5$ pc.

A large and significant asymmetry has been found in the catalog (Larsen and Humphreys 1996). There are more faint stars in the first quadrant ($l = 20° - 45°$) of the Galaxy compared to complementary longitudes on the other side (fourth quadrant) of the center-anticenter line. The stars were chosen to be bluer than $B - V \approx 0.6$ mag and are most likely subdwarfs. Under this assumption, they are on average approximately 3 kpc away in the direction and at the same distance as the stellar bar proposed by Weinberg (1992) *but 1.7 kpc above it.* Possible explanations for this excess include an elliptical thick disk, disk heating from an interaction with the Sagittarius dwarf, or heating of the disk by the bar. Explaining this asymmetry will give important insights into the early Galaxy.

3. Future Work

The APS data has been useful as an exploratory data set. The asymmetry in counts appeared only on the largest spatial scales (across several plates). It was also detected in a region of the sky where our current understanding of galactic structure told us we should not find an asymmetric distribution of stars. The presence of such an asymmetry should serve as ample warning that fundamental galactic structure parameters cannot be reliably determined from only a handful of directions.

In analyzing the full set of APS Galactic Structure Fields, a new methodology is being applied which can be applied to an arbitrary number of directions. A fast, optimized galaxy model suited for multi-field work was created and a robust global optimization routine (the genetic algorithm) was applied for the first time to modeling the Galaxy. From these tools, a set of parameters matching the APS star counts better than extant models was self-consistently derived from the data. These initial studies not only recover the asymmetry, but also several other interesting features.

At the *Formation of the Galactic Halo....Inside and Out* conference it became quite obvious that all-sky surveys are necessary for an understanding of the structure of our Galaxy. The current consensus seems to be that both the Eggen, Lynden-Bell, and Sandage (1962) rapid collapse and the Searle and Zinn (1978) merger processes probably played a role in forming the galactic halo. If true, "pencil-beam" surveys can no longer be relied on to provide an understanding of any more than the local environment of the halo. All-sky surveys are essential to providing the context in which the more detailed small surveys can be interpreted.

References

Eggen, O.J., Lynden-Bell, D., and Sandage, A.R. 1962 Astrophys.J., 136, 748
Humphreys, R.M. and Larsen, J.A. 1995 Astron.J. 110, 2183
Larsen, J.A. and Humphreys, R.M. 1994 Astrophys.J.Lett., 436, L149
Larsen, J.A. and Humphreys, R.M. 1996 Astrophys.J.Lett., *in press*
Searle, L. and Zinn, R. 1978 Astrophys.J. 225, 357

GALACTIC STRUCTURE WITH GSC-II MATERIAL

The North Galactic Pole and the Stock-2 cluster regions

A. SPAGNA[1], M.G. LATTANZI[1], G. MASSONE[1],
B.J. McLEAN[2] AND B.M. LASKER[2]
[1] *Osservatorio Astron. di Torino*
[2] *Space Telescope Science Institute*

1. Introduction

Faint surveys of accurate colors and proper motions are fundamental for studying the physical and kinematical properties of the stellar populations in the Galaxy. A program designed to address such issues has been initiated on selected fields utilizing prototype material from the GSC-II project (see McLean *et al.* in this volume, p. 431). Photographic photometry V, B–V, V–R_c, is derived down to $V = 18.5$ with a precision of about 0.1 mag, together with absolute proper motions to $\sigma_\mu \simeq 3$ mas/yr.

Newly derived results are presented for a galactic plane region, containing the open cluster Stock 2, and for a high latitude field close to the North Galactic Pole.

2. Galactic Models at the North Galactic Pole

A new magnitude limited survey ($\sim 12,000$ stars with $V \leq 18.5$) of colors (BVR_c) and proper motions in a new field at the North Galactic Pole (NGP), with extent $4° \times 5°$ and centered at $b \simeq 87°$, has been presented in Spagna *et al.* (1996). Here we give comparisons with theoretical starcounts *and* kinematics provided by two integrated galactic models, the IASG Galaxy model (Ratnatunga *et al.* 1989), and the Besançon model of stellar population synthesis (Robin & Crézé 1986, Bienaymé *et al.* 1987). Besançon synthetic data are based on the version of 16 November 1994, while the IASG model adopts the default parameters for thin disk and halo, includes a thick disk with $h_z = 750$ pc, $\rho/\rho_0 = 10\%$ and uses the 47 Tuc luminosity function and M3 color-magnitude relation.

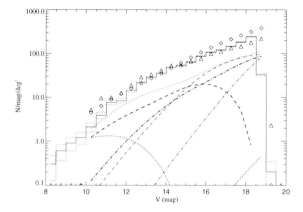

Figure 1. NGP starcounts (*solid* line and ±1σ *dotted* line histograms) shown with Besançon (*triangles*) and IASG (*diamonds*) predicted counts. IASG individual components are also shown: thin disk (*dotted* lines), thick disk (*dashed* lines) and halo (*dotted-dashed* lines); thin and thickened symbols identify main sequence and giant stars (RG, AGB, HGB) respectively.

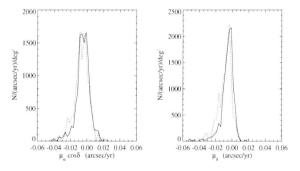

Figure 2. $17 < V \leq 17.5$ and $0.5 < B - V \leq 0.9$. Distribution of $\mu_\alpha \cos\delta$ and μ_δ from NGP data (*solid* line) and Besançon model (*dotted* line).

At intermediate magnitudes ($V \approx 14$–15) both models provide a good match to the observed photometric and kinematical distributions, while a few discrepancies are noticeable at fainter magnitudes ($V > 17$).

NGP starcounts are shown in Figure 1, where systematic differences at faint magnitudes are apparent. These are more pronounced for IASG, and adjusting only for the thick disk scale-height and density normalization does not improve the fit. Color and proper motion distributions help identifying an excess of slow halo giants in IASG.

Besançon kinematics provide better fit to the data, also at faint magnitudes. This is shown in Figure 2 for (μ_α, μ_δ) of stars in the range $17 < V \leq 17.5$ belonging to the blue peak. (Here, a random error of $\sigma = 3$ mas/yr is included in Besançon proper motions.) Besides a small

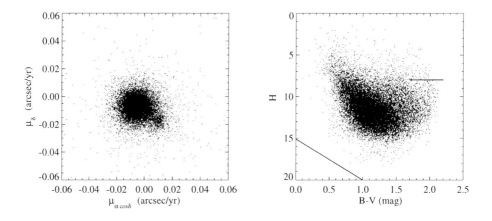

Figure 3. Stock 2. $\vec{\mu}$ VPD (*left*) and H (*right*) diagrams.

excess of stars at intermediate asymmetric drift ($\mu \simeq 20$ mas/yr) and an offset of ~ 1 mas/yr in μ_δ, the agreement of the two distributions is good.

3. Stock 2 Region

We have derived colors (BVR_c) and fundamental (FK5) proper motions for 20,000 stars down to $V = 18$ in a $1.8° \times 1.8°$ field towards the galactic anticenter ($b = 133.2°$, $l = -1.6°$). The field contains the open cluster Stock 2 (Krzeminski & Serkowski 1967) which is outstanding in the vector point diagram (Figure 3 left).

Although a detailed analysis is awaiting for a better modeling of the strong and patchy interstellar extinction in this line-of-sight, in Figure 3 (right) the reduced proper motions, $H = V + 5\log \mu + 5$, already show the scientific potential of our data. The H-constant line represents the loci of red giants at 1 kpc, while the red line on the bottom left traces the theoretical region of POP I WD at 200 pc. Both lines have been appropriately corrected for mean interstellar reddening and extinction. Early type stars belonging to the Per OB1 association at about 2 kpc would appear in the region at $H \leq 10.5$ and $B - V \geq 1.3\text{-}1.5$.

References

Bahcall, J. N., Casertano, S., and Ratnatunga K. U. 1987. Astrophys. J., 320, 515.
Bienaymé O., Robin A.C. and Créze M., 1987. Astron. Astrophys., 180, 94.
Krzeminsky W. and Serkowski K., 1967. Astron. J., 147, 988.
Ratnatunga K. U., Bahcall, J. N., and Casertano, S. 1989. Astrophys. J., 339, 106.
Robin A.C. and Créze M., 1986. Astron. Astrophys., 157, 71
Spagna, A., *et al.* 1996. Astron. Astrophys, 311, 758.

INVESTIGATION OF GALACTIC STRUCTURE WITH DENIS STAR COUNTS

S. RUPHY
Observatoire de Paris

1. Introduction

Thanks to the DENIS program (Deep Near-Infrared Survey of the Southern Sky), relatively deep near-infrared star counts are now available for the first time on a large scale. The basic method to interpret star counts in terms of galactic structure is to compare them with predictions given by models of the point source sky. Of particular promise are studies with DENIS of the spatial distribution of evolved stars in our Galaxy, thanks to its high sensitivity to red giant and to the much lower interstellar extinction that hampers visual observations of far-away stars in the disc of our Galaxy. In this paper, I present a sample of extensive comparisons between two models of the Galaxy and DENIS star counts (Ruphy 1996). I will focus on the analysis of star counts in the anticenter direction, that leads to new values for the distance of the cutoff and the radial scale length of the stellar disc.

2. Observations

The observations presented here are part of the first data released by the DENIS survey which will provide, before the end of the century, the first digitized survey of the southern sky, with a 3σ detection limit of 18, 16 and 14, in the I (0.8 μm), J (1.25 μm) and K_s (2.15 μm) bands respectively (Epchtein, this volume, p. 106). The observations were made in February and November 1995, using the DENIS camera attached at the focus of the ESO 1m telescope at La Silla, Chile.

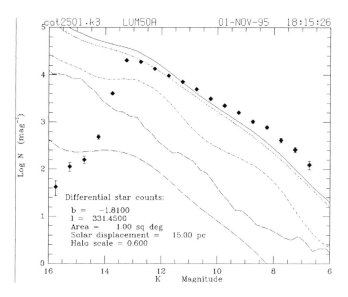

Figure 1. Comparison of SKY predictions and DENIS differential star counts in K_s, at l=331°, b=-1.81°, on 1 deg^2. Diamonds, observed star counts with poissonian error bars; solid line, total model prediction; dots, disc; long dashes, spiral arms; long dash-dot, halo; short dash-dot, ring.

3. Comparisons with Models of the Point Source Sky

3.1. STUDY OF THE INHOMOGENEITIES OF THE GALACTIC PLANE WITH THE 'SKY' MODEL

The SKY model is based on a realistic representation of the Galaxy, including features such as disc, spiral arms, local spur, molecular ring, bulge and halo, and delivers star counts in many filters lying within a large spectral range, from far-UV to far-infrared. The basic model is described by Wainscoat *et al.* (1992), and was considerably enhanced by Cohen (1994a,b). Comparisons with DENIS star counts in J and K at different longitudes along the galactic plane, such as the one presented in Figure 1, allow to improve the description of the spiral arms and the molecular ring in SKY : the lack of bright stars in the model is likely to be due to an inaccurate luminosity function of the spiral arms, and the excess of faint stars can be reduced by changing the geometry of the molecular ring and making it elliptical (Ruphy 1996).

3.2. NEW DETERMINATION OF THE DISC SCALE LENGTH AND RADIAL CUTOFF IN THE ANTICENTER WITH THE BESANÇON MODEL

Comparisons in the anticenter direction have revealed a systematic excess of stars predicted by both SKY and the Besançon model. The Besançon model is a synthetic stellar population model developped by Robin & Crézé (1986). In the outer part of the galactic plane, the most influential parameters are the disc scale length h_R and the distance of the edge of the disc R_{max}. We compared J−K colour distributions in two strips (a strip covers $\simeq$ 5.75 deg^2) crossing the galactic plane at l = 217° and l = 239° respectively, with predictions given by the Besançon model for a large range of h_R and R_{max} values. The best values are obtained by using a maximum likelihood method applied on a set of bins of K and J−K, in different declination ranges (see Ruphy et al. 1996, for details of the procedure and further discussions). For the two strips, the maximum likelihood is obtained for the same values, namely R_{max} = 15±2 kpc and h_R = 2.3±0.1 kpc.

4. Conclusion

Thanks to the very first DENIS data, we confirmed the existence of the cutoff in our Galaxy and obtained a rather short scale length for the disc. Our value is compatible with previous determinations based on star counts in the visible, in the anticenter direction (Robin et al. 1992) or in central parts of the Galaxy (Ojha et al. 1996). It is also in agreement with the recent value found by Fux & Martinet (1994) from kinematical data. Our determination benefits from a small influence of the extinction and the large size of the samples ($\simeq$ 16,000 stars per strip). Furthermore, it is based not only on the dwarf population, as it is the case for analysis of star counts in the visible, but also on the giant population.

Acknowledgements

I thank M. Cohen and A. C. Robin for their close collaboration in the interpretation of the DENIS counts with their models. The DENIS team is also warmly thanked for making the observations available for scientific analysis.

References

Cohen M. 1994a, Astron.J., 107, 582
Cohen M. 1994b, Astrophys.J., 427, 848
Fux R., Martinet L., 1994, Astron.Astrophys., 287, L21
Ojha D.K, Bienaymé O., Robin A.C., Crézé M., Mohan V., 1996, Astron.Astrophys., 311, 456
Robin A. C. and Crézé M., 1986, Astron.Astrophys., 157, 71
Robin A. C., Crézé M., Mohan V., 1992, Astrophys.J., 400, L25
Ruphy S., Robin A. C., N. Epchtein, Copet E., Bertin E., Fouqué P., Guglielmo F., 1996, Astron.Astrophys., 313, L21
Ruphy S., 1996, PhD. dissertation, Paris
Wainscoat R. J., Cohen M., Volk K., Walker H. J., Schwartz D E., 1992, Astrophys.J.Suppl., 83, 111

STARCOUNTS IN THE HUBBLE DEEP FIELD: FEWER THAN EXPECTED, MORE THAN EXPECTED

The Halo main-sequence & Halo white-dwarf luminosity functions

R. A. MÉNDEZ[1], G. DE MARCHI[1], D. MINNITI[2],
A. BAKER[3] AND W. J. COUCH[4]
[1] *European Southern Observatory*
[2] *Lawrence Livermore National Laboratory*
[3] *Institute of Astronomy*
[4] *School of Physics, University of New South Wales*

1. Introduction: The Hubble Deep Field

The Hubble Deep Field (HDF, Williams *et al.* 1996), recently acquired with the Hubble Space Telescope, is the deepest, most detailed optical view of the Universe. Even though intended primarily for the study of the high-redshift Universe, the HDF provides a unique opportunity to find faint stellar objects in our Galaxy, set constraints on number of low-mass Halo stars, estimate the contribution of baryonic matter to the dark Halo, investigate the nature of microlensing sources, and to calibrate Galactic Structure models. The HDF has been an STScI initiative to provide the deepest exposures yet acquired with HST on a non-proprietary basis. Observations of an 'anonymous field' located at $l = 125.9°$, $b = 54.8°$ were performed on the continuous viewing zone, for a total of 150 orbits (10 consecutive days) in the four HST passbands F300W, F450W, F606W and F814W. These observations reached 5σ magnitude limits (for Galaxies) of roughly (STMAG) 30th mag in all these passbands (except F300W which reaches 27th mag). We have used a novel software detection & classification algorithm to create a sample of point-like objects in the HDF. We have also compared the observed stellar counts to constrain the faint end of the Halo field luminosity function by using a recent Galactic Structure model. The unexpected appearance of a faint-blue group of very compact objects is discussed as well.

2. Analysis of the Hubble Deep Field: A Tale of a Few

As a result of combining groups of exposures with slight telescope offsets, the final Wide-Field camera images have a resolution of 0.04 arcsec pixel^{-1} and, therefore, stellar images are actually *subsampled* (Williams et al. 1996). In our strategy, the F606W and F814W images were coadded for detections at the faintest levels allowed by these data, totalling 70h of combined exposure time. The field of view is 4.69 arcmin^2.

To create our point-like sample we have used a recently developed source extraction algorithm (SExtractor: Bertin & Arnouts 1996). This software uses a neural network classifier, trained with observed and simulated stars and galaxies. In addition to providing astrometry and photometry, the software also gives a 'Stellarity Index' (or CLASS). The CLASS parameter can be viewed as the probability of an object being a star; it goes from 0 for extended objects to 1 for point-like objects. We have found that objects with $CLASS < 0.85$ are clearly extended on the frames. On this basis, we have compiled a sample of 16 objects satisfying the criteria that their $CLASS \geq 0.85$, these objects do not seem to exhibit any structure around them.

We are quite confident that we are not missing stars down to $F606W \sim F814W \sim 30$ mag, while the inclusion of some compact objects could be a problem for magnitudes fainter than $F814W \sim 27.5$. More details about the catalogue creation and completeness can be found in Méndez et al. (1996)

3. Model Comparison and Conclusions

A detailed comparison of the observed counts on the HDF and the counts predicted by a recent model by Méndez and van Altena (1996) is given in Méndez et al. (1996). The main result is that the luminosity function for faint ($12 \leq M_V \leq 14$) Halo stars is depressed by, at least, a factor of two (but most likely a factor of four) with respect to the luminosity function for nearby stars by Wielen et al. (1983). These results are coincident with recent findings from deep HST observations of Globular Clusters (De Marchi & Paresce 1995, and references therein) that indicate a sharp decrease in stars fainter than $M_v \sim 12$. Also, recent determinations of the disk luminosity functions from HST observations indicate that the faint end of the Wielen et al. (1983) luminosity function is indeed overpopulated, and that the disk luminosity function *also* has a maximum at $M_v \sim 12$ and a steep decrease at fainter magnitudes (Santiago et al. 1996). This would seem to indicate that the luminosity function of the different stellar populations in our Galaxy behave in a similar way, and that they are not strongly dependent on metallicity.

A group of faint-blue objects (with $25 \leq V \leq 28$, $-0.5 < B - V < 0.5$ and $0 < V - I < 1.2$) is found to be too blue at these magnitudes to be QSOs. If they are assumed to be stellar objects they would be Halo white-dwarfs (WDs). Their numbers would indicate an extreme overdensity of Halo WDs with respect to theoretical expectations. The mass locked in these WDs would be on the order of 10^{-4} $M_\odot pc^{-3}$, and their local number density would be as high as 40% as that of disk WDs. An alternative to scaling the Halo WD luminosity function would be to assume a sharp increase in the number of Halo WDs for $M_v \geq 11$. The colors of these putative WDs roughly agree with the theoretical colors computed by Bergeron et al. (1995), indicating temperatures in the range of $T_{eff} \sim 7-8 \times 10^3$ K. Based on luminosities derived from distances computed from the Galactic models (roughly 8 to 19 kpc), we estimate cooling ages of less than 1 Gyr, similar to the WDs found by Richer et al. (1995) in the globular cluster M4.

The faint-blue objects lay at the boundary of our star-galaxy boundary ($CLASS = 0.85$) and therefore we can not rule out the (more likely) possibility that this objects are distant star-forming regions, perhaps globular clusters or dwarf ellipticals in the process of forming, as discussed by Elson et al. (1996).

References

Bergeron, P., Wesemael, F., Beauchamp, A., 1995, Publ.Astron.Soc.Pacific, 107, 1047
Bertin, E., and Arnouts, S., 1996, Astron.Astrophys.Suppl., 117, 393
De Marchi, G., & Paresce, F. 1995, Astron.Astrophys., 304, 211
Elson, R.A.W., Santiago, B.X., and Gilmore, G.F., 1996, New Astronomy, in press
Méndez, R.A. and van Altena, W.F., 1996, Astron.J., 112, 655
Méndez, R.A., Minniti, D., De Marchi, G., Baker, A., Couch, W.J., 1996, Mon.Not.R.astron.Soc., in press
Richer, H.B., Fahlmann, G.G., Ibata, R.A., Stetson, P.B., Bell, R.A., Bolte, M., Bond, H.E., Harris, W.E., Hesser, J.E., Mandushev, G., Pryor, C., and VandenBerg, D.A., 1995, Astrophys.J., 451, L17
Santiago, B.X., Gilmore, G., and Elson, R.A.W., 1996, Mon.Not.R.astron.Soc., 281, 871
Wielen, R., Jahreiβ, H., Krüger, R., 1983, in The Nearby Stars and the Stellar Luminosity Function, IAU Colloq. No. 76, eds. A.G. Davis Philip and A.R. Upgren (L. Davis Press, Schenectady), p. 163
Williams, R.E., et al., 1996, Astron.J., in press

MULTIWAVELENGTH MILKY WAY:
AN EDUCATIONAL POSTER

D. LEISAWITZ[1], S.W. DIGEL[2] AND S. GEITZ[3,4]
[1]*Astrophysics Data Facility, NASA/GSFC*
[2]*Hughes STX Corporation, NASA/GSFC*
[3]*Technical Graphics Department, Purdue University*
[4]*NASA Visiting Faculty Fellow*

Abstract. The Astrophysics Data Facility at NASA Goddard Space Flight Center supports the processing, management, and dissemination of data obtained by past, current, and future NASA and international astrophysics missions, and promotes the effective use of those data by the astrophysics community, educators, and the public. Our *Multiwavelength Milky Way* poster was printed for broad distribution. It depicts the Galaxy at radio, infrared, optical, X-ray, and gamma-ray wavelengths. In particular, the poster contains images of the Galactic 21-cm and CO ($J = 1 \to 0$) line emission, and *IRAS* 12, 60, and 100 μm, *COBE/DIRBE* 1.25, 2.2, and 3.5 μm, Digitized Sky Survey optical wavelength, *ROSAT/PSPC* 0.25, 0.75, and 1.5 keV X-ray, and *CGRO/EGRET* $E > 100$ MeV gamma ray broadband emission. All of the data sets are publicly available. Captions describe the Milky Way and what can be learned about the Galaxy from measurements made in each segment of the electromagnetic spectrum. The poster is intended to be an educational tool, one that will stimulate heightened awareness by laypersons of NASA's contribution to modern astronomy.

Through an interface available on the World Wide Web at http://adf.gsfc.nasa.gov/adf/adf.html one may view the images that appear on the poster, read the poster captions, and locate the archived data and references.

Part 5. Extra-Galactic Astronomy

WEAK GRAVITATIONAL LENSING—THE NEED FOR SURVEYS

P. SCHNEIDER
Max-Planck-Institut für Astrophysik, Garching

1. What is Weak Lensing, and What is it Good For?

Light rays from distant sources are deflected if they pass near an intervening matter inhomogeneity. This gravitational lens effect is responsible for the well-established lens systems like multiple-imaged QSOs, (radio) 'Einstein' rings, the giant luminous arcs in clusters of galaxies, and the flux variations of stars in the LMC and the Galactic bulge seen in the searches for compact objects in our Galaxy. These types of lensing events are nowadays called 'strong lensing,' to distinguish it from the effects discussed here: light bundles are not only deflected as a whole, but distorted by the tidal gravitational field of the deflector. This image distortion can be quite weak and can then not be detected in individual images. However, since we are lucky to live in a Universe where the sky is full of faint distant galaxies, this distortion effect can be discovered statistically. This immediately implies that weak lensing requires excellent and deep images so that image shapes (and sizes) can be accurately measured and the number density be as high as possible to reduce statistical uncertainties. Weak gravitational lensing can be defined as using the faint galaxy population to measure the mass and/or mass distribution of individual intervening cosmic structures, or the statistical properties of their mass distribution, or to detect them in the first place, independent of the physical state or nature of the matter, or the luminosity of these mass concentrations. In addition, weak lensing can be used to infer the redshift distribution of the faintest galaxies. After introducing the necessary concepts, I will list the main applications of weak lensing and discuss some of them in slightly more detail, stressing the need for very deep and wide-field images of the sky taken with instruments of excellent image quality.

2. The Mapping of Small Sources

A gravitational lens provides a map from the observer's sky to the undistorted sky. The properties of this map are determined by the surface mass density of the deflector (see, *e.g.*, Schneider, Ehlers & Falco 1992). Provided the angular size of a source is very much smaller than the typical angular scale of the deflector, the lens mapping can be linearized locally. This linearized map is then characterized by an isotropic focussing term and an anisotropic distortion, due to the tidal gravitational field. This term is also called shear; it causes a circular source to be mapped into an elliptical image.

From the basic assumption that the intrinsic orientations of an ensemble of galaxies taken from a large cosmic volume are randomly distributed one can statistically infer the local distortion from an ensemble of galaxy images.

The mapping of sources as described above also affects the flux of a galaxy image. This magnification effect can be observed either by its effect on the local number counts of galaxies, an effect called magnification bias (Broadhurst *et al.* 1995), or by effect on the mean image size at fixed surface brightness (Bartelmann & Narayan 1995).

The 'traditional' method to determine the shear and magnification uses the properties of (isolated) galaxy images; for each one has to determine a center and the tensor of second brightness moments, and the rest of the CCD is unused. Alternatively, one can use the two-point auto-correlation function (ACF) of the light distribution on the CCD, $\xi(\vec{\theta})$. This is related to the unlensed ACF $\xi^s(\vec{\theta})$ by $\xi(\vec{\theta}) = \xi^s(A\vec{\theta})$. Since the unlensed ACF can be assumed to be isotropic, the anisotropy of the observed ACF immediately yields the (reduced) shear g. One can calculate the ACF locally and determine g locally. In addition, since the ACF is caused by very many faint galaxies per solid angle, one might suppose that it is a universal function (which can be determined from deep HST exposures); in that case, also the magnification can be determined locally. To avoid being dominated by just the brighter objects on the frame, they can be cut out, so that one works on a field with the topology of a Swiss Cheese. Eventually, if all objects are cut out which are significantly detected, one works in the noise limit. If the ACF of the noise is caused by faint high-redshift galaxies, the value of g determined from the noise should agree with that determined from the images, but gives independent information. This method was proposed and successfully tested both on synthetic images as well as on real data; in the latter case, the shear field obtained from individual galaxy images has been reproduced by the ACF of the noise (van Waerbeke *et al.* 1996). The ACF method is also a sensitive diagnostics for testing image quality; improper data reduction shows up immediately as artificial features in the ACF.

3. Main Applications of Weak Lensing

In this section I will outline the main applications of weak gravitational lensing as currently known.

3.1. RECONSTRUCTION OF CLUSTER MASS PROFILES

Though historically not the first application of weak lensing, the reconstruction of the two-dimensional mass distribution of clusters has been the major application of weak lensing up to now. Tyson *et al.* (1990) discovered a shear field in two clusters and determined the radial mass profiles from that. Kochanek (1990) and Miralda-Escudé (1991) investigated how shear data can be used to constrain the mass profiles of clusters. The pioneering paper by Kaiser & Squires (1993) paved the way for a non-parametric two-dimensional mass reconstruction: As in Newtonian gravity, the shear is given as a linear functional of the (surface) mass density. As was first shown by Kaiser & Squires, this relation can be inverted to express the surface mass density (up to an additive constant) in terms of the shear. Thus, if the shear can be measured, the surface mass density of the lens can be reconstructed.

In the weak lensing regime, the shear can be obtained from the local image ellipticities, as described above, and thus from an ensemble of images, the surface mass density can be evaluated. This method was first applied by Fahlman *et al.* (1994) to the cluster MS1224, and they obtained quite a large lower limit for the mass-to-light ratio of this cluster. Since then, several more clusters have been investigated with that method. The KS method has been modified to allow the inclusion of strong lensing (Schneider & Seitz 1995; Seitz & Schneider 1995; Kaiser 1995), to account for a finite region (*e.g.*, a CCD) on which observational data are given (Seitz & Schneider 1996a; see also Squires & Kaiser 1996), and to account for a broad redshift distribution of the galaxies (Seitz & Schneider 1996b). Using these generalizations, Seitz *et al.* (1996) have reconstructed the mass profile of the inner part of the cluster Cl 0939+4713 from a deep image taken with the HST. The resulting detailed two-dimensional mass map, when compared with the distribution of bright cluster galaxies, shows that the light traces the mass very well in this cluster. Also, the number density effect caused by the magnification has been discovered in this cluster. The mass-to-light ratio is only moderate (~ 200, depending on the mean redshift of the galaxies), but that should be no surprise: Cl 0939 is the highest-redshift cluster in the Abell catalog (A851) and therefore expected to have a very high optical luminosity. A low-resolution X-ray map (Schindler & Wambsganss 1996) indicates that also the X-ray emission traces the (dark) mass; this will be checked in more detail once a HRI map of this cluster becomes available.

The prospects of this method are simply excellent: deep images taking under good conditions will allow to study the dark mass distribution in clusters (*e.g.*, the radial density profile, detection of substructure and ellipticity), independent of assumptions about symmetries or dynamical or thermal equilibrium of the matter. It therefore provides the least prejudiced mass distributions, and can be used to calibrate other methods, *e.g.*, those using the X-ray profile and temperature (for example, see Squires at al. 1996). As stressed before, the accuracy of this method depends sensitively on the data quality, and on the available number density of galaxy images—thus on the depth of the observations. The combination of distortion and magnification effects, using maximum-likelihood techniques (Bartelmann *et al.* 1996), will increase the efficiency and accuracy of the reconstructions.

3.2. STATISTICAL PROPERTIES OF THE (DARK) MASS DISTRIBUTION IN GALAXIES

Individual galaxies are not massive enough to produce a significant shear signal, but statistically combining the signals from many (foreground) galaxies can yield a detectable 'relative alignment' of background images relative to the direction of the nearest foreground galaxy. First attempted by Tyson *et al.* (1984), this effect has now been discovered by Brainerd, Blandford & Smail (1996). Fitting a parametrized model to the alignment data, they have shown that the characteristic velocity dispersion (or rotational velocity) of galaxies is in the range expected from other investigations. In addition, they were able to obtain an interesting lower bound on the spatial extent of the dark halos in galaxies. This study was carried out with a relatively small number of galaxies; Schneider & Rix (1996) have shown that even with moderately-sized samples of galaxies, one can obtain very accurate determinations of model parameters such as σ_* or the characteristic size s_* of an L_* galaxy. In addition, the Tully-Fischer exponent can be probed, as well as the evolution of the mean redshift with apparent magnitude. All that is needed is a collection of wide-field images taking in excellent seeing conditions. Galaxy-galaxy lensing has also been detected in the HST MDS (Griffiths *et al.* 1996) and the HDF (Dell' Antonio & Tyson 1996).

3.3. DETECTION OF 'DARK' MASS CONCENTRATIONS

On wide-field images, one can search for (dark) mass concentrations by looking for statistically significant alignments of faint galaxy images. Based on the aperture densitometry developed by Kaiser (1995), I have investigated the statistical properties of the appropriately-defined aperture mass calculated from the image ellipticities in annular regions (Schneider 1996). The expectation is to detect isothermal halos with velocity dispersion in excess of ~ 600 km/s, without any reference to the optical or X-ray luminosity of these halos. Depending on the cosmological model, one expects about 10 such halos per square degree for a standard CDM model, increasing by a factor of order 10 in a COBE-normalized CDM model. This method will thus allow for the first time to investigate the statistics of dark halos without any assumption about bias factors, so that these results can be directly compared to numerical LSS simulations. In fact, dark halos have already been discovered by their shear effects: the 'dark' lens in the double QSO 2345+007 was discovered by the shear field it creates (Bonnet *et al.* 1993), and significant shear fields have been discovered around several high-redshift radio-loud quasars (Fort *et al.* 1996), supporting the magnification bias hypothesis for the associations of these QSOs with foreground galaxies (*e.g.*, Bartelmann & Schneider 1994).

3.4. CONSTRAINTS ON THE REDSHIFT DISTRIBUTION OF VERY FAINT GALAXIES

The lensing strength of a given deflector increases with increasing source redshift. This yields the possibility to obtain information about the redshift distribution of the faintest detectable galaxies, as proposed by Smail, Ellis & Fitchett (1994), Bartelmann & Narayan (1995) and others. In particular, the fact that significant shear was observed in the high-redshift ($z_d = 0.83$) cluster MS 1054−03 (Luppino & Kaiser 1996) shows that a large fraction of the galaxies used in this study ($21.5 < I < 25.5$) must have a redshift significantly larger than 1. For a different study of source redshifts from weak lensing, using the magnification effect, see Fort, Mellier & Dantel-Fort (1996).

3.5. DETERMINATION OF THE POWER SPECTRUM OF COSMIC DENSITY FLUCTUATIONS

The density fluctuations of the mass inhomogeneities in the Universe distort light bundles from distant sources and can produce an observable effect. It has been shown in several papers (see, *e.g.*, Villumsen 1996 for references) that the statistical properties of the distortion field are directly related to the power spectrum of the density fluctuations. For example, the two-point correlation function of the image ellipticity caused by the LSS is obtained by a convolution of its power spectrum with a known kernel function. Whereas the expected magnitude of the shear is quite small (of order 1%; however, including the non-linear evolution of the density field, the expected rms shear increases to about 4%; B. Jain, private communication), its detection and quantitative investigation will allow to study the statistical properties of the density field in the Universe, on (co-moving) scales much smaller than those achievable with CMB experiments, again without any assumption about bias factors.

3.6. PRACTICAL CONSIDERATIONS

In order to obtain good angular resolution and/or high accuracy on the local determination of the shear and the magnification, one has to take very deep exposures to be able to work with a high number density of galaxy images. The images are affected by any residual anisotropic PSF which mimics a shear. In order to correct for these instrumental effects, a stable PSF is needed, and good sampling of the PSF is required. It is also obvious that the seeing is crucial in this game: seeing circularizes small elliptical images and thus significantly reduces the shear signal. In order to regain the image ellipticities 'before seeing,' correction factors have to be applied, which are determined by simulating images with the same PSF and comparing the input shear with that estimated from the convolved image with pixelization and noise added. These correction factors can be quite large and reduce the accuracy with which the shear can be measured significantly (for a detailed discussion on these methods, see Bonnet & Mellier 1996; Kaiser, Squires & Broadhurst 1995).

4. The Need for Surveys

Weak gravitational lensing—because it is 'weak'—requires statistical ensembles of galaxy images to infer properties of the intervening mass distribution. For relatively strong deflectors, such as a massive cluster of galaxies, a single large CCD frame provides sufficient area for the reconstruction of the central mass distribution. However, to investigate the outskirts of clusters, either wide-field images or mosaics have to be taken. For the search of dark mass concentrations, a deep, high-resolution map of a large consecutive area is needed—a few square degrees mapped to magnitudes of $R \sim 26$ with seeing less than $0.8''$ will certainly reveal a sample of mass concentration selected only by their mass properties. To study the large-scale structure shear, the requirements on image quality are tremendous—the residual anisotropies of the PSF have to be understood at a percent level!

However, it seems that such accuracy can be achieved with the SUSI camera at NTT (Fort *et al.* 1996). For an analysis of galaxy-galaxy lensing, one can compromise between depth of a survey and solid angle; in the latter case, the mass properties of relatively nearby galaxies are probed, and the relatively large angular separations of such potential lenses allows to investigate the spatial extent of the galaxy halos to large distances, whereas for deeper images, the cosmological evolution of galaxy halos can be probed. Concerning the first strategy, the imaging part of the SDSS will almost certainly allow the by far most detailed study of the statistical properties of the mass distribution of (low-to-medium-redshift) galaxies.

Acknowledgements

This work was supported by the "Sonderforschungsbereich 375-95 für Astro-Teilchenphysik" der Deutschen Forschungsgemeinschaft.

References

Bartelmann, M. & Narayan, R. 1995, Astrophys.J. 451, 60.
Bartelmann, M., Narayan, R., Seitz, S. & Schneider, P. 1996, Astrophys.J. 464, L115.
Bartelmann, M. & Schneider, P. 1994, Astron.Astrophys. 284, 1.
Bonnet, H., Fort, B., Kneib, J.-P., Mellier, Y. & Soucail, G. 1993, Astron.Astrophys. 280, L7.
Bonnet, H. & Mellier, Y. 1995, Astron.Astrophys. 303, 331.
Brainerd, T.G., Blandford, R.D. & Smail, I. 1996 Astrophys.J. 466, 623.
Broadhurst, T.J., Taylor, A.N. & Peacock, J.A. 1995, Astrophys.J. 438, 49.
Dell' Antonio, I.P. & Tyson, J.A. 1996, astro-ph/9608043.
Fahlman, G., Kaiser, N., Squires, G. & Woods, D. 1994, Astrophys.J. 437, 56.
Fort, B., Mellier, Y. & Dantel-Fort, M. 1996, astro-ph/9606039.
Fort, B., Mellier, Y., Dantel-Fort, M., Bonnet, H. & Kneib, J.-P. 1996 Astron.Astrophys. 310, 705.
Kaiser, N. 1995, Astrophys.J. 439, L1.
Kaiser, N. & Squires, G. 1993, Astrophys.J. 404, 441.
Kaiser, N., Squires, G. & Broadhurst, T. 1995, Astrophys.J. 449, 460.
Kochanek, C.S. 1990, Mon.Not.R.astron.Soc. 247, 135.
Luppino, G. & Kaiser, N. 1996, astro-ph/9601194.
Miralda-Escudé, J. 1991, Astrophys.J. 370, 1.
Schindler, S. & Wambsganss, J. 1996, preprint.
Schneider, P. 1996, Mon.Not.R.astron.Soc., in press.
Schneider, P., Ehlers, J. & Falco, E.E. 1992, "Gravitational lenses," Springer: New York.
Schneider, P. & Rix, H.-W. 1996, Astrophys.J., in press.
Schneider, P. & Seitz, C. 1995, Astron.Astrophys. 294, 411.
Seitz, C., Kneib, J.-P., Schneider, P. & Seitz, S. 1996, Astron.Astrophys. (in press).
Seitz, C. & Schneider, P. 1995, Astron.Astrophys. 297, 287.
Seitz, S. & Schneider, P. 1996a, Astron.Astrophys. 305, 383.
Seitz, C. & Schneider, P. 1996b, Astron.Astrophys., in press.
Smail, I., Ellis, R.S. & Fitchett, M.J. 1994, Mon.Not.R.astron.Soc. 270, 245.
Squires, G. & Kaiser, N. 1996, preprint.
Squires, G. *et al.* 1996, Astrophys.J. 461, 572.
Tyson, J.A., Valdes, F., Jarvis, J.F. & Mills Jr., A.P. 1984, Astrophys.J. 281, L59.

Tyson, J.A., Valdes, F. & Wenk, R.A. 1990, Astrophys.J. 349, L1.
Van Waerbeke, L., Mellier, Y., Schneider, P., Fort, B. & Mathez, G. 1996, Astron.Astrophys., in press.
Villumsen, J.V. 1996, Mon.Not.R.astron.Soc. 281, 369.

THE EVOLUTION OF QUASARS AND THEIR CLUSTERING

P.S. OSMER
Astronomy Department, The Ohio State University

1. Introduction

George Djorgovski's message to me on behalf of the organizing committee suggested that I discuss 1) Evolution of the Quasar Luminosity Function; 2) Possible Differences among Radio-Loud, Radio-Quiet, and X-Ray Selected Samples; 3) Quasars as Probes of Clustering and Large-Scale Structure at High Redshift; 4) Searches for Extremely Distant Quasars; 5) Theoretical Implications; 6) Major Surveys in Progress and Planned; and 7) What Surveys Should We Do in the Future. To cover all these topics in a 25-minute talk and 7-page article is a challenge, but I am happy to try.

As we approach the end of the century and indeed the millennium, it has been fashionable to write books and articles about the end of subjects such as history and science. What about the end of quasars? That is a more legitimate question, because we have not extended the high-redshift limit at $z = 4.9$ since 1991 and there is quantitative evidence that the space density is falling steeply for $z > 4$. However, in reviewing the literature, I can assure you that we have not reached the end of publications nor work in the field. The ADS abstract service lists 498 articles on quasars in 1995 alone. The number of known quasars is of order 10,000 and will grow by another order of magnitude with planned surveys. Advances in technology now enable surveys for quasars over large fractions of the entire sky to significantly fainter flux limits than could be reached before at X-ray, optical, IR, and radio wavelengths. Furthermore, there is the new and very positive trend of making the data sets from such surveys publicly available, in many cases as soon as the data can be processed. I predict that such public access will greatly increase the scientific results from the surveys by enabling more studies of the data on a broader range of topics than any single person or group could accomplish, to say nothing of permitting the cross-checking of results, which is a basic part of our science.

The plan of this talk is to cover in the sections below the questions posed in the first paragraph (except that question 2 will be discussed only briefly in §6).

2. Evolution of the Quasar Luminosity Function

In discussing the evolution of quasars, we should remember that the fundamental physical question is how quasars themselves form and evolve. However, what we derive from observations is the quasar luminosity function (QLF). And, while we have made progress in characterizing how the QLF evolves with redshift, I do not think we have yet established that the parameterization of the QLF as showing luminosity evolution, for example, means that the quasars themselves are undergoing luminosity evolution in the sense of a long-lived population of objects that are bright at $z \approx 2$ and then fade toward the present day. Rather, I think that establishing how quasars evolve physically should be one of the main goals of future surveys. Boyle (1991, Figure 2) gives a nice description of the observational issues on this matter.

A second important point is understanding what it takes to determine the evolution of the QLF, namely, large, calibrated, quantitative surveys whose detection efficiency is well characterized (see Hewett & Foltz 1994 for an excellent discussion of this topic). Suppose, for example, that we were to require 10 quasars per magnitude bin and 10 magnitude bins per redshift over 10 redshift intervals. This would mean 1,000 quasars overall. The challenge in assembling such a sample, which we have yet to do, is not the total number, but filling the bins at the extremes in redshift and absolute magnitude. High-luminosity quasars are rare, as are quasars at $z > 4$, and one must cover large areas of the sky. Low-luminosity quasars are much more common, but their apparent magnitudes become increasingly faint with increasing redshift. No single survey to date has provided all the data we need. However, we have been able to make progress with relatively small samples because the evolution of the QLF is so pronounced: the space density of quasars increases by a factor of 1000 between redshifts 0 and 2.

At present, determination of the QLF for $z < 2.2$ is based primarily on the Durham/AAT UVX survey (Boyle 1991) and the Large Bright Quasar Survey (Hewett, Foltz, and Chaffee 1993). For $z > 3$, there are three calibrated surveys (Warren, Hewett, & Osmer 1994 [WHO], Schmidt, Schneider, & Gunn 1995 [SSG], and Kennefick, Djorgovski, & de Carvalho 1995 [KDC]). The KDC results provide confirmation that the observed space density of quasars at $z > 4$ does decline as indicated by the earlier work of WHO and SSG. A linear plot of space density against lookback time shows a remarkable peak when the universe was approximately 15% of its present age. This striking feature suggests that quasar activity was very concentrated at redshifts near 2 to 3.

Of course there still are challenges and questions for this picture. Have bright quasars with $z > 4$ been missed (Irwin, McMahon, & Hazard 1991 [IMH])? Are significant numbers of quasars obscured by dust (Webster *et al.* 1995)? Are there other significant populations of quasars that have been missed by surveys to date, such as optically quiet quasars or ones found only by variability (*e.g.*, Hawkins & Veron 1995)? While I agree that some classes of quasars are missed by optical surveys based on color and emission lines, I am not yet persuaded that the overall picture of quasar evolution needs to be changed. For example, Boyle and di Matteo (1995) argue from a *ROSAT*-selected sample of quasars that a significant number of quasars is not being missed. But, this is a scientific question that is resolvable with current technology. Quantitative surveys that follow the criteria of Hewett & Foltz (1994) can provide the answers.

3. Quasars as Probes of Clustering & Large-Scale Structure at High Redshifts

Because of their great luminosity, quasars are in principle the best tracers of clustering and large-scale (1000 Mpc) structure at $z > 1$. But, they are rare objects, which presumably originated in high-density peaks in the primordial fluctuations, and they yield sparse sampling of large-scale structure. Shanks & Boyle (1994) give a good account of previous work in the field and the current state of the subject. Using a compilation of available surveys, they present convincing evidence that quasar clustering on scales up to 10 h^{-1} Mpc is detected at the 4σ level of significance and furthermore argue in favor of stable clustering for both quasars and galaxies out to $z \approx 2$.

The situation at larger scales has been hampered by lack of surveys over sufficiently large areas on the sky, but there have been discoveries of supercluster-sized groups of quasars. It has been difficult to establish the statistical significance of such groups because their sizes have been comparable to the surveys in which they were found. Most recently, Graham, Clowes, & Campusano (1995) apply a minimum spanning-tree analysis to the data and conclude that the several $\sim 100 h^{-1}$ Mpc groups found by Crampton *et al.*(1989) and themselves are real.

Looking ahead, this is a subject that the Sloan Digital Sky Survey and the 2dF quasar survey should definitely settle. With large, connected areas of up to 10,000 deg^2 and sample sizes up to 100,000 objects, we can expect an improvement of the S/N ratio for clustering of more than a factor of 10 at scales up to 10 h^{-1} Mpc. At the $100h^{-1}$ Mpc scales we will finally have adequate control fields to establish the nature of the supercluster-sized groups.

4. Searches for Extremely Distant Quasars ($z > 5$)

The results of WHO, SSG, KDC, and IMH indicate that the expected number of quasars with $z > 5$ is small. Future efforts will need a combination of area on the sky and faint limiting magnitude. Good sensitivity at $\lambda > 0.8\mu$m will also be required because Lyman α absorption along the line of sight will make quasars faint at shorter wavelengths.

The Sloan Digital Sky Survey is expected to make a very important contribution because of the large area on the sky it will cover. At present several groups are making targeted searches for extremely distant quasars. Kennefick and Smith are leading a survey that makes use of the second Palomar Survey in combination with z-band (0.9μm) images from the Burrell and Curtis Schmidts at NOAO to cover 2000 deg^2. Irwin & McMahon and Hewett & Warren are two of the groups in the U.K. making use of the La Palma Observatory and U.K. Schmidt plates to search for $z > 5$ quasars.

Current expectations, based on the searches for $z > 4$ quasars to date and the QLFs derived by WHO, SSG, and KDC, for example, indicate that at I = 18 mag, more than 1000 deg^2 will probably be needed to find one quasar at $z > 4.8$. At I = 20, the limit of the Sloan Survey, one quasar is expected in 15 to 300 deg^2 (SSG to WHO form of the decline with increasing redshift). Thus, the Sloan Survey, with its planned coverage of 10,000 deg^2 has an excellent chance of finding a significant number of quasars with $z > 5$.

5. What Do the Observations Tell Us about the Origins of Nuclear Activity? About the Formation of Large Scale Structure and Galaxies?

The strong peak in the space density of quasars near redshift 2–3 begs the question, 'Are we seeing the epoch of quasar formation?' Recall that in the standard model for quasars, the observed luminosity is a function of the black-hole mass and the fueling rate of material being accreted. Thus, it is very likely that the peak in space density is related to galaxy formation and interactions at high redshift. Furthermore, if quasars represent high-density peaks in structure formation, then the clustering and large-scale distribution of quasars should yield valuable information for models of structure formation.

TOWARDS A BETTER UNDERSTANDING OF ACTIVE GALACTIC NUCLEI

P. PADOVANI
Dipartimento di Fisica, II Università di Roma "Tor Vergata"

Abstract. Active Galactic Nuclei (AGN) are ideal sources for multi-wavelength studies as their emission can cover almost 20 orders of magnitude in frequency from the radio to the γ-ray band. After reviewing their basic properties, I will assess how well we know the multifrequency spectra of AGN as a class. I will then briefly illustrate how currently available and forthcoming sky surveys will help in addressing some of the open questions of AGN studies. Finally, an analysis of the problem of the missing Type 2 QSO will exemplify the dangers of monochromatic sky surveys for AGN.

1. AGN and Unified Schemes

The large number of classes and subclasses which appear in AGN literature might disorientate astronomers working in other fields. A simplified classification, however, can be made based on radio loudness and the width of the emission lines (*e.g.*, Urry and Padovani 1995). Radio-loud (RL) sources, that is objects with radio to optical (B-band) flux ratio $f_\mathrm{r}/f_\mathrm{B} \gtrsim 10$ (*e.g.*, Stocke *et al.* 1992), make up only $\sim 10\text{--}15\%$ of AGN and have, for the same optical luminosity, radio powers 3 to 4 orders of magnitude larger than those typical of radio-quiet (RQ) sources. AGN are also divided in Type 1 (broad-lined) and Type 2 (narrow-lined) objects according to their line-widths, with 1000 km/s (full width half maximum) being the dividing value. We then have RL Type 1 AGN, that is radio quasars and broadline radio galaxies, and RL Type 2 AGN, that is radio galaxies, with their corresponding RQ counterparts, that is Seyfert 1 galaxies and QSO, and Seyfert 2 galaxies respectively. Some objects exist with unusual emission line properties, such as BL Lacs, which have very weak emission lines with typical equivalent widths < 5 Å.

In recent years we have come to understand that classes of apparently different AGN might actually be intrinsically similar, only seen at different angles with respect to the line of sight. The basic idea, based on a variety of observations and summarized in Figure 1 of (Urry and Padovani 1995), is that emission in the inner parts of AGN is highly anisotropic. The current paradigm for AGN includes a central engine, surrounded by an accretion disk and by fast-moving clouds, probably under the influence of the strong gravitational field, emitting Doppler-broadened lines. More distant clouds emit narrower lines. Absorbing material in some flattened configuration (usually idealized as a torus) obscures the central parts, so that for transverse lines of sight only the narrow-line emitting clouds are seen (Type 2 AGN), whereas the near-IR to soft-X-ray nuclear continuum and broad-lines are visible only when viewed face-on (Type 1 AGN). In RL objects we have the additional presence of a relativistic jet, roughly perpendicular to the disk, which produces strong anisotropy and amplification of the continuum emission ("relativistic beaming"). In general, different components are dominant at different wavelengths. Namely, the jet dominates at radio and γ-ray frequencies (although it does contribute to the emission in other bands as well), the accretion disk is thought to be a strong optical/UV/soft X-ray emitter, while the absorbing material will emit predominantly in the IR.

This axisymmetric model of AGN implies widely different observational properties (and therefore classifications) at different aspect angles. Hence the need for "Unified Schemes" which look at intrinsic, isotropic properties, to unify fundamentally identical (but apparently different) classes of AGN. Seyfert 2 galaxies have been "unified" with Seyfert 1 galaxies, whilst low-luminosity and high-luminosity radio galaxies have been unified with BL Lacs and radio quasars respectively (see Alloin 1993 and Urry and Padovani 1995 and references therein).

2. The Multiwavelength Spectrum of AGN

The property that makes AGN ideal sources for multiwavelength studies is their broad-band emission, which covers basically the whole observable electromagnetic spectrum from the radio to the γ-ray band (almost 20 orders of magnitude in frequency). Multifrequency coverage at this level, however, is rare for a single object: see for example Alloin 1993 and Lichti *et al.* 1995 which give simultaneous multifrequency data for NGC 3783, a Seyfert 1 galaxy, and 3C 273, a radio-loud quasar, respectively. Also, the objects studied are not necessarily representative of their class and are usually relatively local.

Single-object studies are certainly important, as specific models can be fitted to their multifrequency spectra to constrain the emission processes. However, one would also like to learn about the general properties of the AGN population and study statistically their emission in various bands, for example to constrain unified schemes. In other words, use multifrequency data to address some of the open questions of AGN research.

Before I address this point, we first have to assess how well we know the multifrequency spectrum of AGN as a class. I have then taken the latest Véron-Cetty & Véron AGN catalogue (1996), which includes 11,662 AGN and gives redshift, optical (U, B, and V) magnitudes, and some radio information (6 and 11 cm fluxes), and cross-correlated it with radio, far/mid-IR, X-ray and γ-ray catalogs available in machine readable form (using BROWSE as implemented at SAX/SDC). The list includes the FIRST and NVSS (respectively about 15% and 40% completed), the NORTH 20 cm, PKS, PMN, GB6, S4, S5, 1 Jy catalogs (radio: 20, 11, and 6 cm), the IRAS PSC and FSC (infrared: 12–100 μ), the *EXOSAT* CMA, the *ROSAT* RASS-BSC and WGA, the *Einstein* Slew, EMSS, IPC catalogs (soft X-ray: ~ 0.05–3.5 keV), a hard X-ray compilation (Malizia and Bassani 1996), and the *GRO* EGRET catalog (γ-ray: 30 MeV–20 GeV). Many of these surveys have been discussed at this meeting by various contributions, where detailed references can be found.

Figure 1a, which summarizes more than 30 years of observations, shows the percentage of AGN with data in a given band as a function of frequency for all AGN and for the RL and RQ subclasses. It can be seen that the multifrequency spectrum of AGN is not that well known. We are not doing too badly in the radio (at a few GHz) and soft X-rays, where about 30% of AGN are detected, while basically all AGN have optical data (due to the fact that the eye is peaked in the optical and we always require an "optical identification"). However, we have major "holes" in our knowledge in the far/mid-IR, hard X-ray, and γ-ray bands. Note also that RL AGN fare better than RQ ones at all frequencies, apart from the IRAS band. If we restrict ourselves to the radio/optical/soft X-ray bands, where we have the best coverage, only about 10% of AGN have data in all three bands.

The fractions of objects shown in Figure 1a are the convolution of two effects: 1. the intrinsic spectral shape of AGN; 2. the sensitivity of our detectors in the various bands. The latter is shown in Figure 1b, which displays the limiting sensitivities of the deepest all-sky (or large-area) surveys available at various frequencies, which were used to construct Figure 1a (deeper, small-area surveys exist in some bands). The POSS I limit, as discussed above, is only indicative as many optical data come from dedi-

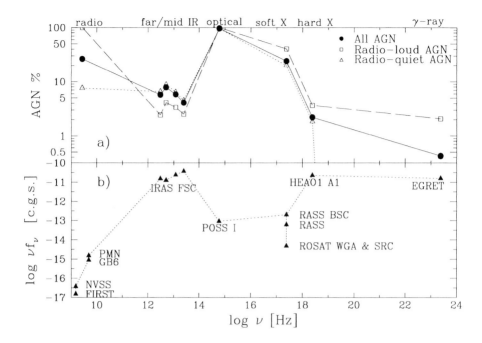

Figure 1. a) (top) The percentage of AGN with detections in a given band vs. frequency for all objects (*solid points*), radio-loud (*open squares*), and radio-quiet (*open triangles*) AGN. Sources without radio data are included in the latter class. b) (bottom) The sensitivity of the deepest all-sky (or very large area) surveys as a function of frequency.

cated observations. The main point here is that the bands where we have less information, *i.e.*, far/mid-IR, hard X-ray and γ-ray, are exactly those where our surveys are less sensitive. It then follows that our poor knowledge of the multifrequency spectra of AGN is due, more than to their intrinsic faintness in some bands, to the limitations of our detectors. (The radio band is an exception: radio surveys are very deep but most AGN are weak radio sources.)

3. The Role of Multiwavelength Sky Surveys in AGN Research

Given the broad-band emission of AGN, it is clear that multiwavelength sky surveys are extremely important to further our understanding of AGN physics, at a zero order simply by providing more data for more AGN.

This is not as trivial as it might sound: AGN are rare, making up only about 1% of all bright galaxies (although low luminosity AGN [*e.g.*, LINERs] could be relatively more numerous). The latest AGN catalog includes about 12,000 sources, while at this meeting catalogs with tens and even hundreds of millions of entries have been discussed (*e.g.*, Canzian,

p. 422). AGN are also hard to find, especially in the optical. Some of the complete samples, which are needed to study AGN evolution, are still quite small, especially the radio-selected ones, and only less than 300 BL Lacs are known over the whole sky (Padovani and Giommi 1995). The cross-correlation between surveys will take advantage of the broad-band emission of AGN, enhancing the efficiency of AGN detection, especially important for rare classes of objects (Nass *et al.* 1997 and Perlman *et al.* 1997).

Specific examples of open questions in AGN research which will benefit from existing and future sky surveys include the following:

- The radio-loud/radio-quiet dichotomy. Despite years of effort, we still do not know what makes an AGN radio-loud nor why the f_r/f_B distribution is bimodal for optically-selected samples. Deep radio surveys, like the NVSS and the FIRST, will allow us to study for the first time large numbers of radio-selected *radio-quiet* objects (Gregg *et al.* 1996), to check if the f_r/f_B distribution is still bimodal when the selection is done in the radio band, and even to study the radio evolution of RQ quasars.
- Thermal versus non-thermal emission. While we have strong evidence in favour of a dominance of non-thermal (synchrotron and inverse Compton) emission in radio-loud AGN, the situation is not that clear for radio-quiet sources, although the optical/UV might be dominated by thermal emission. Multifrequency data for large numbers of AGN are needed to address this question on a statistical basis.
- AGN Evolution. The driving force behind the strong evolution observed in AGN is still not understood. Also, although the form of the evolution seems to be similar at radio, optical, and X-ray frequencies, this should be checked against larger and deeper samples.

This list is no doubt incomplete but it is only meant to give a flavor of the possibilities multiwavelength surveys provide us with. I will now concentrate on a particular problem, which is a text-book example of how monochromatic surveys can be misleading: the missing Type 2 QSO.

4. The Mystery of the Missing Type 2 QSO

According to unified schemes, Type 2 objects are Type 1's seen edge-on, *i.e.*, with their central parts strongly obscured by dust. Seyfert 2's have been identified as the Type 2 equivalent of Seyfert 1's, and radio galaxies as the Type 2 equivalent of radio quasars. We also think that Seyfert 1 galaxies are simply low-luminosity, low-redshift versions of RQ quasars. There should then be Type 2, *i.e.*, narrow-lined, RQ quasars: but where are they?

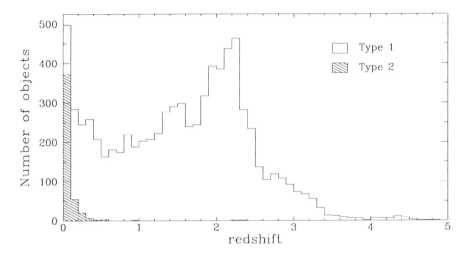

Figure 2. The redshift distribution of the $\sim 8,000$ radio-quiet AGN in the Véron-Cetty & Véron catalog for Type 1 (broad-lined) and Type 2 (narrow-lined) objects. Only 3 Type 2 radio-quiet AGN are known at $z > 0.9$.

Figure 2 shows the redshift distribution of all RQ Type 1 and Type 2 AGN in Véron-Cetty & Véron. Most known Type 2's are local, $\sim 80\%$ of them at $z < 0.1$, and there seem to be no high-z counterparts to the local Seyfert 2's. Note that for semi-opening angles of the obscuring torus between 30° and 45° (derived from various methods), Type 2 AGN should be intrinsically more numerous than Type 1's by factors between 6.5 and 2.4, that is *most RQ AGN should be of Type 2*. Even for angles as large as 60° the two classes should be equally numerous. The point is that most quasars are still identified by optical searches, tuned to find AGN with strong non-stellar continua, *i.e.*, Type 1 quasars. Normally, in fact, candidates are first selected on the basis of their non-stellar colors, then observed spectroscopically to see if they show broad lines. The inferred extinction in Type 2 objects can be larger than 10 magnitudes, reaching even values as high as a few hundred (*e.g.*, Goodrich *et al.* 1994), so in these objects the AGN continuum is swamped by the host galaxy light. Type 2 QSO, then, do not even make it to the candidate level, and even if they did, they would be discarded because they do not show broad lines! Type 2 radio sources, of course, are selected because of their radio emission and, in fact, we have examples of radio galaxies at redshifts almost as high as those of radio quasars. Seyfert 2's, on the other hand, are relatively easy to find locally where the selection is based on host galaxy properties. For example, a complete sample of Seyferts can be derived by taking spectra of all galaxies down to a given magnitude limit. This is of course only feasible for relatively bright (and therefore local) galaxies (*e.g.*, $B \leq 14.5$: Huchra and Burg 1992).

How can we find the high-redshift Type 2 QSO? The energy absorbed by the obscuring material in the optical/UV/soft X-ray bands will have to be re-emitted in the IR band. Type 2 AGN should then be strong IR emitters. In fact, optical identification of IRAS sources has produced a large number of Type 2 AGN, approximately 30% of known objects, and IRAS selected Type 2's make up about 50% of the few sources at $z > 0.1$. Some of these objects have also, as expected, bolometric luminosities typical of quasars. However, the redshift distribution in Figure 2 includes all these IRAS selected Type 2 AGN, which leads to the next question: Why is IRAS not finding more of them? The problem is that the IRAS survey, as illustrated in Figure 1b, is not very sensitive: IRAS has detected only about 8% of all known AGN at 60μ (and an even smaller fraction in the other bands) and these are mostly local sources, $\sim 80\%$ at $z < 0.1$. IRAS is then sampling only the local universe and the few high-z Type 2 AGN detected are probably atypically luminous, like IRAS F10214+4724 ($z = 2.286$), which has recently been shown to be gravitationally lensed (Broadhurst and Lehar 1995).

The torus becomes transparent at hard X-ray energies but even the surveys in this band, made with the HEAO1 observatory, are not very sensitive: only about 2% of known AGN have hard X-ray data (~ 10 keV) and again most of them are local sources ($z < 0.1$).

5. Towards an Unbiased Survey

If our current understanding of AGN is correct, the example I have just given is not isolated. Most available surveys, in fact, give us a "biased" view of the AGN population, not only because different components emit at different frequencies, but mainly because most surveys preferentially detect some particular classes of AGN. Namely, high-frequency radio surveys are dominated by objects with beamed radio emission, the so-called blazars (BL Lacs and core-dominated quasars). Although this is less of a problem at the faint fluxes reached by the FIRST and NVSS surveys, it can be shown that even these surveys will not detect the bulk of the RQ population, so that radio surveys still preferentially detect RL AGN. (Low-frequency radio surveys are less biased in terms of selecting beamed objects but still detect mostly RL sources). The region between near-IR to soft X-ray frequencies is the most biased of all: here in fact the torus is optically thick and only the nuclear continuum emission of Type 1 AGN is detected. Radiation from the central parts manages to escape only at $\lambda \gtrsim 50$–100μ and $E \gtrsim 10$–20 keV, the precise values depending on the density and geometry of the absorbing column. (A likely distribution of these parameters also implies a distribution in the energies at which we can see through the torus.) The γ-ray EGRET sky survey is also biased, as it has only detected blazars. Far-IR surveys, as discussed above, are unbiased, but current surveys are not deep enough (also, they suffer from strong contamination by non-AGN sources, mostly star-forming regions, both galactic and extragalactic). Finally, hard X-ray surveys are also unbiased, but still not deep enough.

What about future far-IR and hard X-ray surveys? As discussed at this meeting (Beichmann, p. 27), our best bet for a *large area*, far-IR survey is the COBRAS/ SAMBA mission, which will perform all-sky surveys in the 30–900 GHz range. In fact, as regards other IR surveys, they will either be barely more sensitive than IRAS (*e.g.*, the ISO 200μ serendipitous survey) or will only sample the mid-IR range (*e.g.*, WIRE) or be limited to small areas of the sky (*e.g.*, ISO ELAIS Oliver *et al.* 1997, SIRTF). As regards hard X-rays, there will be serendipitous surveys covering the 2–30 keV (SAX WFC) and 50–100 keV (*GRO* BATSE) ranges but as these instruments were made for other purposes they will not be much more sensitive than the HEAO1 surveys. On the other hand, the SAX LECS and MECS will provide a deep serendipitous survey in the 1–10 keV band, while deep surveys will also be performed in the 0.25–12 keV and 2–12 keV ranges respectively by XMM (Stewert 1997) and ABRIXAS (Trumper 1997), the former in serendipitous and slew mode, the latter as a proper all-sky survey. Although these energy ranges are not optimized to see through the torus in all sources, the rest-frame frequency increases as (1+z) so that at $z = 1$, for example, ABRIXAS will sample the 4–24 keV range. These surveys will certainly represent a major step forward in our quest to obtain an unbiased view of AGN, independent of orientation.

In summary, the main conclusions are as follows: 1) both because different AGN components emit in different bands and because most sky surveys are biased towards particular AGN classes, no single survey can give us the broad view of AGN we need to understand them. Therefore, multiwavelength sky surveys are *vital* for AGN research; 2) If our understanding of AGN is correct, we may be missing between 70% to 90% of (radio-quiet) AGN. Deep far-IR and hard X-ray surveys are needed to see through the absorbing material we think surrounds the central parts of AGN and identify the large numbers of AGN which have so far escaped detection.

Acknowledgements

By way of acknowledgement, I offer my thanks to various participants in the IAU 179 Symposium for useful discussions. I also acknowledge helpful comments from two long-standing collaborators, Meg Urry and Paolo

Giommi, the latter being also responsible for introducing me to the world of database management.

References

Alloin, D., *et al..*, 1995, Astron.Astrophys., 293, 293
Antonucci, R., 1993, Ann.Rev.Astron.Astrophys., 31, 473
Beichman, C., this volume, 27
Broadhurst, T. and Lehar, J., 1995, Astrophys.J., 450, L41
Canzian, B., this volume, 422
Goodrich, R.W., Veilleux, S. and Hill, G.J., 1994, Astrophys.J., 422, 521
Gregg, M.D., *et al..*, 1996, Astron.J., 112, 407
Huchra, J. and Burg, R., 1992, Astrophys.J., 393, 90
Lichti, G.G., *et al..*, 1995, Astron.Astrophys., 298, 711
Malizia, A. and Bassani, L., 1996, private communication
Nass, P., *et al..*, this volume, 305
Oliver, S., *et al..*, this volume, 112
Padovani, P. and Giommi, P., 1995, Mon.Not.R.astron.Soc., 277, 1477
Perlman, E., *et al..*, this volume, 310
Stocke, J.T., Morris, S.L., Weymann, R.J. and Foltz, C.B., 1992, Astrophys.J., 396, 487
Trümper, J., these proceedings
Urry, C.M. and Padovani, P., 1995, Publ.Astron.Soc.Pacific, 107, 803
Véron-Cetty, M.-P., and Véron, P., 1996, ESO Scientific Report No. 17

MEASURES OF GALACTIC AND INTERGALACTIC MASS IN CLUSTERS

D. WINDRIDGE, S. PHILLIPPS, M. BIRKINSHAW
Astrophysics Group, University of Bristol

1. Estimation of the Electron Gas Mass

If a galaxy cluster's X-ray gas distribution follows an isothermal polytropic β model, we may write the electron radial density distribution as; $n_e = n_{e0}(1 + r^2/r_c^2)^{-3/2\beta}$, r_c being the core radius and n_{e0} the central electron density. This may be related to both an X-ray surface brightness distribution and a Sunyaev-Zel'dovich effect distribution (Sarazin 1986). Fitting to observational data then enables us to constrain the value of β. The normalisation value, n_{e0}, to obtain a total mass estimate is calculated via the relationship between the X-ray and S-Z distribution normalisation constants, and the gas temperature and spectral emissivity parameters from fits to the X-ray spectrum. We are then in a position to evaluate $n_e(r)$ and its integral; the total electron gas mass. If we can further assume that there exists a simple ratio between the electron and proton number densities within the gas, we may straightforwardly posit a value for the total gas mass. An additional method of determining the polytropic gas index exists, with optical constraints on the galactic velocity dispersion, through the relation; $\beta = \mu m_H \sigma_z^2 / k_B T_e$. Studies at optical, as well as X-ray and radio wavelengths are thus useful as a corroborative measure in determining the total gas mass.

2. Estimation of the Baryonic Galactic Mass

In formulating the total baryonic galactic mass, we shall follow Persic & Salucci (1992) in treating the galactic mass-to-light ratio as a power-law with respect to luminosity; i.e., $\Gamma(L)=A(L/L_\star)^\eta$. Combining this with the Schechter (1976) luminosity function, we may derive a class-specific estimate of the total field density across the entire luminous range:

$\rho_b = \phi_\star L_\star A \int_{x_{min}}^{x_{max}} x^{1+\alpha+\eta} e^{-x} dx$, with x$\equiv \frac{L}{L_\star}$, and $\phi_\star$ & $L_\star$; Schechter normalisation parameters. The value η is not subject to rigorous observational constraint, but we shall argue for a value of ~ -0.1 by noting that the galactic stellar M/L ratio is $\approx 1 \frac{M_\odot}{L_\odot}$, practically across the entire observational range (Longmore 1982); whilst the HI mass remains roughly constant. The *baryonic* M/L ratio can then only rise modestly with decreasing luminosity, the excess baryons being supplied by the increasing HI gas mass fraction. Since the density equation is critically sensitive to the sum $\alpha+\eta$ for values around -1.8 (*i.e.*, $\alpha \sim -1.7$), below which the integral diverges with decreasing x_{min}, and since this is so close to dwarf galaxy LF slope indicated by deep cluster surveys, we shall treat the dwarf and giant galaxies as discrete Schechter populations (Driver *et al.* 1994). Evaluating the mass integral separately for the two classes, then, we find a field dwarf/giant mass density ratio of 2.5:1. In a flux-limited cluster survey, in which only the giants are reliably detected, a total baryonic mass estimate may then be obtained through the introduction of a simple multiplying factor.

3. Estimation of the Total Galactic Mass in Clusters

Any estimate of the *total* galactic mass in clusters proceeding in the manner set out above must incorporate some consideration of the variation of dark matter mass with luminosity, paying particular regard to the mass-to-light ratios at the dwarf luminosities, where most of the matter would appear to be concentrated. Perhaps the best indication of dwarf mass-to-light ratios at very faint luminosities comes from the recent survey of the local group dSphs by Irwin & Hatzidimitriou (1995), suggesting a $L^{-.965}$ power-law operating over the survey range. Since it would be unphysical for this power-law to extend much beyond the survey luminosity limits, we envisage a composite power-law for the *general* dwarf Γ function, with a baryonic ($L^{-0.1}$) power-law taking-over from the $\eta = -.965$ dSph behaviour at luminosities greater than $10^{6.7} L_\odot$. Performing the density integration within each of the two domains of behavior then gives a dwarf/giant density ratio of 5.5:1 allowing an estimate of the total galactic mass in clusters from the giant galaxy number counts alone.

References

Birkinshaw, M. & Hughes, J. P., 1994, Astrophys. J. 420, 33
Driver S. P. *et al.*, 1994, Mon. Not. R. astron. Soc. 266, 155
Irwin M. & Hatzidimitriou D., 1995, Mon. Not. R. astron. Soc. 277, 1354
Longmore A. J. *et al.*, 1982, Mon. Not. R. astron. Soc. 200, 325
Persic M. & Salucci P., 1992, Mon. Not. R. astron. Soc. 258, 14P
Sarazin C. L., 1986, Rev. Mod. Phys. 58, 1
Schechter P. L., 1976, Astrophys. J. 203, 297

THE SCALING RELATIONS FOR CLUSTERS OF GALAXIES

Fundamental Planes, Cluster Structure, and Cluster Evolution

J. ANNIS
Experimental Astrophysics Group, Fermilab

1. Introduction

The dominant baryonic component of clusters of galaxies is their X-ray emitting atmosphere. X-ray surveys have made it possible to examine this component in large samples of clusters. David *et al.* (1993) provided a catalog of X-ray temperatures. I have measured 79 clusters from that list using Einstein IPC data (Harris *et al.* 1990), determining several measures of the radius. The most useful turns out to be the Petrosian radius. Given a radius, one can measure the surface brightness at that position.

2. Fundamental Planes

The set of observables r_p-I_p-T_x are interrelated in such a way that given any two I can predict the third. This is at root the result of the virial theorem implying that $r \propto T\eta^{-1}$, and the bremsstrahlung expression for surface brightness, $I_p \propto r_p^{-1} T_x^{1/2} \eta^2$ (where η is the mass surface density). The relationship I measure is

$$r_p \propto T_x^{1.30} I^{-0.42}$$

This relation differs from the virial in a way that can be understood as a variation of M_{lum} with M_{tot}.

In practice, any mass indicator may be used in place of T_x, including L_x and N_A. The fundamental plane relationships of r_p-I_p-T_x and L_x-I_p-T_x can be used as distance indicators. The width of both are consistent with the measurement error on T_x, so they can probably be improved with better measurements of T_x.

3. Cluster Structure

I have found two exotic distance indicators, relations that seem *not* to be the result of the virial theorem, but rather related to cluster structure. These are the r_p-I_m/I_p and L_r-I_m relations. (I_m is the surface brightness at 250 kpc.) The L_x-I_m relation in particular seems to suggest a quite stunning thing. It can be understood if the dominant contributor to the luminosity in all clusters is distributed with the same functional form and similar scale radii. And indeed, the four very different clusters of Jones and Forman have profiles that, while differing in detail, are roughly straight lines of constant slope.

4. Cluster Evolution

We can use distance measuring to look for evolution. The methodology is the same as if I was trying to measure q_0. The distance indicator relations are linear in log-log plots. I assume evolution occurs in the form $r \propto r_\star(1+z)^a$, $L \propto L_\star(1+z)^b$, and $T \propto T_\star(1+z)^c$. Evolution then appears as a shift in the zeropoint of the distance indicator relation as a function of z.

Each distance indicator has a different sensitivity to evolution. The fundamental plane indicators are primarily sensitive to T_x evolution, though there is also sensitivity to L_x. The "exotic" indicators are primarily sensitive to r_p evolution. The data are consistent with no evolution in T_x, but they are also consistent with evolution in both L_x and T_x that balance each other. The data are consistent with little or no evolution in r_p.

The Einstein sample reported here cannot place too strong a constraint on cluster evolution for two reasons: a) there are only a handful of clusters at $z > 0.15$, and b) the selection criteria for the $z > 0.15$ clusters was haphazard at best. A further analysis with ROSAT and ASCA data promises to place stringent limits.

5. Summary

Studying clusters in the X-ray reveals a very powerful set of scaling relations. A full report on the study sketched here can be found in Annis 1996.

References

Annis, J. 1996 Astrophys.J., submitted.
David, L. P., Slyz, A., Jones, C., Forman, W., Vrtilek, S. D. and Arnaud, K. A. 1993 Astrophys.J. 412 479
Harris, D. E. *et al.* 1990, The Einstein Observatory Catalog of IPC X-ray Sources (Smithsonian Institution Astrophysical Observatory)

THE RADIO PROPERTIES OF X-RAY SELECTED EXTRAGALACTIC OBJECTS

L. RAMÍREZ-CASTRO,
I. PÉREZ-FOURNON AND F. CABRERA-GUERRA
Instituto de Astrofísica de Canarias

1. The WGACAT Catalog of ROSAT Point Sources

We have used the 1st revision of the WGACAT catalogue, generated from all ROSAT PSPC pointed observations from February 1991 to March 1994, available at the HEASARC public archive. This catalogue, described by White *et al.* (1994), contains 68,907 detections, with more than 62,000 individual sources.

The ROSAT PSPC field was divided in two parts: the first includes the center of the field roughly up to the inner support structure at 19 arcmin radius. The second is the region outside the inner support and runs from 18 to 55 arcmin. We have only considered the sources detected in the inner part of the PSPC images.

2. The VLA FIRST Survey

FIRST (Faint Images of the Radio Sky at Twenty-cm) is a project designed to produce the radio equivalent of the Palomar Observatory Sky Survey (Becker *et al.* 1995, White *et al.* 1997) using the NRAO Very Large Array.

The catalogue contains ∼138,000 sources produced from the 1993 and 1994 images. The astrometric reference frame of the maps is accurate to 0.05″. Approximately 15% of the sources have optical counterparts at the limit of the POSS I plates; unambiguous optical identifications are achievable to m_v ∼24. The survey area has been chosen to coincide with that of the Sloan Digital Sky Survey; at its m_v ∼24 limit, ∼50% of the optical counterparts to FIRST sources will be detected.

3. The FIRST Radio Counterparts of WGACAT Sources

2,300 objects, all the WGACAT detections in the inner part of the PSPC images, within the area were selected. 151 of these sources have a FIRST counterpart and only 68 of them have optical identifications. Most of the FIRST sources are not detected in the ROSAT images. Redshifts are available for 44 associations. The classification of the 68 WGACAT/FIRST associations with optical identifications is given in Table 1.

TABLE 1. Classification of the 68 WGACAT/FIRST associations with available optical identifications

Galaxies	QSOs	BL-Lacs	AGN	Clust. of galax.	Stars	Radio Galaxies
28	17	8	8	4	2	1

4. Radio Morphology

The radio counterparts of 111 of the WGACAT sources are single sources, while 23 are double, 11 are triple and 6 have four or more radio components.

Figure 1 shows as an example the optical DSS, the FIRST radio, and the ROSAT images. The optical counterpart is the faint object close to the X-ray position and to the central radio component. This is probably a distant radio galaxy.

5. Optical Follow-up of the WGACAT/FIRST Sources

Since the WGACAT catalogue contains sources from a large number of ROSAT images, and most of these lack complete optical identifications, the analysis of the WGACAT/FIRST associations is strongly limited. We plan to complete the optical identification of complete samples. The objects identified previously come from relatively bright sources, by studying the fainter population we expect to find larger and complete samples of X-ray objects with radio counterparts at higher redshifts. The current highest redshift in our WGACA/FIRST sample is 2.334

References

Becker, R. et al. 1995, Astrophys.J. 450, 559
White, N.E. et al. 1994, IAU Circ. No.6100
White, R.L. et al. 1997, Astrophys.J. in press

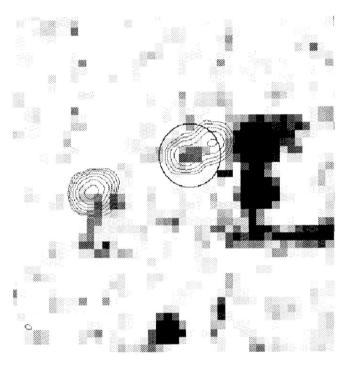

Figure 1. Optical (image), radio (contours), and X-ray data (error circle of radius~6.5 arcsec) of one of the radio-loud X-ray selected sources of the ROSAT sample.

UNCOVERING ULTRA-LUMINOUS GALAXIES IN THE IRAS FSC THROUGH RADIO AND OPTICAL CROSS-IDENTIFICATION

G. ALDERING
University of Minnesota, Department of Astronomy

1. Introduction

Ultraluminous infrared galaxies (ULIRGs) have luminosities ($10^{12} L_\odot$) once exclusive to QSOs. This suggests they might be the early, dust-enshrouded stages of QSOs. ULIRGs have $\sim 3.5\times$ the space density of QSOs at the present epoch. Quasars reached their peak space density at $z \sim 2$, so if ULIRGs are QSO precursors, there should be a dramatic increase in their space density up to $z \sim 2$. The small number of known ULIRGs makes it difficult to explore links between ULIRGs and QSOs, much less measure their evolution. To do so, a large sample of ULIRG candidates must be identified. The *IRAS* FSC contains $\sim 60,000$ probable galaxies, of which 1%–3% should have $L_{FIR} > 10^{12} L_\odot$. We discuss an efficient and reliable method which uses *IRAS*-VLA-*APS* cross-identification and flux ratios to mine the FSC for likely ULIRGs.

2. Exploiting the L_{FIR} versus F_{FIR}/F_{opt} Correlation

Likely ULIRGs can be identified by using the strong correlation between F_{FIR}/F_{opt} and L_{FIR}, as shown in Figure 1. Since the *IRAS* positions are too crude for a direct cross-identification, the radio-FIR correlation (see below) was exploited using the *NVSS* (Condon, this volume, p. 19) to refine the positions. Fits to the forward and inverse relation are shown as dashed lines. The inverse relation has $\sigma \sim 0.22$ dex once outliers are rejected. Among the sub-sample of 64 *IRAS* galaxies with $F_{FIR}/F_{opt} > 10^{1.2}$, 18 of 22 ULIRGs would be found. This translates to an efficiency of $\sim 30\%$ and a completeness of $\sim 80\%$.

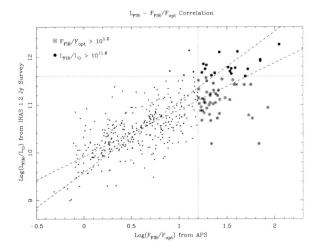

Figure 1. The F_{FIR}/F_{opt}–L_{FIR} relation constructed from sources matched between the *IRAS* 1.2 Jy survey, the *NVSS* 1.49 GHz radio catalog, and the available *APS* catalogs. The forward and inverse unweighted least-squares fits are shown. The solid symbols highlight those galaxies with $L_{FIR} > 10^{11.6}$ $L_\odot$. The concentric circles highlight ULIRG candidates selected using $F_{FIR}/F_{opt} \geq 10^{1.2}$. These cuts have been used to estimate the efficiency and completeness of the F_{FIR}/F_{opt}-excess selection technique.

3. Increasing Reliability using the Radio-FIR Correlation

Determining reliable optical identifications for sources with large F_{FIR}/F_{opt} is plagued by the large size of the *IRAS* error ellipse at the limit of the FSC ($\sim 7'' \times 25''$) relative to the surface density of faint optical candidates. The presence of IR cirrus further complicates the identification of faint 60 μm *IRAS* sources. The strong correlation between radio continuum emission and FIR flux, $L_{1.49GHz} = 10^{11.379\pm0.004}(L_{FIR}/L_\odot)^{1.030\pm0.014}$ with $\sigma \sim 0.18$ dex determined from the Figure 1 dataset, enables confirmation and positional refinement of faint FIR sources.

4. Application

We are performing cross-identification of sources from the *IRAS* FSC as the corresponding radio (NVSS & FIRST) and optical (POSS I & POSS II) data become available. The entire sky with $\delta > -33°$ and $|b| > 20°$ is being processed. ULIRGs are out there waiting to be found; in related work, we have recently discovered one ULIRG at $z = 1.1$, and two at $z = 1.3$ from spectroscopic follow-up of sources in the *IRAS* VFSS. These were found in relatively small samples, suggesting an excess of high redshift ULIRGs, as expected if ULIRGs evolve like QSOs.

EARLY RESULTS FROM AN HST IMAGING SURVEY OF THE ULTRALUMINOUS IR GALAXIES

K.D. BORNE[1,2], H. BUSHOUSE[3], L. COLINA[3] AND R.A. LUCAS[3]
[1] *Hughes STX*
[2] *NASA-GSFC*
[3] *STScI*

1. Why Study the Ultraluminous IR Galaxies?

The intense study of interacting galaxies originated in large part with the discovery by IRAS that the most IR-luminous galaxies are nearly all products of collisions and may be the missing link in the chain of evolution from quasars to normal quiescent galaxies (Sanders *et al.* 1988a,b). These galaxies (with $L_{IR} > 10^{12} L_\odot$) are considered to be the most strongly starbursting of all galaxies in the local universe, have a higher space density than quasars, emit >90% of their power in the IR, are rich in the raw materials of star formation, and to a large extent owe their peculiar morphologies to encounters with other galaxies. The particular importance of IR-luminous galaxies in the grand scheme of cosmology and galaxy evolution has been underscored by the luminosity function studies of Soifer *et al.* (1986), which indicated that most galaxies have gone through a high-IR luminosity stage. We are using the Hubble Space Telescope to survey the fine-scale features that are associated with the interaction- and activity-related processes that are at work within the Ultraluminous IR Galaxy Sample. It is widely believed that these galaxies are undergoing star formation at a prodigious rate and are abnormally dust-enshrouded. An alternative to the starburst hypothesis is that these galaxies' IR luminosity is powered by a dust-hidden quasar at its center (Sanders *et al.* 1988a). *It is important for our understanding of the evolution of galaxies and quasars to determine which of these two hypotheses is valid, or in which objects they are separately valid.*

2. The Ultraluminous IR Galaxy Sample: $L_{IR} > 10^{12} L_\odot$

We have created our survey sample from a combination of sources. The first "bright" sample of 10 galaxies that satisfied the $L_{IR} > 10^{12} L_\odot$ constraint was compiled by Sanders et al. (1988a). A second, partially overlapping, "warm" sample of 12 galaxies (9 new ones) was also identified at that time by Sanders et al. (1988b). An additional, partially overlapping, "bright" sample of 17 galaxies in the south was later compiled by Melnick & Mirabel (1990). Lawrence et al. (1996) found another 126 low-flux objects in the QDOT all-sky redshift survey of IRAS galaxies, while Kim et al. (1995) and Clements et al. (1996) have added to the numbers at low flux levels.

3. Early Results from the HST Survey

To date, we have received HST WFPC2 I-band (F814W) images for about 40 galaxies from our total combined sample of 160 ultraluminous IR galaxies. With the help of such high-resolution imaging, the properties of this class of objects are now being better defined, including a clarification of the nature of the energy source. *Fine structure is seen within a radius < 2″ for each galaxy.* In ∼20% of the galaxies, the structure is smooth and centrally concentrated, suggestive of a bright nuclear energy source (AGN?). *In the other cases, the sub-arcsecond morphology is chaotic and extended, suggestive of strong starburst activity.* The peculiar, disturbed morphologies that are seen on large (kiloparsec) scales among this sample of galaxies are continued down to the smallest scales in the cores of these strongly starbursting systems (Figure 1). A rich variety of morphological features are seen; these are probably related to the recent interaction-induced starburst episode. These starburst-related features, (*e.g.*, numerous bright clumps of star formation, shells, and bubbles) are similar to those seen in previous HST imaging observations of strongly interacting and merging galaxies.

References

Clements, D. L., et al. (1996) Mon.Not.R.astron.Soc., 279, 459.
Kim, D.-C., et al. (1995) Astrophys.J.Supp., 98, 129.
Lawrence, A., et al. (1996) Mon.Not.R.astron.Soc., submitted.
Melnick, J., & Mirabel, F. (1990) Astron.Astrophys., 231, L19.
Sanders, D. B., et al. (1988a) Astrophys.J., 325, 74.
Sanders, D. B., et al. (1988b) Astrophys.J., 328, L35.
Soifer, B. T., et al. (1986) Astrophys.J., 303, L41.

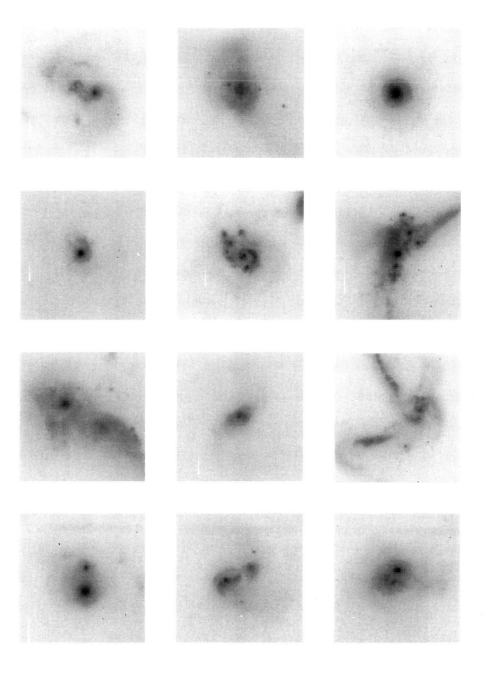

Figure 1. Selected HST WFPC2 I-band images (10″ square) for 12 ultraluminous IRAS galaxies. Note the clear interaction/merger morphology for many of the galaxies, but also note the AGN-like appearance of at least 2 of them.

THE K-BAND LUMINOSITY FUNCTION OF GALAXIES

J. P. GARDNER[1,2], R. M. SHARPLES[2],
C. S. FRENK[2] AND B. E. CARRASCO[3]
[1] *NASA – GSFC*
[2] *University of Durham, Physics Dept.*
[3] *INAOE*

1. Introduction

The luminosity function of galaxies is central to many problems in cosmology, including the interpretation of faint number counts. The near-infrared provides several advantages over the optical for statistical studies of galaxies, including smooth and well-understood K-corrections and expected luminosity evolution. The K–band is dominated by near-solar mass stars which make up the bulk of the galaxy. The absolute K magnitude is a measure of the visible mass in a galaxy, and thus the K–band luminosity function is an observational counterpart of the mass function of galaxies.

2. Data

Previously the K–band luminosity function has remained poorly determined, relying on the results of small-area surveys, (Glazebrook *et al.* 1995), or samples selected in other bands (Mobasher, Sharples & Ellis 1993). We have conducted a photometric survey of 10 square degrees in the B, V, I and K–bands, (Gardner *et al.* 1996; Baugh *et al.* 1996), and obtained spectra of 564 galaxies selected at $K < 15$, achieving a 90% redshift identification rate.

3. Results

We present the first determination of the near-infrared K–band luminosity function of field galaxies from a wide field K–selected redshift survey. The best fit Schechter function parameters are $M^* = -23.12 + 5\log(h)$, $\alpha = -0.91$, and $\phi^* = 1.66 \times 10^{-2} h^3$ Mpc^{-3}. Based on extensive Monte

THE K-BAND LUMINOSITY FUNCTION OF GALAXIES

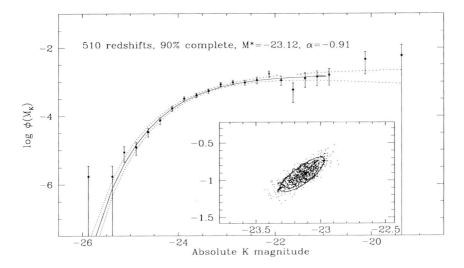

Figure 1. The differential K–band luminosity function of galaxies. The points, and their errors were determined from our data using the SWML method of Efstathiou, Ellis & Peterson (1988). The solid line is the best fit Schechter (1976) function determined using the maximum likelihood method. The dashed lines are the 1σ errors on this fit determined from the error ellipse, which is plotted in the inset figure. Also plotted are the results of 1000 Monte Carlo simulations of our survey parameters, and the 68% contour of the simulations.

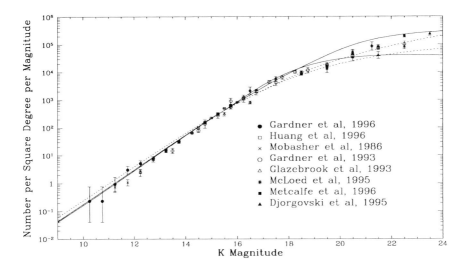

Figure 2. The K–band number counts with models based upon our luminosity functions. The solid lines include the effects of passive evolution, the dotted lines are pure K-correction models. The higher line in each case is for $q_0 = 0.02$, while the lower lines are for $q_0 = 0.5$. References for the data are given in Gardner et al. (1996).

Carlo modelling, we estimate that systematics are no more than 0.1 mag in M^* and 0.1 in α, which is comparable to the statistical errors on this measurement.

References

Baugh, C. M., *et al.* 1996, Mon.Not.R.astron.Soc., in press
Efstathiou, G., Ellis, R. S., & Peterson, B. A. 1988, Mon.Not.R.astron.Soc., 232, 431
Gardner, J. P., *et al.* 1996, Mon.Not.R.astron.Soc., 282, L1
Glazebrook, K., *et al.* 1995, Mon.Not.R.astron.Soc., 275, 169
Mobasher, B. , Sharples, R. M., Ellis, R. S. 1993, Mon.Not.R.astron.Soc., 263, 560
Schechter, P. 1976, Astrophys.J. 203, 297

THE K-BAND WIDE FIELD SURVEY: UNDERSTANDING THE LOCAL GALAXIES

J. HUANG
Institute for Astronomy, University of Hawaii

1. The Bright K-band Imaging

We have conducted a wide-field K-selected galaxy survey with complementary optical I- and B-band imaging in six fields with a total coverage of 9.8 square degrees (Huang *et al.* 1997). The observations were carried out on the UH 0.6m and the UH 2.2m telescopes. The purpose of this survey is to study the properties of the local galaxies and explore the evolution of K-selected galaxies at low redshifts. Star-galaxy discrimination is performed using both galaxy color properties and object morphologies, and 6264 galaxies are found. This survey establishes the bright-end K-band galaxy number counts in the magnitude range $13 < K < 16$ with high precision. We find that our bright-end counts have a significantly steeper slope than the prediction of a no-evolution model, which cannot be accounted for by observational or theoretical error. Since it is very unlikely that there is sufficient evolution at such low redshifts to account for this effect, we argue that there is a local deficiency of galaxies by a factor of 2 on scale sizes of around 300 Mpc. This would imply that local measurements of Ω_0 underestimate the true value of the cosmological mass density by this factor and that local measurements of H_0 could be high by as much as 33%.

2. The Hawaii-Australia K-band Redshift Survey

We have been granted AAT time to conduct the redshift measurement on our K-selected sample (Huang *et al.* 1997) by using the 2DF. Actually, one of our fields was observed during the 2DF commissioning time. There are three primary scientific goals which will be achieved by this redshift survey.

First, we will construct the local K-band luminosity functions at different redshifts up to z=0.2. By comparing these local luminosity functions, we can determine whether the steep slope of the local K-band counts is caused by a local galaxy deficiency or by galaxy evolution at low redshifts (Huang et al. 1997).

Second, we will be able to determine the evolution of the K-selected galaxies at redshifts > 0.5. Cowie et al. (1996) derived the K-band luminosity functions at redshifts > 0.5. The change in the luminosity function can be determined by comparing the K-band luminosity function of Cowie et al. (1996) with the local one.

Finally, we aim to investigate galaxy clustering from a cleanly selected galaxy sample. Previous studies have been based on optical samples (e.g., Loveday et al., 1992) or far-infrared IRAS galaxies (e.g., Dekel et al., 1993), both of which are inherently biased to star-forming systems and are already known to avoid rich clusters. This has led to much dispute about b, the biasing parameter of light relative to mass. A K-selected sample is inherently superior in that the K band light is a close tracer of underlying stellar mass rather than star-formation. This has long been recognized as desirable and has led to large all-sky K surveys such as 2MASS and DENIS; these surveys, however, are part of longer term efforts. My large contiguous areas will allow investigation of galaxy clustering on scale $\leq 20h^{-1}$ Mpc. Although the volume surveyed is smaller then some of the $> 10^5$ optically selected redshift surveys, it will provide a valuable check on their results and allow their biases as a function of morphology and color to be assessed. Uniquely, this wide field K-band survey will also allow us to investigate the evolution of clustering to z=0.2. Not only do we have an unchanging morphological mix with redshift, we also see the same population as at z=0. Again, this contrasts with optically selected surveys—at z=0.2 they are dominated by small starburst systems, a different population whose clustering properties may well be intrinsically different (Efstathiou et al. 1991).

3. The Morphology Classification and Surface Photometry of the K-selected galaxies

The morphological mix of the K-selected galaxy sample has yet to be determined. Huang et al. (1997) obtained the morphological mix from the mean color-magnitude relation. By fitting the mean color-magnitude relation, they conclude that a K-selected galaxy sample contains mainly elliptical and early type spiral galaxies, but very few irregular galaxies. This conclusion needs to be confirmed by the morphological classification.

The morphology mix is a key factor in understanding the galaxy evolution and modeling the galaxy number counts. With the morphological classification and redshifts measured in the 2DF redshift survey, we will be also able to construct the local type-dependent luminosity functions. These type-dependent luminosity functions will allows us to model the number counts more precisely to understand galaxy evolution.

Large format CCDs have provided an effective way for us to obtain high S/N images for a sample with wide coverage. An 8k×8k CCD has been built by UH astronomers, and is now available on the CFHT and the UH 88-inch telescopes. The field of view is about 0.5° with a scale of 0.22″ pixel^{-1}. One of our fields has been observed in R-band on the CFHT with the 8k×8k CCD. The exposure time is one hour, and the seeing is 0.″7. The image can allow us to clearly classify the morphologies of our K-selected sample to K=15.

Surface photometry of the K-selected sample is another goal of the 8k CCD imaging. Recently surface photometry (Schade et al. 1996) of optically selected samples has led some authurs to conclude there is luminosity evolution of the spiral galaxies. However, the redshift distributions from other redshift surveys do not support this conclusion. We will be able to reinvestigate this issue by using the K-selected sample. Our advantages are the high S/N, larger sample and insensitivity of the K-selected sample to star-formation.

4. The QUick Infrared Galaxy Survey

The K-band galaxy number counts below K=13 remain controversial. Two current K-band surveys (Huang et al. 1997, Gardner et al. 1996) have too limited sky coverage (both cover about 10 square degrees) to obtain precise galaxy number counts below K=13. Though Mobasher et al. (1986) conducted a substantial K-band survey below K=12, their sample was selected from the B-selected sample and might suffer from some systematic errors. A K-band survey with a sky coverage of 300 square degrees is required to solve this problem.

Though there will be large near-infrared surveys such as the 2MASS and DENIS, The QUick Infrared Galaxy survey is designed to attack this problem easily and effectively. The University of Hawaii has built the QUick Infrared Survey Telescope (QUIST). QUIST is a 25-cm f/10 infrared optimized telescope which mounts on top of QUIRC (QUick InfraRed Camera). QUIRC has a 1024×1024 HgCdTe infrared array sensitive from 0.7 to 2.5 microns. QUIST with QUIRC are mounted in binocular fashion with the University of Hawaii's 61-cm telescope located on Mauna Kea. QUIST has a field of view of 0.43×0.43 degrees and pixels are 1.5 arcseconds across.

The observation on the QUIST is totally remote-controlled. Please check the IfA web page for more detailed information and the QUIST images (http://www.ifa.hawaii.edu/images/quist/).

This survey has been in operation since May, 1996. We find that a total exposure of 6 minutes can allow us to detect a galaxy with K=14. So far about 30 square degrees of sky has been observed. We expect to finish this survey before the end of this year.

References

Cowie, L. L., Songaila, A., & Hu, E. M. 1996, Astron.J., In press
Dekel, A., *et al.* 1993, Astrophys.J. 412, 1
Efstathious, G., *et al.* 1991, Astrophys.J. 380, L47
Gardner, J. P., *et al.* 1996, Mon.Not.R.astron.Soc., in press.
Huang, J-S., *et al.* 1997, in press
Loveday, J., *et al.* 1992, Mon.Not.R.astron.Soc., 247, 1P
Mobasher, B., Ellis, R. S., & Sharples, R. M. 1986, Mon.Not.R.astron.Soc., 223, 11

MULTI-COLOR SURFACE PHOTOMETRY OF NEARBY GALAXIES

T. ICHIKAWA[1], N. ITOH[2] AND K. YANAGISAWA[2]
[1] *Kiso Observatory, The University of Tokyo*
[2] *Department of Astronomy, The University of Tokyo*

1. Introduction

Near-infrared (NIR) emission in galaxies is mainly radiated by old population low temperature stars, which construct the basic stellar structure and keep the trails of past galaxy evolution. On the other hand, optical observations show recent star formation activity, especially in spiral galaxies. Therefore multi-color observations from optical to near-infrared wavelengths are very important to understand the past and recent star-formation history. Nearby large galaxies are well studied not only in optical but also in mid- and far-infrared by IRAS, CO and HI radio observations. However, the study in the near-infrared is still limited because large format arrays are not common. Here we show a wide-field, near-infrared imaging of nearby elliptical and spiral galaxies and discuss their star-formation history.

2. Observations and Results

Target galaxies were selected in terms of size (larger than $5'$ in diameter), inclination (more edge-on than $65°$), and morphological types (from E to Sc). In JHK' bands, the observations were made with the Kiso Observatory Near-Infrared Camera (Yanagisawa *et al.* 1996) attached to the prime focus (F/3.1) of the 105cm Schmidt telescope (Ichikawa *et al.* 1996). The camera is equipped with a 1040×1040 PtSi near-infrared array (supplied by Mitsubishi Electric Co.) with a field of view of $18.4' \times 18.4'$. The pixel size of $17\ \mu m \times 17\ \mu m$ corresponds to $1.1'' \times 1.1''$. Although the low quantum efficiency is a disadvantage of the PtSi array, its good uniformity, stability of performance, and lack of memory effect allow us to conduct accurate surface photometry. Optical BVR_cI_c observations were made with

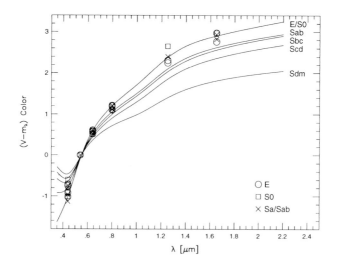

Figure 1.

the same Schmidt telescope and a CCD camera equipped with a TI 1000 × 1018 array having a field of view of $12.5' \times 12.5'$. The large field of view is very important for accurate sky subtraction especially for NIR, where sky background overwhelmed the galaxy emission. Small errors in sky background estimates tend to lead not only wrong magnitude colors but also wrong scale lengths of the bulge and the disk.

3. Comparison with Models

A preliminary result for elliptical and early spirals is shown in Figure 1 which compares the total magnitude of the galaxies with the evolutionary model from Yoshii and Takahara (1988). The deviation of some galaxies from the model is significant in the B band. The model was generally constructed by comparing aperture photometry and spectroscopic data observed in the galaxy center region, where the bulge and galactic center emission dominate. The disk and bulge components should be separately discussed in galaxy evolution models because their history is supposed to be different. These detailed discussions will be made elsewhere.

References

Ichikawa, T., Yanagisawa, K. and Itoh, N., 1996, SPIE 2744, 104
Yanagisawa, K., Ichikawa, T. and Itoh, N., 1996, SPIE 2744, 92
Yoshii, Y. and Takahara, F. 1988, Astrophys.J. 326, 1

MUTI COLOR IMAGING OF CLUSTERS OF GALAXIES WITH MOSAIC CCD CAMERAS

S. OKAMURA[1], M. DOI[1], N. KASHIKAWA[2], W. KAWASAKI[1],
Y. KOMIYAMA[1], M. SEKIGUCHI[2], K. SHIMASAKU[1],
M. YAGI[1] AND N. YASUDA[1]
[1] *Department of Astronomy and Research Center for the Early Universe, University of Tokyo, Japan*
[2] *National Astronomical Obvservatory of Japan*

1. Introduction

At present, the photometric data for clusters at $z \lesssim 0.2$ mainly come from photographic photometry. The lack of CCD data for such clusters is simply due to the fact that no CCD camera had been available until recently that covers the wide extension of clusters within a reasonable amount of observing time. We have developed a large mosaic CCD camera and conducted multicolor imaging observations of $z \lesssim 0.2$ clusters using the 40-inch Swope telescope at Las Campanas Observatory.

2. 5K×8K Mosaic CCD Camera

Our mosaic CCD camera consists of forty 1K×1K CCDs located on a grid of the 8×5 array. The CCDs are TC-215 with 12-micron pixels manufactured by Texas Instruments Japan. The spacing between adjacent CCDs is slightly smaller than twice the side of the CCD. The image of a contiguous field, which consists of 14,500×9,280 pixels, is composed of four offset exposures. Figure 1 shows the camera as of April 1996. A brief description of the camera is given by Kashikawa *et al.* (1995a)

3. Observations and Data Reductions

We observed more than a dozen nearby ($z \lesssim 0.2$) clusters of galaxies in the B, V, and R bands in 1994–5. The camera evolved during this period from the 4×7 array to the 5×8. At the Cassegrain focus of the Swope telescope,

Figure 1. The 5K×8K mosaic CCD camera

the camera (5×8 array) gave a field of view of about $1°\!\!.41 \times 0°\!\!.90 = 1.27$ square degrees with the resolution of 0.35 arcsec pixel^{-1}.

The data are processed with a software package dedicated to our camera. In addition to the pre-processings common to CCD data such as bias/dark subtraction and flat fielding, we need to do the frame mosaicking and PSF equalization, which are pre-processings characteristic of our camera. The former is to establish a consistent coordinate system and a consistent flux scale while the latter is to assure the uniformity of object detection and extracted parameters, over a contiguous field composed of four separate offset exposures taken by 40 different chips. The mosaicked and PSF-equalized data of a contiguous field is subjected to the more or less usual image analysis consisting of object detection, measurement of photometric and shape parameters, and star/galaxy discrimination.

4. Results

Based on the R-band data of four clusters, A1656 (Coma; $z = 0.02$), A1367 ($z = 0.02$), A1644 ($z = 0.05$), and A1631 ($z = 0.05$), we detected a significant cluster-to-cluster variation in the faint part ($M_R \gtrsim -18$mag) of the luminosity function of early-type galaxies (Kashikawa et al. 1995b). The data for A1656 and A1367 were obtained with our old mosaic CCD camera attached to the Kiso Schmidt telescope (Sekiguchi et al. 1992).

Our morphological classification (Doi et al. 1993) is based solely on the shape of the luminosity distribution in a single color band (R band), which is subject to the seeing effect. In order to examine the reliability of our

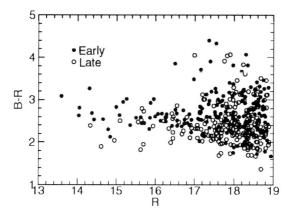

Figure 2. $B - R$ colors (zero point arbitrary) of galaxies in A1631 as a function of R magnitude

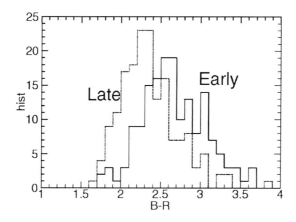

Figure 3. $B - R$ colors (zero point arbitrary) of galaxies in A1631 as a function of $B - R$ colors

classification we analyzed the B-band data of A1631 and obtained the $B-R$ color for galaxies. The colors are computed from the aperture magnitudes. Aperture radius is set to be $1.5 r_e$, where r_e is the *isophotal equivalent radius* measured in the R band, i.e., $r_e = \{(isophotal\ image\ area)/\pi\}^{1/2}$. The internal photometric accuracy is estimated as $\lesssim 0.08$ mag in r.m.s. in $m_R < 18.5$.

Figure 2 shows the $B - R$ *relative* color as a function of R magnitude while Figure 3 shows the histograms for the two types. The histogram of the early-type is shifted to the red with respect to that of the late-type by about 0.3 mag. This systematic difference is taken to be evidence that our morphological classification is broadly valid at $z \sim 0.05$.

References

Doi, M., Fukugita, M., & Okamura, S. 1993, Mon.Not.R.astron.Soc., 264, 832.
Kashikawa, N. et al. 1995a, in "Scientific and Engineering Frontiers for 8–10m Telescopes," eds. M.Iye and T.Nishimura(Tokyo: Universal Academy Press), pp.105–110.
Kashikawa, N. et al. 1995b, Astrophys.J., 452, L99.
Sekiguchi, M. et al. 1992, Publ.Astron.Soc.Pacific, 104, 744.

QSO COLOR SELECTION IN THE SDSS

H. J. NEWBERG AND B. YANNY
Fermi National Accelerator Laboratory

1. Introduction

The Sloan Digital Sky Survey (SDSS) will image 10,000 square degrees in the north galactic cap in five filters. We hope to identify and obtain spectra for about 100,000 quasars brighter than 20th magnitude in this area. The selection will be primarily on the basis of point spread function and colors, but we will also identify quasars from a catalog of FIRST radio sources. The selection areas in color space must be determined during the testing period prior to the official start of the survey. This task may determine the length of the test period. In anticipation of this becoming the critical path, we have written a body of software that will allow us to quickly analyze a set of multicolor data and make a first cut at the selection limits.

2. Parameterization of the Stellar Locus

We model the stellar locus with a set of locus points, each with an ellipse fit to the stellar locus cross section at that point. Here, *stellar locus* will refer to the data points in color space, and *locus points* will refer to the model. Each locus point is iteratively moved to the centroid of its associated stars, which are (roughly) those that are closer to that locus point than they are to any other. Ellipse fits to the stellar locus are measured at each locus point from the projection of the associated stars onto a plane perpendicular to the stellar locus. With each iteration, new locus points are added in between the existing locus points, but only in places along the locus where the distance between locus points is larger than a factor times the local width of the stellar locus. This allows points to be more closely spaced when the stars are concentrated in a narrow line, but keeps the locus from wandering freely when the stellar locus is broad. This algorithm takes as input parameters the two approximate endpoints of the stellar locus, the

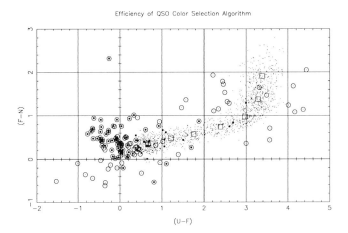

Figure 1. This plot shows the stellar locus fit (open squares) to the Trevese *et al.* data. By selecting sources outside a three sigma ellipse around the fit locus points, we found 63 of the 80 QSOs (79%), with 53% of the selected targets yielding QSOs. Stars are represented by dots, and quasars by filled squares. Sources that were chosen as quasar candidates are circled. We expect the Sloan survey to have approximately the same completeness and efficiency. Although this data set goes 1 or 2 magnitudes fainter (which makes it easier to find QSOs), the photometric errors are about five times larger than we expect the SDSS to achieve.

maximum distance from the stellar locus that a star will still be considered as part of the locus, the number of iterations of the algorithm, and the factor which determines how closely spaced the locus points can be. We generally iterate until the number of locus points is limited by the width of the locus itself, and then iterate several more times until the locus has stabilized.

3. QSO Color Selection

We have applied a simple target selection algorithm to a set of UJFN point source data from Trevese *et al.* (1994), and used their QSO identifications (from color selection and proper motion studies) to check the completeness and efficiency of algorithm. A three sigma cut from the model stellar locus yields 79% completeness with 47% contamination (Figure 1).

References

Gaidos, E. J., *et al.* 1993 Publ.Astron.Soc.Pacific, 105, 1294
Trevese, D., *et al.* 1994, Astrophys.J., 433, 494

THE CALAR ALTO DEEP IMAGING SURVEY

H. HIPPELEIN, S. BECKWITH, R. FOCKENBROCK, J. FRIED,
U. HOPP, C. LEINERT, K. MEISENHEIMER, H.-J. RÖSER,
E. THOMMES AND C. WOLF
Max-Planck-Institut für Astronomie, Heidelberg

1. Introduction

The **C**alar **A**lto **D**eep **I**maging **S**urvey (CADIS) combines deep multicolor imaging in broad and narrow band filters with deep probing in several selected wavelength intervals through a Fabry-Pérot (FP) interferometer with a spectral resolution of ∼450.

Though the survey is primarily designed to detect emission line galaxies such as primeval galaxies at high redshift and faint blue galaxies, its multi-filter technique also allows the classification and redshift determination of early type galaxies, of faint QSOs beyond $z = 3$, and of faint M stars. In the case of emission line galaxies, the accurate redshift is obtained from the FP observations, which allows to derive the faint end of the luminosity function of galaxies at $z \lesssim 1$, and their three dimensional correlation function.

2. Survey Strategy

The limiting flux is dictated by the search for primeval galaxies. In order to detect a representative number of primeval galaxies the sensitivity for the deep FP observation of the redshifted Lyα line needs to be $F_{lim} = 3 \times 10^{-20}\,\mathrm{W\,m^{-2}}$ (Thommes & Meisenheimer 1994). With this sensitivity we should be able to detect bright spiral galaxies out to $z > 1$, and the first burst of star formation in primeval galaxies at $z \lesssim 5$ down to star forming rates of $\sim 30\,\mathrm{M_\odot\,yr^{-1}}$. The filter observations should reach a limiting (5σ) magnitude of $R = 24.3\,\mathrm{mag}$.

Since we expect most primeval galaxies to be located at $z \gtrsim 5$ we have selected for the Lyα line search three atmospheric windows centered at 700, 820, and 918 nm. In each window we observe at 10 wavelength settings at intervals of 1.5 nm. In order to discriminate foreground galaxies that light

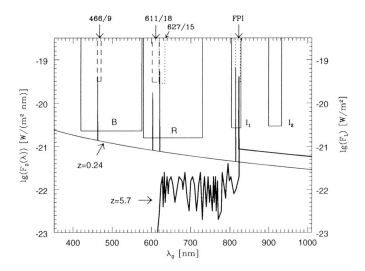

Figure 1. Demonstration of the veto-filter strategy used in CADIS. Solid boxes indicate the bandwidths and 5σ limits for the deepest broad and medium band filters and the dotted and dashed boxes the same for the Fabry-Pérot scans and for the veto-filters 466/9, 611/18, and 627/15 nm. Whilst a Lyα galaxy at $z = 5.7$ will be only detected in the FPI band and (perhaps) the I-bands, for example a foreground galaxy at $z = 0.24$, which has its Hα line at the FPI band at 820 nm exhibits detectable emission lines in both the 466/9 and 627/15 nm filters.

up with emission lines in the same FP bands we have introduced several so-called *veto*-filters, which are selected in such a way that for every prominent emission line appearing in the FP window a second or third line should show up in them. Figure 1 demonstrates this strategy for a galaxy detected in Hα.

We added a complete set of broad band (B, R, K') and of medium band filters ranging from 400 to 930 nm, and with bandwidths between 10 and 30 nm. These are optimized both for the continuum determination and the identification of contaminants such as faint galactic M stars. We thus get a global view of the spectral energy distribution (SED) for every object in the field, without biasing towards a fixed color. With these SEDs we should be able to classify most of the objects and determine the redshift of early type galaxies with an accuracy of $\Delta z \lesssim 0.05$, in the redshift range $0.5 < z < 1.2$.

3. Survey Fields

CADIS surveys 10 empty $11' \times 11'$ fields. The selection criteria are: no stars brighter than $R \sim 16$ mag, and a low FIR(100μm) flux of < 2.0 MJy sr^{-1}. The total area covered is thus 0.33 deg^2. The observations are done at the 2.2 m and 3.5 m telescopes on Calar Alto, Spain. The total survey will consist of ~4500 CCD exposures and yield a list of approximately 50,000 objects. We expect 20 to 200 primeval galaxies at $z \gtrsim 5$, more than 10,000 emission line galaxies at redshift 0.24 to 0.9, several 100 early type galaxies, more than 300 QSOs, and about 1500 low mass stars (possibly brown dwarfs). The first two CADIS fields will be completed at the beginning of 1997.

References

Thommes & Meisenheimer 1994, Galaxies in the Young Universe, p.242.

THE CALAR ALTO DEEP IMAGING SURVEY: FIRST RESULTS

E. THOMMES, K. MEISENHEIMER,
R. FOCKENBROCK, H. HIPPELEIN AND H.-J. RÖSER
Max-Planck-Institut für Astronomie, Heidelberg, Germany

1. Introduction

The **C**alar **A**lto **D**eep **I**maging **S**urvey (CADIS) is a very deep emission line survey using a Fabry-Pérot (FP), combined with deep broad- and medium-band photometry (for an overview see Hippelein *et al.* 1996). This survey is specifically designed to detect primeval galaxies, but it will in addition produce a large data base for investigations of faint galaxies at intermediate redshifts ($0.2 < z < 1.2$). We present some first results from the initial data recorded with the CADIS strategy.

2. Data

These data were taken with the 2.2m telescope at Calar Alto in the CADIS field 9H. Due to delays in getting the 2k× 2k CCDs, we employed a 1k×1k CCD (field of view $8 \times 8'$). We got four FP settings in the wavelength region 814nm to 818.5nm (resolution=1.8nm). Every setting consists of 7 individual exposures of 1500 s integration. We reached a 5σ detection limit of $S_{lim}(5\sigma) \approx 5 \times 10^{-20} \text{W/m}^2$. To get an estimate of the continuum near the emission lines, we did exposures with a filter $\lambda/\Delta\lambda$=812/17 nm ($F_{lim}(5\sigma) \approx 5.8 \times 10^{-21}\text{W}/(\text{m}^2\text{nm})$). The FP exposures were supplemented by broad band exposures with the filters BV (centered at 500 nm, 5σ limit $\approx 25.^m8$), R_c (5σ limit $\approx 25.^m0$) and I (5σ limit $\approx 23.^m1$). Further narrow band exposures with the filters 466/9 ($F_{lim}(5\sigma) \approx 11 \times 10^{-21}\text{W}/(\text{m}^2\text{nm})$), 612/10 ($F_{lim}(5\sigma) \approx 11 \times 10^{-21}\text{W}/(\text{m}^2\text{nm})$) and 614/28 ($F_{lim}(5\sigma) \approx 7 \times 10^{-21}\text{W}/(\text{m}^2\text{nm})$) enable to detect further emission lines of foreground objects. Emission line objects are selected by the requirement, that they have a 5 σ detection in at least one FP wavelength setting and that the line flux exceeds the continuum by more than 3.5σ.

3. Results

In this exploratory CADIS data set we already found 147 emission line galaxies which satisfy the conditions mentioned above. 104 show at least a marginal detection in the BV band and are therefore classified as faint blue galaxies in the foreground. 74 of these emission line objects with a blue detection also show signals in the additional narrow band exposures. These allowed us to identify them as being either at $z \approx 0.24$ or $z \approx 0.63/0.68$. Of the 43 emission line objects without detection in the blue band, 35 show at least a marginal detection in one of the additional narrow band filters and therefore also could be identified as galaxies at these redshifts. From the remaining 8 galaxies 7 are promising candidates for Ly-α emitting PGs at $z \approx 5.7$. Three of these 7 candidates have no continuum flux shortward of the emission line at $\lambda \approx 816$ nm and 4 candidates showed a marginal detection in our red filter ($R \leq 26.0$). Five of the candidates are clearly resolved ($\geq 2''$ FWHM), while two objects may be unresolved.

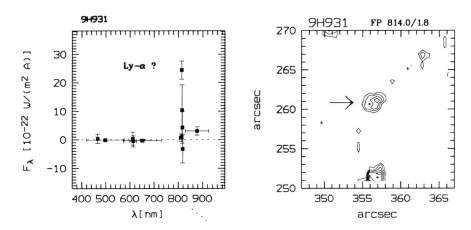

Figure 1. Photometric spectrum and conture plot of one of the 7 promising candidates for Ly-α emitting PGs at z=5.75.

About 30% of the foreground emission line objects have no detectable continuum in the BV and R band. These objects are a new class of objects which are overlooked by broad band selected redshift surveys. We derived lower limits to the luminosity function from our data for the pure emission line objects in the redshift bins 0.24 and 0.64. Although the covered spatial angle of the data presented here ($\approx 60 \;\square'$) is yet too small to draw definite conclusions these are in good agreement with the results of the AUTOFIB redshift survey (Colles et al. 1996) We expect that our sample of 7 candidates for primeval Ly-α emitting galaxies at $z \approx 5.7$ is still contaminated by several types of foreground objects. Statistical considerations indicate, that artifacts and emission line galaxies at $z = 0.24$ and $z = 0.63$ could hardly account for more than two of the candidates. The unknown fraction of galaxies at $z = 1.2$ with strong [OII] line but undetectable blue continuum could however well make up for half of the candidates. Therefore, high S/N ratio slit-spectroscopy at medium resolution ($\Delta\lambda \approx 0.5$ nm) and/or deeper continuum I, R and B band images are required to identify the true Ly-α galaxies among our candidates. We are pretty optimistic that at least some of the objects will, in fact, turn out to be at $z \approx 5.7$.

References

Colles *et al.*, 1996, to appear in The Early Universe with the VLT, proceedings of ESO-workshop held in Garching bei München, Springer, Berlin, Heidelberg, New York, pp. 87.
Hippelein *et al.* 1996, this volume, 293

HAMBURG/SAO SURVEY OF EMISSION-LINE GALAXIES

V. LIPOVETSKY[1], D. ENGELS[2], A. UGRYUMOV[1], U. HOPP[3*],
G. RICHTER[4], Y. IZOTOV[5], A. KNIAZEV[1] AND C. POPESCU[6]
[1] *Special Astrophysical Observatory, Russia*
[2] *Hamburger Sternwarte, Hamburg, Germany*
[3] *Universitätssternwarte München, Munich, Germany*
[4] *AIP, Potsdam-Babelsberg, Germany*
[5] *Main Astronomical Observatory, Goloseevo, Kiev, Ukraina*
[6] *MPI für Astronomy, Heidelberg, Germany*

1. Introduction

We present first results of the Hamburg/SAO Survey of emission-line galaxies (hereafter HSS, SAO—Special Astrophysical Observatory, Russia) initiated to search for extremely metal-deficient ($Z < Z_\odot/10$) galaxies and to create a large sample of Blue Compact Galaxies (BCG). This "Northern BCG Sample," will be assembled by merging the HSS with samples from the Second Byurakan Survey (SBS) (Stepanian et al. 1987) and the Case Low-Dispersion Northern Sky Survey (Pesch et al. 1991).

The strongly metal-deficient galaxies are important for the estimation of primordial helium abundance—a crucial parameter in Standard Big Bang nucleosynthesis theory—for the understanding of basic processes determining the chemical evolution of galaxies, for constraints to models of stellar nucleosynthesis, and for the determination of the initial stellar mass function. During the last three decades only few galaxies of extremely low-metallicity such as I Zw 18 or SBS 0335-052 were found, indicating that they are rare. New surveys are therefore necessary to increase the volume sampled significantly to improve the search. One way is to decrease the limiting [OIII] $\lambda\lambda 4959, 5007$ Å emission line fluxes as it is done by the KISS project (Kniazev et al. 1996). The other way is to increase the sky area surveyed, keeping the sensitivity to the detection of the [OIII] emission

---

[1] Visiting astronomer Calar Alto Observatory.

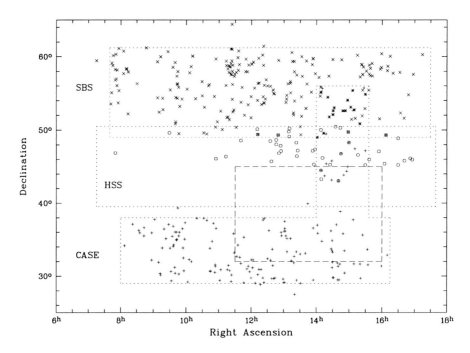

Figure 1. Sky positions of known BCGs from the Northern Sample, drawn from the SBS, HSS and Case surveys.

lines on a level similar to those of previous surveys. Both these strategies should be followed as they are complementary to each other.

BCGs, having strong emission lines in their optical spectra give the opportunity to map the low-mass galaxy population at distances well beyond the Local Supercluster (Salzer 1989; Salzer *et al.* 1993; Hopp *et al.* 1995; Pustil'nik *et al.* 1995). They may help trace the spatial distribution of low-mass galaxies and test cosmological theories of the origin and of the development of large-scale structures in the Universe. However all the most representative samples of BCGs from the SBS, Case and UM surveys are located in narrow bands and yield significant boundary effects for any statistical analysis. The aim of the Hamburg/SAO Survey is not only to provide a new large BCG sample, but also to fill the gap between the areas covered by the SBS and Case surveys.

The three surveys together will constitute the "Northern BCG sample" with a total area on the sky of ≈ 3000 deg^2. Overlap between the surveys will be used to study selection effects. The current HSS region covers about 1100 deg^2 within the boundaries: $\alpha = 7^h20^m$–17^h40^m and $\delta = +40°$–$+50°$. We plan to extend it down to $\delta = +35°$.

2. Selection of Candidates and Follow-Up Spectroscopy

The selection of candidates of emission-line galaxies was made on digitized objective-prism photoplates of the Hamburg Quasar Survey (HQS; Hagen *et al.* 1995, see also Popescu *et al.* 1996 for a related work on emission-line galaxies) during October–December 1994. Spectra with apparent [OIII] emission-lines were selected from the digitized database of low-resolution spectra in semi-automatic mode and then rescanned with full resolution. The high-resolution spectra were classified according to the prominence of the emission lines. Presently, the candidate selection is finished in the region: $\alpha = 8^h-17^h30^m, \delta = 45°-50°$ and $\alpha = 8^h-15^h30^m, \delta = 40°-45°$. First follow-up spectroscopy was obtained in 1995 with the 2.2 m Calar Alto (German-Spanish Observatory, Spain) and the 6 m SAO (Russian) telescopes in snap-shot mode. Standard spectral data reduction was performed at the SAO under MIDAS.

3. Preliminary Results

Follow-up spectroscopy was performed for 218 candidates, confirming 75 new emission-line objects. 46 of them appear to be BCGs, 4 are QSOs, and the others are different types of galaxies including AGN and starburst. Their magnitude range is $16^m \leq B \leq 20^m$. About 20 BCGs have low oxygen abundance in the range 12+log(O/H) = 7.7–8.0, but no new galaxies with extremely low metallicity were discovered. Comparing with the SBS and Case surveys, we find that the HSS BCG have similar distributions of absolute magnitudes and redshifts. It appears that the galaxies in all three samples have similar physical properties.

References

Hagen, H.-J., *et al.* 1995. Astron.Astrophys. Suppl., 11, 195.
Hopp, U., *et al.* 1995. Astron.Astrophys. Suppl.,109, 537.
Kniazev, A., *et al.* 1996, this volume, 302
Pesch, P., Sanduleak, N. and Stephenson C.B., 1991. Astrophys.J. Suppl., 76, 1043.
Pustil'nik, S.A., *et al.* 1995. Astrophys.J., 443, 499.
Popescu, C.C., *et al.* 1996. Astron.Astrophys. Suppl., 116, 43.
Salzer, J., 1989. Astrophys.J., 347, 152.
Salzer, J. and Rosenberg, J., 1993. In ESO/OHP Workshop on Dwarf galaxies, ed. G. Meylan & P. Prugniel, p. 129.
Stepanian, J.A., Lipovetsky, V.A., Erastova, L.K. and Shapovalova, A.I., 1987. In "Observational Evidence of Activity of Galaxies," IAU Symp. 128, p. 17.

KISS: A NEW DIGITAL SURVEY FOR EMISSION-LINE OBJECTS

A. KNIAZEV[1], J. SALZER[2], V. LIPOVETSKY[1], T. BOROSON[3], J. MOODY[4], T. THUAN[5], Yu. IZOTOV[6], J. HERRERO[2,7] AND L. FRATTARE[2,8]

[1] *Special Astrophysical Observatory, Russia*
[2] *Wesleyan University, USA*
[3] *NOAO, Gemini Project, USA*
[4] *Brigham Young University, USA*
[5] *University of Virginia, USA*
[6] *Main Astronomical Observatory, Ukraine*
[7] *Center for Astrophysics, USA*
[8] *Space Telescope Science Institute, USA*

1. Introduction

We have initiated a major new survey for emission-line galaxies (ELGs) which we call the KPNO International Spectroscopic Survey (KISS). Survey observations began in March 1994 with the 0.61-m Burrell Schmidt telescope[1]. The technique we employ combines the benefits of a traditional photographic objective-prism survey with the advantages of using a CCD detector. The field of view of our CCD is 1.1° square, and the prism employed provides a dispersion of 19 Å/pixel at 5000 Å. The spectral range covered (4800–5500 Å) is restricted by a specially designed filter that transmits from rest-frame Hβ to just shortward of the strong night-sky line at 5577 Å; this greatly reduces the sky background. We expect KISS to be sensitive to galaxies with magnitudes as faint as B = 20^m–21^m, much deeper than existing photographic surveys. Our initial pilot project covers 100 square degrees (Salzer *et al.* 1994) and overlaps the CfA/Dartmouth Century Redshift Survey ($\alpha = 8^h30^m$–16^h45^m and $\delta = 29°$–$30°$) in the North Galactic cap.

---

[1] Observations made with the Burrell Schmidt of the Warner and Swasey Observatory, Case Western Reserve University.

We are able to detect in our survey: galaxies with redshifts up to z = 0.10 showing strong [OIII] λ5007 line, high redshift galaxies via their [OII] λ3727 emission in the range z = 0.23 – 0.48, QSOs with redshifts in the range 2.9 – 3.5 via Lyα.

Among the goals of the survey are: obtain a deep sample of ELGs, complete to 2–3 magnitudes fainter than previous surveys; study the large-scale distribution of emission-line galaxies and their relation with normal galaxies; search for new extremely low metallicity dwarf galaxies like I Zw 18 or SBS 0335–052; determine the luminosity function for star-forming galaxies, with special emphasis on the faint end.

The initial strip has been completely observed and partly processed. Follow-up spectra of some newly discovered candidates have been obtained with the Russian 6-m telescope. Future plans include: (1) survey additional areas of the sky; (2) obtain follow-up optical spectroscopy of all candidates with a multifiber spectrograph like HYDRA; (3) obtain objective-prism images in other spectral ranges (*e.g.*, around Hα); (4) application of our survey method to larger telescopes.

2. Survey Data Reduction and Some Preliminary Results

The survey input data consist of deep direct images taken using two filters, B and V, images with low-resolution prism spectra and short exposure direct B and V images taken on photometric nights in order to calibrate the magnitudes of the sources.

For the reduction of this complicated data set new software was developed in both the MIDAS and IRAF environments. The results presented here were obtained with the former package, developed principally by AK. The reduction procedure consists of the following steps: (1) preliminary reductions, (2) alignment and combining of direct and spectral images, (3) 2D sky fitting (Shergin *et al.* 1996), (4) adaptive filtration for the direct images, (5) transformation of the direct image positions to the spectral image, (6) direct image inventory, (7) astrometry using the direct images, (8) B and V photometry using the direct images, (9) 1D spectra extraction, (10) searching for emission features in spectra, (11) detection and tabulation of candidates with emission features.

The Output KISS Database contains accurate positions (X, Y, α(1950), δ(1950)), magnitudes, colors and Star/Galaxy classification for all objects found during the inventory of the combined direct image plus additional information (position of emission peak, signal-to-noise ratio (SNR) for that peak, *etc.*) for all objects detected in the spectral image. In addition, the Database contains 1D extracted spectra, brightness profiles in the spatial direction, noise vectors (noise value for each pixel of spectrum), and spectra of continuum and continuum subtracted spectra for all extracted objects.

All ELG candidates selected in each KISS field are copied into a separate Output KISS Database. Typically 40–50 candidates are selected in each field. Final inspection of selected candidates above the threshold SNR value is made visually. It allows us to remove all complicated cases (artifacts, patterns of saturated bright stars, overlapping spectra, *etc.*) which cannot be filtered out with our current algorithms.

To estimate the quality of our photometry, astrometry and Star/Galaxy classification we compare the results of our reductions with results from Willmer *et al.* (1996), Moreau *et al.* (1995) and data from the APM machine. For this comparison we have identified all objects from those papers which have counterparts in the KISS fields $\alpha=12^h55^m$, 13^h00^m, 13^h05^m, 13^h10^m and have examined the differences in the photometry and positions of these objects between the mentioned papers and the KISS derived values. We summarize the results of this comparison below.

3. Conclusions

1. A new method of automatic search for emission-line candidates has been developed and applied to our deep survey data.
2. The new software for the search for emission-line candidates has been written in both the MIDAS and IRAF environments.
3. The comparison of our data with other surveys shows:

    (a) In spite of our data undersampling (1 pixel = 2.07″) the astrometric precision after reduction is about 0.5″, good enough for follow-up spectroscopy of candidates with a multifiber spectrograph like HYDRA.
    (b) KISS has reliable photometry to at least V=18.5^m when compared to these other surveys (absolute error is ≤ 0.15 mag.). Internal photometric errors are ≤ 0.10 mag. at V=20^m.

4. The follow-up spectroscopy with the 6-m telescope proved that there is no noticeable bias between candidates selected visually and automatically. Candidates above the SNR threshold are shown to be real emission-line objects of various types.

References

Moreau O. and Reboul H., 1995. Astron. Astrophys. Suppl., 111, 169
Salzer, J., *et al.* B.A.A.S., 26, 916
Shergin, V.S., Kniazev A.Y. and Lipovetsky, V.A. 1996. Astr. Nachrichten, 2, 95
Willmer, C.N.A., *et al.* Astrophys. J. Suppl., 104, 199

HOW TO FIND BL LAC OBJECTS IN THE RASS

P. NASS
Max-Planck-Institut für Extraterrestrische Physik, Germany

1. Introduction

The correlation of an X-ray survey like the ROSAT All-Sky Survey (RASS) with optical surveys like the Hamburg Quasar Survey (HQS) can significantly increase the number of known BL Lac objects.

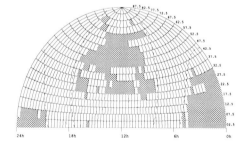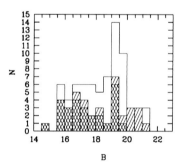

Figure 1. Left: In the studied area (shaded) we identified 35 new BL Lac objects (Nass *et al.* 1996) ; 43 BL Lac objects were previously known, from which we excluded 4 star-like objects. The ROSAT All-Sky Survey (RASS) was correlated with these 216 fields from the Hamburg Quasar Survey (HQS) during an ongoing AGN identification program. To select BL Lac candidates we used the criterion $\log f_X/f_B > 1.3$. During the follow-up spectroscopy of the first subsample of $\log f_X/f_B > 1.3$ candidates, 40 % of them proved to be BL Lac objects. The efficiency rate will be higher after the complete identification, because the remaining candidate objects have higher $\log f_X/f_B$ and for a given X-ray flux BL Lac objects tend to be optically fainter than other types of AGN. Right: Among the newly identified BL Lac objects (blank) are more optically faint ones than among the previously known BL Lac objects of which 27 (crossed) were and 12 (striped) were not detected in the RASS. The decrease beyond B=19 is probably caused by the platelimit and the difficulty to detect a featureless spectrum of such a weak object. Follow-up observations of BL Lac candidates with 'empty' RASS error circles on HQS plates could enhace the number of BL Lac objects fainter than $B = 19.5$.

X-ray sources with the highest ratios of X-ray to optical flux are primarily BL Lac objects. Therefore, by selecting high $\log(f_x/f_B)$ objects, very good BL Lac candidates can be chosen. This is a *radio independent* search technique which is very important for the weaker BL Lac objects which do not enter common radio catalogues with thresholds of 25 to 40 mJy. However, cross-correlation with more sensitive radio catalogs helps to optimize the candidate list and saves optical observing time at large telescopes (very important for faint objects). RBL-like objects cannot be discovered with this technique but with the combined X-ray/radio/optical method which consists of the calculation of the energy indices of X-ray sources with a radio counterpart. By taking spectra of candidates which populate the typical BL Lac area in the $\alpha_{RO} - \alpha_{OX}$ diagram one can simply identify new BL Lac objects.

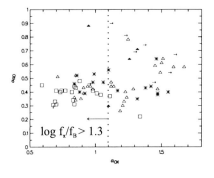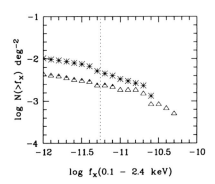

Figure 2. Colour-Colour-Diagram(left): The 39 previously known BL Lac objects can be found in the typical areas: XBLs populate a "box" in the lower left corner and RBLs are aligned along a diagonal at higher α_{OX} and higher α_{RO}. 27 objects (open triangles) were detected in the RASS, 12 not. 4 (filled triangles) of these 12 objects were detected in pointed PSPC observations. For the remaining 8 objects (arrows), we used the upper limits from the RASS. Most new objects (asterisks) populate the XBL region, those without entry in radio catalogues(empty boxes) at lower α_{RO}. Log N - log S(right): The obtained surface density is in accordance with previous studies. However, we suspect more BL Lac objects at lower X-ray fluxes which remained undiscovered due to previously used search techniques.

2. Conclusions

The combination of X-ray, optical, and radio surveys is a very efficient means for the discovery of new BL Lac objects in X-ray surveys like the RASS. Our results indicate that about 200 BL Lac objects can be found among the high $\log f_x/f_B$ BL Lac candidates in the studied area. A larger sample of BL Lac objects is needed to ensure that selection effects did not influence the existing log N - log S distributions and BL Lac models. Particularly, the reason for the seperation of the whole BL Lac class into XBLs and RBLs is important. Will we still see it in a larger BL Lac sample? Additionally, further studies can be made, *e.g.*, our environmental studies of BL Lac objects with possibly extended X-ray emission.

References

Nass P., Bade, N. ,Kollgaard, R. I. *et al.*,1996, Astron.Astrophys.,309,419

THE WARPS X-RAY SURVEY OF GALAXIES, GROUPS, AND CLUSTERS

D. HORNER[1], C.A. SCHARF[1], L.R. JONES[2], H. EBELING[3],
E. PERLMAN[4], M. MALKAN[5] AND G. WEGNER[6]
[1] *Lab. for High Energy Astrophysics, NASA/GSFC*
[2] *School of Physics and Space Research, Univ. of Birmingham*
[3] *Institute for Astronomy, Univ. of Hawaii*
[4] *Space Telescope Science Institute*
[5] *Dept. of Astronomy, UCLA*
[6] *Dept. of Physics & Astronomy, Dartmouth College*

1. Introduction

We have embarked on a survey of ROSAT PSPC archival data searching for all detected surface brightness enhancements due to sources in the innermost $R \leq 15'$ of the PSPC field of view in the energy band 0.5–2.0 keV. This project is part of the Wide Angle ROSAT Pointed Survey (WARPS) and is designed primarily to measure the low luminosity, high redshift, X-ray luminosity function of galaxy clusters and groups. Accurate measurements of the high redshift XLF would allow the form of the XLF evolution to be determined via the position of the Schechter function break. This would help discriminate between luminosity and density evolution, and discriminate between different hierarchical models, *e.g.*, those including a different mix of fundamental particles, a flat power spectrum of the initial fluctuations, and reheating of the intracluster gas at high redshifts.

2. Data Analysis

Our source detection method, Voronoi Tessellation and Percolation or VTP, represents a significant advance over conventional methods and is particularly suited for the detection and correct quantification of extended and/or low surface brightness emission which could otherwise be missed or wrongly interpreted. We also use energy dependent exposure maps to estimate the fluxes of sources which can amount to corrections of as much as 15%. We have nearly complete sky coverage for clusters out to a redshift of $z \simeq 1$, with 50% coverage at a redshift of $z \sim 1.4$. We have 80% coverage for groups (or faint, small clusters) to $z \sim 0.2$ and for galaxies to about $z \sim 0.1$.

In an ongoing optical follow up program, we are obtaining both CCD imaging and spectroscopic data of the extended sources and selected point-like VTP sources (specifically those with galaxy counterparts). The follow-up procedure has been designed to minimize incompleteness and misidentifications of the X-ray source candidates (catalogue in preparation). Classification based solely on X-ray criteria may be in error at low fluxes like those used in WARPS. We have found that only 70–80% of our VTP extended sources are clusters. Imaging will also allow determinations of the richnesses and morphologies of the clusters and correlation of optical and X-ray properties of clusters.

3. Results

The first results for for an initial 91 fields (17.2 deg^2) at fluxes $> 3.5 \times 10^{-14}$ erg s^{-1} cm^{-2} are presented in detail by Scharf et al. (1996). The sky density of extended objects with detected flux $> 3.5 \times 10^{-14}$ erg sec^{-1} cm^{-2} is 2.8–4.0 (± 0.4) deg^{-2}. A comparison with a point source detection algorithm has demonstrated that our VTP approach typically finds 1–2 more objects deg^{-2} to this flux limit, suggesting that the conventional method fails to detect a significant fraction of extended objects. The surface brightness limit of the WARPS cluster survey is $\sim 1 \times 10^{-15}$ erg sec^{-1} cm^{-2} arcmin^{-2}, approximately 6 times lower than the Extended Medium Sensitivity Survey (EMSS). The WARPS LogN-LogS (which currently represents a lower limit) shows a significant excess over previous measurements for $S \gtrsim 8 \times 10^{-14}$ erg sec^{-1} cm^{-2} (0.5–2 keV). We attribute this mainly to a larger measured flux from extended sources as well as new detections of low surface brightness systems in the WARPS.

For further information and preprints of WARPS papers, see the WARPS home page at
   http://lheawww.gsfc.nasa.gov/~caleb/warps/warps.html.

References

Scharf, C.A., et al. Astrophys. J., in press

THE WARPS BLAZAR SURVEY

Searching for the Faintest X-ray Selected Blazars

E.S. PERLMAN[1], P. PADOVANI[2], L. JONES[3], P. GIOMMI[4],
A. TZIOUMIS[5], J. REYNOLDS[5] AND R. SAMBRUNA[6]
[1]*Space Telescope Science Institute*
[2]*II Università di Roma*
[3]*University of Birmingham*
[4]*ASI/SAX Data Center*
[5]*CSIRO*
[6]*LHEA/GSFC*

1. The Survey

The WARPS (Wide-Angle ROSAT Pointed Survey) blazar survey is a deep X-ray search for BL Lac objects and flat-radio-spectrum quasars (FRSQs), drawn from a cross-correlation of serendipitous sources in the ROSAT PSPC database WGACAT (White *et al.* 1994) with the Green Bank 6 cm and 20 cm (Condon *et al.* 1989, Condon & Broderick 1985), the Parkes radio (Bolton *et al.* 1979), and the Parkes-MIT-NRAO (Griffith & Wright 1993, Wright *et al.* 1994, Griffith *et al.* 1994, 1995) catalogs. Our sample contains 165 new blazar candidates and 95 previously known blazars.

As single-dish surveys yield positions no better than those produced by ROSAT (error circles $10''$–$1'$), we used ongoing VLA surveys (FIRST, Becker *et al.* 1995; NVSS, Condon *et al.* 1996) to refine the positions to the arcsecond level for sources north of $-15°$. For southern ($\delta < -15°$) sources, which also lacked spectral index information, we have done a survey at 6 cm and 3.6 cm with the ATCA. We then obtained finder charts using the Digitized Sky Survey. Where there is no candidate at the best position, a deeper image is being obtained at a 1m class telescope.

Because of its depth and breadth (Figure 1), the WARPS blazar survey will yield the very first X-ray selected sample of FRSQs, allowing their X-ray luminosity function to be computed for the first time. This will produce constraints on the opening angle and γ of the X-ray jet, parameters which are currently unconstrained. We will also address the current controversy

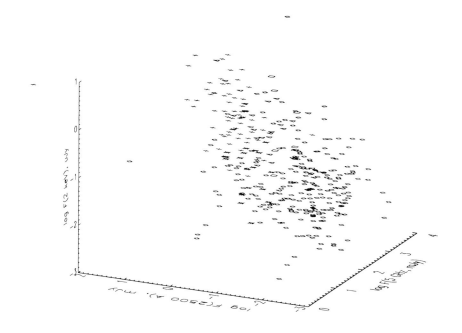

Figure 1. The parameter space covered by the WARPS blazar survey. Diamonds are the WARPS blazars and candidates, crosses are Slew Survey BL Lacs, triangles are 1 Jy BL Lacs, asterisks are EMSS BL Lacs, and squares are S4 FRSQs.

over BL Lac evolution (Perlman *et al.* 1996, Stickel *et al.* 1991). Finally, we willexplore interrelationships betwene the two blazar subclasses.

References

Becker, R. H., White, R. L., & Helfand, D. J., 1995, Astrophys.J. 450, 559.
Bolton *et al.* 1979, Aust J Phys., Astrophys. Suppl., No. 46
Condon, J. J., & Broderick, J. J. 1985, Astron.J. 90, 2540.
Condon, J. J., Broderick, J. J., & Seielstad, G. A. 1989, Astron.J. 97, 1064.
Condon, J. J., *et al.* 1996, preprint.
Griffith, M.R., & Wright, A.E. 1993, Astron.J. 105, 1666.
Griffith, M.R., *et al.* 1994, Astrophys.J.Suppl. 90, 179.
Griffith, M.R., *et al.* 1995, Astrophys.J.Suppl. 97, 347.
Perlman, E. S., *et al.* 1996, Astrophys.J. 456, 451.
Stickel, M., *et al.* 1991, Astrophys.J. 374, 431.
White, N. E., Giommi, P., & Angelini, L., HEAD Meeting 1994
Wright, A.E., *et al.* 1994, Astrophys.J.Supp. 91, 111.

RESULTS FROM ASCA SKY SURVEYS

Y. OGASAKA[1], Y. UEDA[2], Y. ISHISAKI[3], T. KII[2],
T. TAKAHASHI[2], K. MAKISHIMA[4], H. INOUE[2], K. OHTA[5],
T. YAMADA[6], T. MIYAJI[7] AND G. HASINGER[8]
[1] *NASA Goddard Space Flight Center*
[2] *The Institute of Space and Astronautical Science*
[3] *Department of Physics, Tokyo Metropolitan University*
[4] *Department of Physics, University of Tokyo*
[5] *Department of Astronomy, Kyoto University*
[6] *Institute of Physical and Chemical Research (RIKEN)*
[7] *Max-Planck-Institute für Extraterrestrische Physik*
[8] *Astrophysikalisches Institut Potsdam*

1. Introduction

The origin of the Cosmic X-ray Background (CXB) radiation has been investigated extensively by soft X-ray deep survey imaging observations with *Einstein* and *ROSAT*. In contrast, the lack of telescopes capable of detecting hard X-rays has prevented us from extensive study of the nature of the CXB in the energy range above 2 keV before *ASCA*.

ASCA Deep Sky Survey (DSS) and Large Sky Survey (LSS) were intended to carry out unbiased surveys in the wide energy range of 0.5–10 keV. DSS was planned to survey a small sky region with extremely high sensitivity reaching to the source confusion limit of the *ASCA* XRT, while LSS covers a much larger sky area with relatively shallow exposures. These two surveys play complimentary roles in our approaches to the nature of the faint X-ray objects and the origin of the CXB, especially in the 2–10 keV band.

2. DSS Results

The DSS consists of moderately deep pointings of five sky regions and extremely deep pointings of Selected Area 57 (SA57) (Ogasaka 1996). The data reduction and analysis have been done for the Lynx Field, the Lockman Hole and a part of SA57, covering a total solid angle of 0.29 deg^2. The $N(>S)$ in the 2–10 keV band was derived as 55 ± 25 deg^{-2} at the flux limit of 3.80×10^{-14} erg sec^{-1} cm^{-2}. This is consistent with the extrapolation of Log$N$–Log$S$ relations from previous experiments(e.g., Hayashida 1991; Piccinotti et al. 1982). At this flux limit about 40% of the CXB intensity in the 2–10 keV band is resolved into discrete sources. On the other hand, the $N(>S)$ in the 0.5–2 keV band is consistent with more sensitive $ROSAT$ Log$N$–Log$S$ relation derived by Hasinger et al. (1993).

3. LSS Results

The Large Sky Survey, where a continuous sky regions near the North Galactic Pole is systematically surveyed, covered 6 deg^2 up to the present (Ueda 1996). We detected ~ 50 sources in the survey energy band of 2–10 keV whose flux distribute from 1.5×10^{-13} to 2.0×10^{-12} erg sec^{-1} cm^{-2}. The derived Log$N$–Log$S$ relation above 2 keV is consistent with the extrapolation from the previous results. The average spectrum for the sources with flux less than 2.5×10^{-13}erg sec^{-1} cm^{-2} shows a photon index of 1.5±0.2 above 2 keV, which is harder than that of bright AGN observed so far. This result suggests that the hard sources responsible for the CXB above 2 keV appear when the sensitivity drops to $\sim 10^{-13}$erg sec^{-1} cm^{-2}.

4. Source Identifications

We have carried out the optical follow-up observations for X-ray sources detected from the Lynx, SA57 and LSS fields. Significant fraction of hard X-ray dominated sources were identified with narrow-line objects. One of them is the narrow-line quasar at z$\simeq$0.9 discovered from the Lynx Field (Ohta et al. 1996). It is possible that this object is the "type-2 quasar," whose existence has been expected from the unified model of AGNs, and from the "spectral paradox" of the CXB.

Acknowledgements

YO acknowledges the support of the Postdoctoral Fellowships for Research Abroad of the Japan Society for the Promotion of the Science.

References

Hasinger, G. et al., 1993, Astron.Astrophys., 275, 1.
Hayashida, K. 1989, Ph. D dissertation of Univ. of Tokyo, ISAS RN 466.
Ogasaka, Y. 1996, Ph. D dissertation of Gakushuin University
Ohta, K. et al., 1996, Astrophys.J., 458, L57
Piccinotti, G. et al. 1982, Astrophys.J., 253, 485.
Ueda, Y. 1996, Ph. D dissertation of University of Tokyo

Part 6. Large Scale Structure

LARGE SCALE STRUCTURE OF THE UNIVERSE

N.A. BAHCALL
Princeton University Observatory, Princeton

Abstract. How is the universe organized on large scales? How did this structure evolve from the unknown initial conditions of a rather smooth early universe to the present time? The answers to these questions will shed light on the cosmology we live in, the amount, composition and distribution of matter in the universe, the initial spectrum of density fluctuations that gave rise to this structure, and the formation and evolution of galaxies, clusters of galaxies, and larger scale structures.

To address these fundamental questions, large and accurate sky surveys are needed—in various wavelengths and to various depths. In this presentation I review current observational studies of large scale structure, present the constraints these observations place on cosmological models and on the amount of dark matter in the universe, and highlight some of the main unsolved problems in the field of large-scale structure that could be solved over the next decade with the aid of current and future surveys. I briefly discuss some of these surveys, including the Sloan Digital Sky Survey that will provide a complete imaging and spectroscopic survey of the high-latitude northern sky, with redshifts for the brightest $\sim 10^6$ galaxies, 10^5 quasars, and $10^{3.5}$ rich clusters of galaxies. The potentialities of the SDSS survey, as well as of cross-wavelength surveys, for resolving some of the unsolved problems in large-scale structure and cosmology are discussed.

1. Introduction

Studies of the large-scale structure of the universe over the last decade, led by observations of the distribution of galaxies and of clusters of galaxies, have revealed spectacular results, greatly increasing our understanding of this subject. With major surveys currently underway, the next decade will provide new milestones in the study of large-scale structure. I will highlight what we currently know about large-scale structure, emphasizing some of the unsolved problems and what we can hope to learn in the next ten years from new sky surveys.

Why study large-scale structure? In addition to revealing the "skeleton" of our universe, detailed knowledge of the large-scale structure provides constraints on the formation and evolution of galaxies and larger structures, and on the cosmological model of our universe (including the mass density of the universe, the nature and amount of the dark matter, and the initial spectrum of fluctuations that gave rise to the structure seen today).

What have we learned so far, and what are the main unsolved problems in the field of large-scale structure? I discuss these questions in the sections that follow. I first list some of the most interesting unsolved problems on which progress is likely to be made in the next decade using upcoming sky surveys.

- Quantify the measures of large-scale structure. How large are the largest coherent structures? How strong is the clustering on large scales (*e.g.*, as quantified by the power spectrum and the correlation functions of galaxies and other systems)?
- What is the topology of large-scale structure? What are the shapes and morphologies of superclusters, voids, filaments, and their networks?
- How does large-scale structure depend on galaxy type, luminosity, surface brightness? How does the large-scale distribution of galaxies differ from that of other systems (*e.g.*, clusters, quasars)?
- What is the amplitude of the peculiar velocity field as a function of scale?
- What is the amount of mass and the distribution of mass on large scales?
- Does mass trace light on large scales? What is in the "voids?"
- What are the main properties of clusters of galaxies: their mass, mass-function, temperature-function, and dynamical state?
- What is the mass density, $\Omega_m \equiv \rho_m/\rho_{\rm crit}$, of the universe?
- How does the large-scale structure evolve with time?
- What are the implications of the observed large-scale structure for the cosmological model of our universe and for structure formation? (*e.g.*, What is the nature of the dark matter? Does structure form by gravitational instability? What is the initial spectrum of fluctuations that gave rise to the structure we see today? Were the fluctuations Gaussian?)

2. Clustering and Large-Scale Structure

Two-dimensional surveys of the universe analyzed with correlation function statistics (Groth and Peebles 1977, Maddox et al. 1990) reveal structure to scales of at least $\sim 20h^{-1}$ Mpc. Large redshift surveys of the galaxy distribution reveal a considerably more detailed structure of superclusters, voids, and filament network extending to scales of ~ 50–$100h^{-1}$ Mpc (Gregory and Thompson 1978, Gregory et al. 1981, Chincarini et al. 1981, Giovanelli et al. 1986, de Lapparent et al. 1986, de Costa et al. 1988, Geller and Huchra 1989) The most recent and largest redshift survey, the Las Campanas Redshift Survey (Kirshner et al. 1996; see also Landy et al. 1996), with redshifts for $\sim 25 \times 10^3$ galaxies, is presented in Figure 1; it reveals the "cellular" nature of the large-scale galaxy distribution. The upcoming Sloan Digital Sky Survey (SDSS), expected to begin operation in 1997 (see §5), will provide a three dimensional map of the entire high-latitude northern sky to $z \sim 0.2$, with redshifts for approximately 10^6 galaxies. This survey, and others currently planned, will provide the large increase in the survey volume required to resolve some of the unsolved problems listed above. (See contribution by McKay, this volume, p. 49.)

The angular galaxy correlation function was first determined from the 2D Lick survey and inverted into a spatial correlation function by Groth & Peebles. They find $\xi_{gg}(r) \simeq 20r^{-1.8}$ for $r \lesssim 15h^{-1}$ Mpc, with correlations that drop to the level of the noise for larger scales. This observation implies

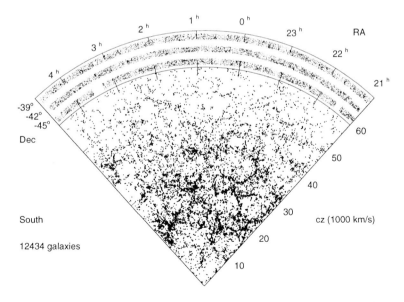

Figure 1. Redshift cone diagram for galaxies in the Las Campanas survey (Kirshner et al. 1996).

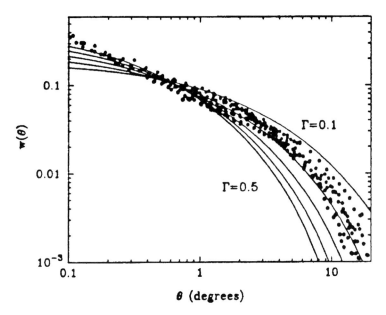

Figure 2. The scaled angular correlation function of galaxies measured from the APM survey plotted against linear theory predictions for CDM models (normalized to $\sigma_8 = 1$ on $8h^{-1}$ Mpc scale) with $\Gamma \equiv \Omega_m h = 0.5, 0.4, 0.3, 0.2$ and 0.1 (Efstathiou *et al.* 1990)

that galaxies are clustered on at least $\lesssim 15h^{-1}$ Mpc scale, with a correlation scale of $r_o(gg) \simeq 5h^{-1}$ Mpc, where $\xi(r) \equiv (r/r_o)^{-1.8} \equiv Ar^{-1.8}$. More recent results support the above conclusions, but show a weak correlation tail to larger scales. The recent two-point angular galaxy correlation function from the APM 2D galaxy survey (Maddox *et al.*, Efstathiou *et al.* 1990) is presented in Figure 2. The observed correlation function is compared with expectations from the cold-dark-matter (CDM) cosmology (using linear theory estimates) for different values of the parameter $\Gamma = \Omega_m h$. Here Ω_m is the mass density of the universe in terms of the critical density and $h \equiv H_0/100$ km s^{-1} Mpc^{-1}. The different $\Omega_m h$ models differ mainly in the large-scale tail of the galaxy correlations: higher values of $\Omega_m h$ predict less structure on large scales (for a given normalization of the initial mass fluctuation spectrum) since the CDM fluctuation spectrum peaks on scales that are inversely proportional to $\Omega_m h$. It is clear from Figure 2, as was first shown from the analysis of galaxy clusters (see below), that the standard CDM model with $\Omega_m = 1$ and $h = 0.5$ does not produce enough large-scale power to match the observations. As Figure 2 shows, the galaxy correlation function requires $\Omega_m h \sim 0.15$–0.2 for a CDM-type spectrum, consistent with other large-scale structure observations.

The power spectrum, $P(k)$, which reflects the initial spectrum of fluctuations that gave rise to galaxies and other structure, is represented by the Fourier transform of the correlation function. One of the recent attempts to determine this fundamental statistic using a variety of tracers is presented in Figure 3 (Peacock and Dodds 1994; see also Landy et al., Vogeley et al. 1992, Fisher et al. 1993, Park et al. 1994). The determination of this composite spectrum assumes different normalizations for the different tracers used (optical galaxies, IR galaxies, clusters of galaxies). The different normalizations imply a different bias parameter b for each of the different tracers [where $b \equiv (\Delta\rho/\rho)_{\text{gal}}/(\Delta\rho/\rho)_m$ represents the overdensity of the galaxy tracer relative to the mass overdensity]. Figure 3 also shows the microwave background radiation (MBR) anisotropy as measured by COBE (Smoot et al. 1992) on the largest scales ($\sim 1000h^{-1}$ Mpc) and compares the data with the mass power spectrum expected for two CDM models: a standard CDM model with $\Omega_m h = 0.5$ ($\Omega_m = 1, h = 0.5$), and a low-density CDM model with $\Omega_m h = 0.25$. The latter model appears to provide the best fit to the data, given the normalizations used by the authors for the different galaxy tracers. The recent Las Campanas redshift survey has reported excess power on $\sim 100H^{-1}$ Mpc scale over that expected from a smooth CDM spectrum (Landy et al. 1996). This is a most important observation that will need to be verified by larger surveys.

The next decade will provide critical advances in the determination of the power spectrum and correlation function. The large redshift surveys now underway, the Sloan and the 2dF surveys, will probe the power spectrum of galaxies to larger scales than currently available and with greater accuracy. These surveys will bridge the gap between the current optical determinations of $P(k)$ of galaxies on scales $\lesssim 100h^{-1}$ Mpc and the MBR anisotropy on scales $\gtrsim 10^3 h^{-1}$ Mpc (see McKay, this volume, p. 49). This bridge will cover the critical range of the spectrum turnover, which reflects the horizon scale at the time of matter-radiation equality. This will enable the determination of the initial spectrum of fluctuations at recombination that gave rise to the structure we see today and will shed light on the cosmological model parameters that may be responsible for that spectrum (such as $\Omega_m h$ and the nature of the dark matter). In the next decade, $P(k)$ will also be determined from the MBR anisotropy surveys on small scales ($\sim 0.1°$ to $\sim 5°$), allowing a most important overlap in the determination of the galaxy $P(k)$ from redshift surveys and the mass $P(k)$ from the MBR anisotropy. These data will place constraints on cosmological parameters including $\Omega (= \Omega_m + \Omega_\Lambda), \Omega_m, \Omega_b, h$, and the nature of the dark matter itself.

Another method that can efficiently quantify the large-scale structure of the universe is the correlation function of clusters of galaxies. Clusters

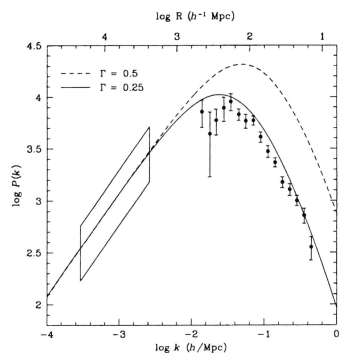

Figure 3. The power spectrum as derived from a variety of tracers and redshift surveys, after correction for non-linear effects, redshift distortions, and relative biases; from Peacock and Dodds 1994. The two curves show the Standard CDM power spectrum ($\Gamma = 0.5$), and that of CDM with $\Gamma = 0.25$. Both are normalized to the COBE fluctuations, shown as the box on the left-hand side of the figure.

are correlated in space more strongly than are individual galaxies, by an order of magnitude, and their correlation extends to considerably larger scales ($\sim 50h^{-1}$ Mpc). The cluster correlation strength increases with richness ($\propto$ luminosity or mass) of the system from single galaxies to the richest clusters (Bahcall and Soneira 1983, Bahcall 1988). The correlation strength also increases with the mean spatial separation of the clusters (Szalay and Schramm 1985, Bahcall and Burgett 1986). This dependence results in a "universal" dimensionless cluster correlation function; the cluster dimensionless correlation scale is constant for all clusters when normalized by the mean cluster separation.

Empirically, the two general relations that satisfy the correlation function of clusters of galaxies, $\xi_i = A_i r^{-1.8}$, are: $A_i \propto N_i$, and $A_i \simeq (0.4 d_i)^{1.8}$ (Bahcall and West 1992). (Here A_i is the amplitude of the cluster correlation function, N_i is the richness of the galaxy clusters of type i, and d_i is the mean separation of the clusters.) These observed relations have been compared with expectations from different cosmological models, yielding powerful constraints on the models (see below).

LARGE SCALE STRUCTURE OF THE UNIVERSE

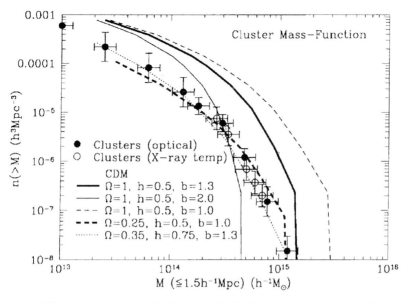

Figure 4. The mass function of clusters of galaxies from observations (points) and cosmological simulations of different $\Omega_m h$ CDM models (Bahcall and Cen 1992, 1993).

The observed mass function (MF), $n(>M)$, of clusters of galaxies, which describes the number density of clusters above a threshold mass M, can also be used as a critical test of theories of structure formation in the universe. The richest, most massive clusters are thought to form from rare high peaks in the initial mass-density fluctuations; poorer clusters and groups form from smaller, more common fluctuations. Bahcall and Cen (1993) determined the MF of clusters of galaxies using both optical and X-ray observations of clusters. Their MF is presented in Figure 4. The function is well fit by the analytic expression

$$n(>M) = 4 \times 10^{-5}(M/M^*)^{-1}\exp(-M/M^*)h^3 \text{ Mpc}^{-3}, \qquad (1)$$

with $M^* = (1.8 \pm 0.3) \times 10^{14} h^{-1} M_\odot$, (where the mass M represents the cluster mass within $1.5 h^{-1}$ Mpc radius).

Bahcall and Cen (1992) compared the observed mass function and correlation function of galaxy clusters with predictions of N-body cosmological simulations of standard ($\Omega_m = 1$) and nonstandard ($\Omega_m < 1$) CDM models. They find that none of the standard $\Omega_m = 1$ CDM models, with any normalization, can reproduce both the observed correlation function and the mass function of clusters. A low-density ($\Omega_m \sim 0.2$–0.3) CDM-type model, however, provides a good fit to both sets of observations (see, *e.g.*, Figure 4).

3. Peculiar Motions on Large Scales

How is the mass distributed in the universe? Does it follow, on the average, the light distribution? To address this important question, peculiar motions on large scales are studied in order to directly trace the mass distribution. It is believed that the peculiar motions (motions relative to a pure Hubble expansion) are caused by the growth of cosmic structures due to gravity. A comparison of the mass-density distribution, as reconstructed from peculiar velocity data, with the light distribution (i.e., galaxies) provides information on how well the mass traces light (Dekel 1994, Strauss and Willick 1995). A formal analysis yields a measure of the parameter $\beta \equiv \Omega_m^{0.6}/b$. Other methods that place constraints on β include the anisotropy in the galaxy distribution in the redshift direction due to peculiar motions (for a review, see Strauss and Willick 1995).

Measuring peculiar motions is difficult. The motions are usually inferred with the aid of measured distances to galaxies or clusters that are obtained using some (moderately-reliable) distance-indicators (such as the Tully-Fisher or $D_n - \sigma$ relations), and the measured galaxy redshift. The peculiar velocity v_p is then determined from the difference between the measured redshift velocity, cz, and the measured Hubble velocity, v_H, of the system (the latter obtained from the distance-indicator): $v_p = cz - v_H$.

The dispersion in the current measurements of β is very large. No strong conclusion can therefore be reached at present regarding the values of β or Ω_m. The larger and more accurate surveys currently underway, including high precision velocity measurements, may lead to the determination of β and possibly its decomposition into Ω_m and b (e.g., Cole et al. 1994).

Clusters of galaxies can also serve as efficient tracers of the large-scale peculiar velocity field in the universe (Bahcall et al. 1994). Measurements of cluster peculiar velocities are likely to be more accurate than measurements of individual galaxies, since cluster distances can be determined by averaging a large number of cluster members as well as by using different distance indicators. Using large-scale cosmological simulations, Bahcall et al. (1994) find that clusters move reasonably fast in all the cosmological models studied, tracing well the underlying matter velocity field on large scales. A comparison of model expectation with the available data of cluster velocities is presented by Bahcall and Oh (1996). The current data suggest consistency with low-density CDM models. Larger velocity surveys are needed to provide more robust comparisons with the models.

4. Dark Matter and Baryons in Clusters of Galaxies

Optical and X-ray observations of rich clusters of galaxies yield cluster masses that range from $\sim 10^{14}$ to $\sim 10^{15} h^{-1} M_\odot$ within $1.5 h^{-1}$ Mpc radius of the cluster center. When normalized by the cluster luminosity, a median value of $M/L_B \simeq 300h$ is observed for rich clusters. This mass-to-light ratio implies a dynamical mass density of $\Omega_{\rm dyn} \sim 0.2$ on $\sim 1.5h^{-1}$ Mpc scale. If, as suggested by theoretical prejudice, the universe has critical density ($\Omega_m = 1$), then most of the mass in the universe *cannot* be concentrated in clusters, groups and galaxies; the mass would have to be distributed more diffusely than the light.

A recent analysis of the mass-to-light ratio of galaxies, groups and clusters (Bahcall *et al.* 1995) suggests that while the M/L ratio of galaxies increases with scale up to radii of $R \sim 0.1$–$0.2h^{-1}$ Mpc, due to the large dark halos around galaxies, this ratio appears to flatten and remain approximately constant for groups and rich clusters, to scales of ~ 1.5 Mpc, and possibly even to the larger scales of superclusters (Figure 5). The flattening occurs at $M/L_B \simeq 200$–$300h$, corresponding to $\Omega_m \sim 0.2$. This observation may suggest that most of the dark matter is associated with the dark halos of galaxies and that clusters do *not* contain a substantial amount of additional dark matter, other than that associated with (or torn-off from) the

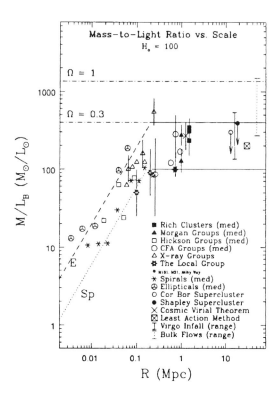

Figure 5. A composite mass-to-light ratio of different systems—galaxies, groups, clusters, and superclusters—as a function of scale. See Bahcall *et al.* 1995 for details.

galaxy halos, and the hot intracluster medium. Unless the distribution of matter is very different from the distribution of light, with large amounts of dark matter in the "voids" or on very large scales, the cluster observations suggest that the mass density in the universe may be low, $\Omega_m \sim 0.2$–0.3.

Clusters of galaxies contain many baryons. Within $1.5h^{-1}$ Mpc of a rich cluster, the X-ray emitting gas contributes ~ 3–$10h^{-1.5}\%$ of the cluster virial mass (or ~ 10–30% for $h = 1/2$) (Briel et al. 1992, White and Fabian 1995). Visible stars contribute only a small additional amount to this value. Standard Big-Bang nucleosynthesis limits the mean baryon density of the universe to $\Omega_b \sim 0.015h^{-2}$ (Walker et al. 1991). This suggests that the baryon fraction in some rich clusters exceeds that of an $\Omega_m = 1$ universe by a large factor (White et al. 1993, Lubin et al. 1995). Detailed hydrodynamic simulations (White et al. 1993, Lubin et al. 1995) suggest that baryons are not preferentially segregated into rich clusters. It is therefore suggested that either the mean density of the universe is considerably smaller, by a factor of ~ 3, than the critical density, or that the baryon density of the universe is much larger than predicted by nucleosynthesis. The observed baryonic mass fraction in rich clusters, when combined with the nucleosynthesis limit, suggests $\Omega_m \sim 0.2$–0.3; this estimate is consistent with the dynamical estimate determined above. Future optical and X-ray sky surveys of clusters of galaxies should help resolve these most interesting problems.

5. The Sloan Digital Sky Survey

A detailed description of the upcoming Sloan Digital Sky Survey (SDSS) is presented in this volume by McKay (p. 49). I will not repeat it here. I only summarize that the SDSS is a complete photometric and spectroscopic survey of π steradians of the northern sky, using 30 2048^2 pixel CCDs in five colors (u', g', r', i', z'), and two spectrographs ($R = 2000$) with 640 total fibers. The 5-color imaging survey will result in a complete sample of $\sim 5 \times 10^7$ galaxies to a limiting magnitude of $r' \sim 23^m$, and the redshift survey will produce a complete sample of $\sim 10^6$ galaxy redshifts to $r' \sim 18^m$ ($z \sim 0.2$), $\sim 10^5$ galaxy redshifts to $r' \sim 19.5^m$ ($z \sim 0.4$) for the reddest brightest galaxies, $\sim 10^5$ quasar redshifts to $g' \sim 20^m$, and $\sim 10^{3.5}$ rich clusters of galaxies.

What are some of the most interesting scientific problems in large-scale structure that the large and accurate Sloan sky survey can address?

- Quantify the clustering (of galaxies, clusters of galaxies, quasars) on large scales using various statistics (power spectrum, correlation function, void-probability distribution, and more).

- Quantify the morphology of large-scale structure (the supercluster, void, filament network).
- Determine the distortion in the redshift space distribution and its implication for the mass-density of the universe.
- Determine the clustering as a function of luminosity, galaxy type, surface brightness, and system type (galaxies, clusters, quasars).
- Determine the clustering properties of clusters (superclustering, correlation function and its richness dependence, power spectrum).
- Study the dynamics of clusters of galaxies. (With the availability of up to hundreds of redshifts per cluster, the mass of clusters can be well determined and compared with X-ray and lensing masses. The cluster mass-function and velocity function will be accurately determined, as well as the M/L and Ω_{dyn} implications).
- Study the evolution of galaxies, clusters, and superclusters to $z \sim 0.5$, and the evolution of quasars to $z \gtrsim 5$. These should provide important new constraints on cosmology.
- Use all the above to place strong constraints on the cosmological model and Ω, as discussed in the previous sections.

6. Important Future Surveys

What are some of the important surveys needed in order to address the main unsolved problems listed in the introduction? I list below such surveys.

- Optical, infrared, and radio redshift surveys (of galaxies, clusters, quasars, AGNs). These will help solve the quantitative description of large-scale structure, its strength and topology, and the relation among the structures described by different objects.
- X-ray surveys of clusters, quasars, and possibly superclusters. These will allow a good determination of the contribution of the hot gas component in the universe, cluster masses and temperature function, baryon fraction in clusters (and superclusters?), and the evolution of clusters and quasars.
- Gravitational lensing surveys. These will allow the most direct determination of the total mass and mass-density distribution in galaxies, clusters, and large-scale structure.
- Peculiar motion surveys of galaxies and clusters should yield most important constraints on Ω_m and b.
- High redshift surveys, using optical ground based telescopes (Keck), HST, X-rays, and radio, should reveal the important but yet unknown time evolution of structure in the universe. This will provide a fundamental clue to models of galaxy formation and cosmology.

- MBR anisotropy surveys, currently underway, will provide the fluctuation spectrum of the microwave background radiation and hopefully determine many of the cosmological parameters such as Ω, Ω_m, Ω_b, H_o, and the initial spectrum of fluctuations.
- All the above surveys will greatly constrain, and possibly determine the cosmological parameters of the universe (H_o; Ω; Ω_b; q_o; λ).

Research support by NSF grant 93-15368 and NASA grant NGT-51295 is gratefully acknowledged.

References

Bahcall, N. A., & Soneira, R. M. 1983, Astrophys.J., 270, 20.
Bahcall, N. A., & Burgett, W. S. 1986, Astrophys.J., 300, L35.
Bahcall, N. A. 1988, Ann.Rev.Astron.Astrophys., 26, 631.
Bahcall, N. A., & West, M. L. 1992, Astrophys.J., 392, 419.
Bahcall, N. A., & Cen, R. Y. 1992, Astrophys.J., 398, L81.
Bahcall, N. A., & Cen, R. Y. 1993, Astrophys.J., 407, L49.
Bahcall, N. A., Gramann, M., & Cen, R. 1994, Astrophys.J., 436, 23.
Bahcall, N. A., Lubin, L., & Dorman, V. 1995, Astrophys.J., 447, L81.
Bahcall, N. A., & Oh, S. P. 1996, Astrophys.J., 462, L49.
Briel, U. G., Henry, J. P., & Boringer, H. 1992, Astron.Astrophys., 259, L31.
Chincarini, G., Rood, H. J., & Thompson, L. A. 1981, Astrophys.J., 249, L47.
Cole, S., Fisher, K. B., & Weinberg, D. H. 1994, Mon.Not.R.astron.Soc., 267, 785.
da Costa, L. N., et al. 1988, Astrophys.J., 327, 544.
Dekel, A. 1994, Ann.Rev.Astron.Astrophys., 32, 371.
de Lapparent, V., Geller, M., & Huchra, J. 1986, Astrophys.J., 302, L1.
Efstathiou, G., Sutherland, W., & Maddox, S. 1990, Nature, 348, 705.
Fisher, K. B., et al., 1993, Astrophys.J., 402, 42.
Geller, M., & Huchra, J. 1989, Science, 246, 897.
Giovanelli, R., Haynes, M., & Chincarini, G. 1986, Astrophys.J., 300, 77.
Gregory, S. A., & Thompson, L. A. 1978, Astrophys.J., 222, 784.
Gregory, S. A., Thompson, L. A., & Tifft, W. 1981, Astrophys.J., 243, 411.
Groth, E., & Peebles, P. J. E. 1977, Astrophys.J., 217, 385.
Kirshner, R., et al., 1996, preprint (to be published in Astrophys.J.).
Landy, S. D., et al., 1996, Astrophys.J., 456, L1.
Lubin, L., Cen, R., Bahcall, N. A., & Ostriker, J. P. 1996, Astrophys.J., 460, 10.
Maddox, S., et al., 1990, Mon.Not.R.astron.Soc., 242, 43p.
Park, C., Vogeley, M. S., Geller, M. J., & Huchra, J. P. 1994, Astrophys.J., 431, 569.
Peacock, J., & Dodds, S. J. 1994, Mon.Not.R.astron.Soc., 267, 1020.
Smoot, G. R., et al., 1992, Astrophys.J., 396, L1.
Strauss, M., & Willick J. 1995, Physics Reports, 261, 271.
Szalay, A., & Schramm, D. 1985, Nature, 314, 718.
Vogeley, M. S., Park, C., Geller, M. J., & Huchra, J. P. 1992, Astrophys.J., 391, L5.
Walker, T. P., et al. 1991, Astrophys.J., 376. 51.
White, S. D. M., Navarro, J. F., Evrard, A., & Frenk, C.S. 1993, Nature, 366, 429.
White, D., & Fabian, A. 1995, Mon.Not.R.astron.Soc., 273, 72.

LARGE SCALE STRUCTURE IN THE LYα FOREST

G. WILLIGER[1,2], A. SMETTE[3], C. HAZARD[4,5],
J. BALDWIN[5,6] AND R. MCMAHON[5]
[1] *MPIA, Heidelberg*
[2] *NASA/GSFC*
[3] *Kapteyn Institute*
[4] *Phys & Astron, U. Pittsburgh*
[5] *Inst of Astronomy*
[6] *CTIO*

1. Introduction

We have conducted a survey of the Lyα forest in the redshift domain $2.15 < z < 3.37$ in front of nine QSOs within a 1° field to probe spatial structure along planes perpendicular to the line-of-sight. We find evidence for correlations of the Lyα absorption line wavelengths in the whole redshift range, and, at $z > 2.8$, of their equivalent widths. Such a correlation is consistent with the emerging picture that Lyα lines arise in filaments or large, flattened structures.

2. Data

We have examined the Lyα forest in front of 9 QSOs in a $\sim 1°$ field. The 2 Å resolution spectra were obtained in the course of a parallel C IV absorber study (Williger *et al.* 1996). We exclude Lyα forest lines within 5000 km s^{-1} from the background QSO to avoid problems with the "proximity effect," as were all Lyα lines corresponding to known metal systems from the C IV survey (Williger *et al.* 1996). 299 Lyα lines with $W_0 \geq 0.3$ Å were detected at $\geq 5\sigma$ confidence level at redshifts $2.15 < z < 3.37$, with 377 lines at a lower (incomplete) detection threshold of $W_0 = 0.1$ Å. We checked that the individual lines-of-sight do not present peculiarities that could undermine our analysis. Indeed, we find no significant variations in the number of Lyα absorbers from a power law distribution in redshift, and no anomalously large voids (Ostriker *et al.* 1988).

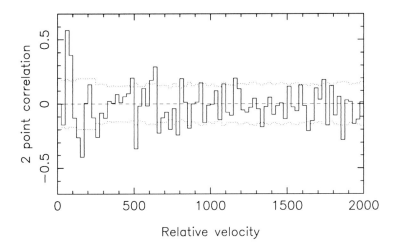

Figure 1. The Lyα forest two point correlation function at $2.6 < z < 3.3$ with rest equivalent width $W_0 \geq 0.1$ Å. Dotted lines show 1σ scatter from 1000 synthetic datasets. The chance probability of a feature arising anywhere at least as significant at $50 < \Delta v < 100$ km s^{-1} is $P = 0.0001$

3. Analysis

Various statistical methods (two-point correlation function, nearest neighbour test, distribution of line separations in velocity space) were then applied to probe for structure in velocity space, ignoring the angular separation between the quasars. This tests for structures in the plane of the sky at various redshifts. Sets of 1000 synthetic datasets using data-shuffling techniques were used to estimate the significance of features in the distribution functions. We find an excess of Lyα absorbers with separations $50 < \Delta v < 100$ km s^{-1}, especially strong over $2.60 < z < 3.25$ (Figure 1). We find no significant difference in the correlations as a function of angular separation. The correlations must come from different lines of sight as the minimum separation between absorbers along any one line of sight is $\Delta v \sim 300$ km s^{-1}. We have examined the difference in rest equivalent width for all line pairs with $50 < \Delta v < 100$ km s^{-1}, where the overabundance is found in the two point correlation function. For the 183 lines with $W_0 \geq 0.1$ Å and $2.80 < z < 3.25$, we find that the pairs of lines with rest equivalent width differences $\Delta W_0 \leq 0.2$ Å are overabundant (Figure 2). This is not simply from an overall normalisation difference; rather, the line pair excess is concentrated toward small values. The correlation appears to be independent of W_0 and angular separation. We have re-analysed data on a smaller angular scale from the literature (*i.e.*, Cotts 1989) as above, using all lines with significance $> 4\sigma$, and find similar effects.

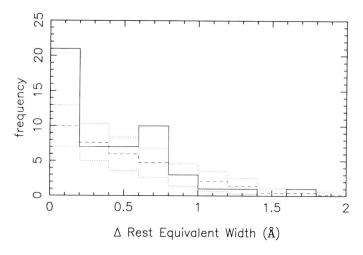

Figure 2. The equivalent width difference distribution at $2.8 < z < 3.25$, $W_0 \geq 0.1$ Å and $50 < \Delta v < 100$ km s^{-1}. The solid line denotes data, the dashed line the mean from 1000 synthetic datasets and the dotted line the $1\sigma\,scatter$. The chance probability of a feature at $0 < \Delta W_0 < 0.2$ Å at least as significant as observed is $P < 0.001$.

References

Crotts, A., 1989, Astrophys.J. 336, 550
Ostriker, J., Bajtlik, S., Duncan, R., 1988, Astrophys.J. 327, L35
Williger, G., Hazard, C., Baldwin, J., McMahon, R., 1996, Astrophys.J.Supp. 104, 145

A SEARCH FOR LARGE VOIDS FROM COMBINED SAMPLES OF GALAXY CLUSTERS AND GALAXIES

K.Y. STAVREV
Institute of Astronomy, Bulgarian Academy of Sciences

1. Introduction

This paper is part of a more extensive study of the large voids in spatial volumes both wide and deep. Three samples of optical tracers with spectroscopically measured redshifts in a volume with $b \geq 30°$ and $z \leq 0.14$ have been extracted from the catalogue of Lebedev and Lebedeva (1986) and the NASA/IPAC Extragalactic Database (NED):

– (1) a homogeneous sample of 277 rich Abell clusters
– (2) an inhomogeneous sample of 969 objects: rich and poor Abell clusters, Zwicky clusters, other clusters, and groups
– (3) a sample of 18 623 NED galaxies.

2. Void Catalog

We have found from the normalized to 1 Mpc^3 distributions of the number of objects by redshift that sample 1 is fairly complete to z=0.8–0.9, while the completeness of the subsamples of Zwicky clusters, other clusters, and groups falls sharply at about twice smaller redshifts. All samples are processed following Stavrev (1990). With the void selection criterion: void diameter $D \geq 80$ Mpc ($H_0 = 100$ Mpc km^{-1} s^{-1}) and exclusion of "open" voids, the procedure generates a catalog of 18 and 22 large voids, respectively for samples 1 and 2.

We have added sample 3 to sample 2 and repeated the void-search on this combined sample. The generated catalog contains 20 large galaxy voids (Table 1). Figure 1 shows the dependence of the void diameters on their distance. The increase of the void sizes in the interval above 300 Mpc is obviously due to sample incompleteness. The dimensions of the voids

TABLE 1. Large voids completely devoid of galaxies

Void No.	α (h)	δ (°)	l (°)	b (°)	R (Mpc)	D (Mpc)	A (°)	MAX (Mpc)	e	Percolating Void(s) No.
1	8.5	49	171	37	330	80	14	80	1.0	2
2	9.1	64	151	39	321	100	18	142	0.7	1, 5
3	9.5	25	204	45	281	92	19	152	0.6	4, 7
4	9.8	14	221	46	306	108	20	180	0.6	3, 7
5	10.6	50	162	56	338	80	13	80	1.0	2
6	10.9	−14	266	40	341	100	17	100	1.0	9
7	11.0	21	220	64	356	100	16	188	0.5	3, 4, 8
8	11.1	29	203	67	391	86	13	86	1.0	7
9	11.2	−6	265	49	369	108	17	188	0.6	6,13,14
10	11.3	37	180	69	279	88	18	88	1.0	
11	12.0	−21	288	40	248	82	19	116	0.7	
12	12.2	36	162	79	346	88	14	88	1.0	
13	12.3	−12	292	50	338	94	16	148	0.6	9,14,15
14	13.2	12	323	74	354	96	15	202	0.5	9,13,18
15	13.7	−9	325	51	387	126	18	126	1.0	13
16	13.8	48	98	66	350	80	13	80	1.0	17
17	13.9	39	79	72	336	80	14	80	1.0	16
18	14.2	12	0	65	331	100	17	169	0.6	14
19	15.6	15	24	49	370	88	14	88	1.0	
20	15.6	60	94	47	383	86	13	131	0.7	

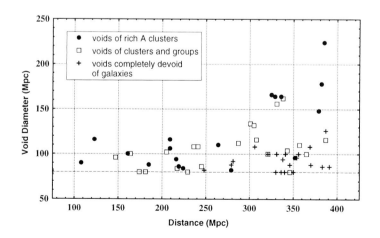

Figure 1. Void diameter versus distance

estimated from the more complete part of the volume are about 100 Mpc, in agreement with Einasto *et al.* (1994). The nearer part of the volume does not contain large, completely empty voids.

Acknowledgements

This research has made use of the NASA/IPAC Extragalactic Database. Thanks are due to A. Mutafov for help in the preparation of the figure. The author thanks the SOC of Symposium 179 for support to attend it.

References

Einasto, M., Einasto, J., Tago, E., and Dalton, G. B. 1994 ESO Sci. Preprint No. 987.
Lebedev, V. S., and Lebedeva, I. A. 1986 Astron. Tsirk. No. 1469, p. 4.
Stavrev, K. Y. 1990 Publ. Astron. Dep. Eötvös Univ., No. 10, p. 115.

LOW LUMINOSITY GALAXY DISTRIBUTION IN LOW DENSITY REGIONS

U. HOPP
Universitätssternwarte München

1. Introduction

We performed two surveys at the Calar Alto Observatory to identify low luminosity galaxies (LLG) in 4 fields towards nearby voids. While the central parts of the voids remain empty, we found about 20 very isolated (nearest neighbour distances of $D_{NN} \geq 4$Mpc) galaxies along the rims of some (but not all) voids (Hopp et al. 1995, Hopp & Kuhn 1995, Kuhn et al. 1996, Popescu et al. 1996). Many of them are dwarfs, a few are giants. CCD surface photometry revealed normal properties (Vennik et al. 1996), HI-observations show a tendency that the isolated dwarfs are overabundant in neutral gas compared to sheet and cluster galaxies of the same luminosity (Huchtmeier et al. 1996).

2. Analysis

Our sample of isolated dwarfs is still too small for most statistical applications. Thus I tried to combine our surveys with the results of similar studies of LLG's (Pustil'nik et al. 1995; Rosenberg et al. 1996; Salzer 1989) which also detected highly isolated galaxies. Naturally, the combined sample is incomplete and inhomogeneous, but it can serve as a test case for very wide angle survey (see Lipovetzki et al., this volume, p. 299) Our comparison sample for giants is the CfA2 catalogue (Huchra et al. 1990, from ZCAT, $B = 15.7^m$). All samples are restricted to $v_r \leq 10^4$ km s^{-1}. All LLG samples follow the same luminosity function with a steep slope at the faint end ($\alpha \sim -1.6$). Especially, the $\sim$100 isolated ($D_{NN} \geq 3$Mpc) galaxies in the combined sample show the same function as sheet members. Most of the dwarfs ($M_B \geq -18^m$) follow the well-known features outlined by the giants.

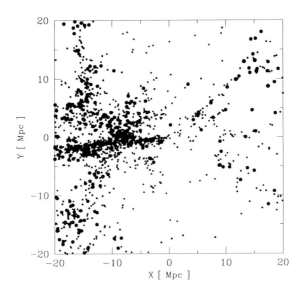

Figure 1. Spatial distribution in cartesian coordinates, Z projected (±20 Mpc). Symbol coding according to M_B. Most of the (absolutely) faint galaxies (small dots, $-18^m \leq M_B \leq -12^m$) follow the distribution outlined by the bright ones (big dots, $-23^m \leq M_B \leq -18^m$), but some additional structures are only occupied by these faint galaxies.

Some of the dwarfs populate structures which are avoided by bright galaxies (Figure 1) while Lyα clouds may even populate sheets which are avoided by dwarfs (see Shull *et al.* 1996). These additional spatial structures may point to hierarchical clustering.

Acknowledgements

The author acknowledges the support of the SFB 375 of the Deutsche Forschungsgemeinschaft.

References

Hopp,U., *et al.*, 1995 Astron.Astrophys.Suppl. 109, 537
Hopp,U., Kuhn,B. 1995 "Reviews in Modern Astronomy," 7, 277
Huchra,J.P., Geller,M.J., de Lapparent,V., Corwin,H.G. 1990 Astrophys.J.Suppl. 72, 433.
Huchtmeier,W.K., Hopp,U., Kuhn,B. 1996 Astron.Astrophys. accepted
Kuhn,B., Hopp,U., Elsässer,H. 1996 Astron.Astrophys. accepted
Popescu,C.C. *et al.*, 1996 Astron.Astrophys.Suppl. 116, 1,
Pustil'nik,S.A. *et al.*, 1995 Astrophys.J. 443, 499
Rosenberg,J.L. *et al.*, 1994 Astron.J. 108, 1557
Salzer,J.J. 1989 Astrophys.J. 347, 152
Shull,M.J. *et al.*, 1996 Astron.J. 111, 72
Vennik,J. *et al.*, 1996 Astron.Astrophys.Suppl. 117, 261

CLUSTERING PROPERTIES OF FAINT BLUE GALAXIES

M.W. KÜMMEL AND S.J. WAGNER
Landessternwarte Heidelberg, Germany

1. Introduction

Log N - log S diagrams are being used as powerful diagnostic tools to probe evolutionary properties of different extragalactic populations. In the optical/near-infrared regime the discrepancy between near-infrared number counts which follow theoretical predictions and counts in the B band which show an excess density revealed a new population of galaxies. The nature of this population of faint blue galaxies is still unknown.

To study the clustering properties of the faint blue galaxies and to investigate whether they are related with excess populations in various other wavelength regimes, *e.g.*, the population of milli-Jansky sources in the radio domain, faint IRAS sources in the far infrared or X-ray sources, we have performed a medium-deep survey at optical and near-infrared wavelengths at the North Ecliptic Pole (NEP). This survey extends homogeneously over an area of one square degree.

2. The Optical/Near-Infrared Survey at the NEP

The optical surveys were carried out in the B_j (460nm) and R (700nm) bands at with a TEK CCD at the prime focus (0.40 arcsec pixel^{-1}) of the 3.5 m telescope at Calar Alto, Spain. The $K'(2.1\mu m)$ data were taken with the MAGIC camera (1.6 arcsec pixel^{-1}) and a NICMOS3 array at the 2.2 m telescope on Calar Alto. To cover the area of $1\deg^2$ in the R band we took a grid of 10×10 fields. In B_j and K' we used a grid of 9×9 exposures with smaller overlap between adjacent fields but still covering $1\deg^2$.

To derive homogeneous photometry over the whole field we took in each band a snapshot survey with small exposure times and large fields in photometric conditions. In K' this was done with the 1.2 m telescope on Calar Alto. In the optical we used the focal reducer CAFOS at the 2.2 m telescope which has a circular field of view with a diameter of 15'. Although the snapshot survey does not cover one square degree, there is sufficient overlap with the frames of the deep survey to transfer the photometry with high accuracy. After standard reduction we identified sources automatically, derived the photometry and classified all objects due to their morphology. In the B_j and the R-band we detected 92,000 and 86,000 sources, respectively. In K' we reach a detection limit of 19.0 mag. Down to this level we detect 25,000 sources.

3. Galaxy Luminosity Function in the Optical

We compared the galaxy luminosity function in B_j with results from other surveys (*e.g.*, Metcalfe *et al.* 1995). The data agree well up to 23.5 mag. At this level incompleteness sets in for the whole survey. Individual, good frames are deeper by up to 0.8 mag. Statistical corrections for incompleteness allow comparisons down to 25 mag. In the R band the comparison of published luminosity functions with our galaxy counts show the completeness limit of our survey at 23 mag with limiting magnitudes (and complete sub-samples from smaller fields) fainter than 24.0 mag.

4. Color-Magnitude Diagram and Correlation Function

Because of the difference between the B_j and the B-band the color-magnitude diagram in B_j-R from our survey shows around 0.2 mag redder colors than the B−R color-magnitude distribution in Metcalfe *et al.* (1995) The limiting magnitudes (25.0 mag and 24.0 mag in B_j and R, respectively) of our survey result in a cut-off at $B_j=25.0$, $B_j-R=1$ with slope 1.

In order to study the nature of the faint populations in the radio, far-infrared and X-ray bands we compared the color-magnitude diagrams of the counterpart—candidates to those of the entire survey and found no statistically significant difference. To study the correlation properties we computed the cross correlation function for the total sample and for various color-selected subsamples. The correlation strength on scales of a few hundred arcseconds is 1.75 in B_j and R. We compared the correlation strengths for four diffent samples in the color-magnitude plane. The red objects show a higher correlation strength than blue objects.

References

Metcalfe N. *et al.* 1995 Mon.Not.R.astron.Soc., 273, 257-276

THE K-BAND HUBBLE DIAGRAM FOR X-RAY SELECTED BRIGHTEST CLUSTER GALAXIES

R.G. MANN[1] AND C.A. COLLINS[2]
[1] *Astrophysics Group, Imperial College*
[2] *Astrophysics Group, Liverpool John Moores University*

1. Introduction

The Hubble (magnitude-redshift) diagram for brightest cluster galaxies (BCGs) is a classic cosmological tool, widely studied because of the remarkably small dispersion (~ 0.3 mag) in the absolute optical magnitudes of low redshift BCGs (Postman and Lauer 1995). Extending the BCG Hubble diagram to higher redshifts would greatly enhance its role as a cosmological probe, but this has been frustrated by several technical problems:

- the conventional means of cluster selection in the optical become increasingly compromised by projection effects at $z > 0.1$
- at higher redshifts the interpretation of optical magnitudes becomes increasingly complicated by the effects of possible star formation.

We have overcomed both of these problems in a recent study (Collins and Mann *in preparation*):

- X-ray cluster selection minimises projection effects, yielding a BCG sample homogeneously selected out to $z \sim 1$
- by using the K-band we are primarily sampling the mature stellar population of the BCGs over that full range, and are insensitive to any possible bursts of star formation.

2. Cluster Sample

Our sample comprises BCGs in 48 X-ray selected clusters, principally drawn from the *Einstein Extended Medium Sensitivity Survey* (EMSS; Gioia and Luppino 1994), and supplemented by two *ROSAT* (Voges *et al.* 1996) clusters. The sample covers the redshift range $0.05 < z < 0.82$, and K-band images for them were obtained with the 3.8m United Kingdom Infrared Telescope (UKIRT).

3. Data Analysis

We measured K-band magnitudes for the sample within a constant metric aperture of diameter 50 kpc ($H_0 = 50$ km s^{-1} Mpc^{-1}, $q_0 = 0.5$), which were converted to absolute magnitudes using the k-correction model of Glazebrook et al. (1995). This yielded an rms dispersion in the absolute magnitudes of 0.50 mags, greatly exceeding that found in previous optical (Postman and Lauer 1995) and K-band (Aragon-Salamanca et al. 1993) studies. Much of the scatter in absolute magnitudes comes from the lower-z half of the sample. The rms dispersion for the higher-z half ($z > 0.225$) is 0.28 mags, and that reduction is highly significant: Monte Carlo tests show that only 0.4% of randomly-selected 24 galaxy subsamples of our data set produce an rms dispersion lower than that. The physical cause of this effect is readily understood. The EMSS clusters were selected above an X-ray flux limit, so the lower redshift clusters tend to have lower X-ray luminosities, and we find a strong correlation between the K-band luminosities of our BCGs and the X-ray luminosities, L_X, of their host clusters. Our results also confirm the existence of the $L_m - \alpha$ relationship for BCGs, [where L_m is the metric luminosity within an aperture of size r_m and α is the logarithmic slope of $L_m(r)$ at $r = r_m$], as found by previous optical studies (Postman and Lauer 1995). If we correct our K-band magnitudes for their correlations with L_X and α, then the rms dispersion of the 48 galaxy sample falls to 0.33 mags.

Furthermore, we find that the low-L_X, low-α clusters at low redshift also tend to have lower numbers of neighbouring galaxies within a projected separation of 100 kpc. We infer, therefore, that, by selecting clusters above an X-ray flux limit, we have included in our sample a low redshift ($z < 0.3$) population of first-ranked ellipticals in groups or poorer clusters which are different in kind from the cD galaxies in rich clusters conventionally envisaged as producing the classic BCG Hubble diagram. This is dramatically illustrated by the results of making an L_X cut to our sample: the 27 BCGs in our sample whose host clusters have $L_X > 2.3 \times 10^{44}$ erg s^{-1} in the 0.3–3.5 keV band yield an rms dispersion in absolute magnitudes of only 0.23 mags after correction for L_X and α, and there is no evidence for a correlation between BCG luminosity and redshift in such a sample. Upon removal of one obvious outlier, which is a full magnitude brighter than the mean of this subsample and whose image comprises at least four very close components, one of which may well be a star, the rms dispersion falls to 0.17 mags, lower than any previously reported figure for BCGs. We are investigating the significance of this result for the study of BCGs and their host clusters, and for their use as cosmological probes.

References

Aragon-Salamanca A., *et al.*, 1993, Mon.Not.R.astron.Soc., 262, 764
Collins C.A., Mann R.G., in preparation
Gioia I.M., Luppino G.A., 1994, Astrophys.J.Suppl., 94, 583
Glazebrook K., *et al.*, 1995, Mon.Not.R.astron.Soc., 275, 169
Postman M., Lauer T.R., 1995, Astrophys.J., 440, 28
Voges W., *et al.*, 1996, Astron.Astrophys., in press

MULTIWAVELENGTH STUDY OF THE SHAPLEY CONCENTRATION

S. BARDELLI[1], E. ZUCCA[2,3], G. ZAMORANI[2,3],
G. VETTOLANI[3] AND R. SCARAMELLA[4]
[1] *Osservatorio Astronomico di Trieste*
[2] *Osservatorio Astronomico di Bologna*
[3] *Istituto di Radioastronomia del CNR*
[4] *Osservatorio Astronomico di Roma*

1. Introduction

The Shapley Concentration is a prominent supercluster in the southern sky. It is interesting not only for its relevance in the peculiar motion problem (it seems to be responsible of $\sim 30\%$ of the acceleration acting on the Local Group of galaxies), but also because it is the most remarkable feature which appears studying the distribution of the Abell-ACO clusters of galaxies: Zucca et al. (1993) found that at every density contrast the Shapley Concentration stands out as the richest supercluster in the sky. In particular, the central part of this supercluster is dominated by a complex containing the three ACO clusters A3556, A3558 and A3562 and the poor cluster SC1329-314, which form a structure elongated $\sim 3°$ along the East-West direction. We are carrying on a long term study of the Shapley Concentration in order to describe its dynamical state and to determine its mass. The project consists of redshift determinations (with the ESO telescopes at La Silla) for galaxies both in the clusters (Bardelli et al. 1994) and in the inter-cluster field of this supercluster, X-ray observations (ROSAT) of the hot gas in the clusters (Bardelli et al. 1996) and radio observations (ATCA, MOST and VLA, Venturi et al. 1996) of the radiogalaxies.

2. The Inter-Cluster Survey

The Shapley Concentration was discovered as a *cluster* structure, but nothing is known about the distribution and the overdensity of *galaxies*. In order to study the global dynamics of this supercluster, we started a redshift survey of galaxies in the central part of this structure and outside clusters or obvious bi-dimensional overdensities. We observed 26 fields obtaining 443 redshifts in the [17–18.8] magnitude range. The number of spectra which present emission lines is 74, corresponding to $\sim 16\%$, similar to the percentage found in clusters. In the field, the percentage is $\sim 44\%$ in the same magnitude bin (Vettolani *et al.*, this volume, p. 346). We found also that this structure is similar to the Great Wall, with a bridge of galaxies connecting the most massive clusters.

3. The Core of the Shapley Concentration

The central regions of rich superclusters could be considered as ideal laboratories for studying the dynamical phenomena related to the formation of rich galaxy clusters. The higher local density excesses lead to high peculiar velocities which, when added to the relative small volume involved, increase the "cross section" for collisions and mergings. The Shapley Concentration shows two main cluster complexes, signatures of ongoing merging processes. We studied the complex dominated by A3558 (Bardelli *et al.* 1994, 1997 in preparation) using a sample of 714 redshifts, finding that it is a single connected structure of strongly interacting clusters. We calculated the dynamical parameters for the clusters, finding the presence of a number of substructures, both in A3558 and A3556. In particular, two groups, located between A3562 and A3558, are very prominent both in the optical (Bardelli *et al.* 1994) and in the X-ray band (Bardelli *et al.* 1996). The galaxy distribution in the core of the Shapley Concentration resembles the results of Burns *et al.* (1993) for the case of a merging between two clusters. The knowledge of the overdensity of inter-cluster galaxies will permit us to estimate the relative impact velocity and to study the evolution of the structure. Moreover, our radio-optical analysis of the brightest objects of this structure will be useful for studying the influence on the physical properties of the galaxy population of a cluster-cluster merging process.

References

Bardelli S., Zucca E., Vettolani, G., *et al.*, 1994, Mon.Not.R.astron.Soc. 267, 665
Bardelli S., Zucca E., Malizia A., *et al.*, 1996, Astron.Astrophys. 305, 435
Burns J.O., Roettiger K., Ledlow M., Klypin A., 1993, Astrophys.J. 427, L87
Venturi, T., *et al.* 1996, Mon.Not.R.astron.Soc., in press
Zucca E., *et al.* 1993, Astrophys.J. 407, 470

A RICH CLUSTER REDSHIFT SURVEY FOR LARGE-SCALE STRUCTURE STUDIES

D. BATUSKI[1], K. SLINGLEND[1], J. M. HILL[2], S. HAASE[1], C. MILLER[1] AND K. MICHAUD[1]
[1]*Department of Physics and Astronomy, University of Maine*
[2]*Steward Observatory, University of Arizona*

1. Introduction

We have used the MX multifiber spectrometer on the Steward 2.3m measuring redshifts of up to 25 galaxies in each of about 90 Abell cluster fields with richness class $R \geq 1$ and $\mathrm{mag}_{10} \leq 16.8$ (estimated $z \leq 0.12$) and no more than one previously-measured redshift. This work has resulted in a deeper, more complete sample for two-point correlation and other studies of large-scale structure. To date, we have collected such data for 110 clusters (a few not in our sample). For most, we have seven or more cluster members with redshifts, enough to add significantly to the available sample of cluster velocity dispersions for other studies of cluster properties.

2. Instrumentation and Observations

The MX Spectrometer velocity data for this survey is a multi-aperture fiber optic device designed and built by John M. Hill of Steward Observatory. Thirty-two fibers, each on a computer-controlled positioning probe, are used to collect galaxy spectra while another 24 fibers collect background sky spectra. The spectrograph utilizes a 300 groove/mm grating tilted to provide a wavelength range of 380–670 nm with a resolution of 1.1 nm (0.37 nm/pixel). The spectra were collected on a Texas Instruments 800 × 800 CCD chip and, in recent observations, on a Loral 800 × 1200 CCD.

The spectroscopic observations took place from 1990 to 1996. One hour integrations are typically required for each cluster. Each integration yields up to 32 quality spectra down to a limiting magnitude of ~ 18. The spectral coverage is ~ 300 nm centered at ~ 530 nm. Before each cluster exposure, comparison lamps were obtained for each fiber for later use in wavelength calibration. Twenty-four sky fibers, positioned next to the galaxy fibers on 24 of the 32 probes, collected simultaneous sky spectra for subsequent sky subtraction.

3. Data Reduction

Once the data were collected, reduction was performed using the NOAO IRAF package. The images were bias-subtracted and flat-field corrected. Individual spectra are located on each image and read into one-dimensional files. The spectra were wavelength-calibrated and the sky spectra that were collected by MX during the integration on the clusters were then averaged and subtracted from each of the galaxy spectra.

The last step in the reduction is the determination of the actual galaxy redshifts by cross-correlation. Nineteen spectra of nearby galaxies of known redshift are cross-correlated with each galaxy spectrum. The resulting average is our best estimate of the galaxy's redshift. We accepted a redshift if it met criteria based on the height of the cross-correlation peak, the number of template spectra that correlated well with the galaxy spectrum, and the dispersion among the estimates of the galaxy's redshift after the spectrum passed a visual inspection.

4. Results

Thus far, 110 clusters have been observed in this program for 99% completion to $m_{10} = 16.8$. For most of the target clusters, 10 or more redshifts per cluster field were obtained, so that average velocities, velocity dispersions, and projection effects can be evaluated accurately. This means that we now have a much-enlarged sample (from 104 for the $D \leq 4$ sample of Bahcall and Soniera (1983) to 242 now) of $R \geq 1$ Abell clusters with at least two measured redshifts per cluster. Combining this sample with the recently completed survey of ACO clusters by den Hartog (1995, PhD Thesis, Leiden University), we have not only been able to calculate a refined two-point correlation function for the northern clusters, but also one for a sample covering the whole unobscured sky. For the northern cluster case, we obtain a power law two-point spatial correlation function with $\gamma = -2.26 \pm 0.14$ and $r_0 = 21.9 \pm 1.5$ Mpc. For the all-sky case, we found $\gamma = -2.53 \pm 0.15$ and $r_0 = 21.0 \pm 1.5$ Mpc.

References

Bahcall, N. and Soneira, R.M. 1983 Astrophys.J, 270, 20.

THE ESO SLICE PROJECT (ESP) REDSHIFT SURVEY

G. VETTOLANI[1], E. ZUCCA[2,1], A. CAPPI[2], R. MERIGHI[2],
M. MIGNOLI[2], G. STIRPE[2], G. ZAMORANI[2,1],
H. MacGILLIVRAY[3], C. COLLINS[4], C. BALKOWSKI[5],
V. CAYATTE[5], S. MAUROGORDATO[5], D. PROUST[5],
G. CHINCARINI[6], L. GUZZO[6], D. MACCAGNI[7],
R. SCARAMELLA[8], A. BLANCHARD[9] AND M. RAMELLA[10]
[1] *Istituto di Radioastronomia del CNR, Italy*
[2] *Osservatorio Astronomico di Bologna, Italy*
[3] *Royal Observatory Edinburgh, United Kingdom*
[4] *Liverpool John Moores University, United Kingdom*
[5] *DAEC, Observatoire de Paris-Meudon, France*
[6] *Osservatorio Astronomico di Brera-Merate, Italy*
[7] *Istituto di Fisica Cosmica e Tecnologie Relative, Italy*
[8] *Osservatorio Astronomico di Roma, Italy*
[9] *Universitè Louis Pasteur, France*
[10] *Osservatorio Astronomico di Trieste, Italy*

1. The ESO Slice Project Redshift Survey

The ESO Slice Project (ESP) is a galaxy redshift survey we have recently completed as an ESO Key-Project over about 30 square degrees, in a region near the South Galactic Pole (Vettolani *et al.*, submitted to A&A). The survey is nearly complete to the limiting magnitude $b_J = 19.4$ and consists of more than three thousands galaxies with reliable redshift determination.

The ESP survey is intermediate between shallow, wide angle samples and very deep, monodimensional pencil beams: spanning a volume of $\sim 10^5$ h^{-3} Mpc^3 at the sensitivity peak ($z \sim 0.1$) it can provide an accurate determination of the "local" luminosity function and the mean galaxy density (Zucca *et al.*, submitted to A&A). Moreover, it can allow clustering analyses not biased anymore by nearby structures. Finally, this uniform set of spectra will allow us interesting studies about the K-correction and the galaxy evolutionary properties, based on a large homogeneous sample.

Here we report some results about the luminosity function.

2. The Luminosity Function

We find that, although a Schechter function (with $\alpha = -1.22$, $M^*_{b_J} = -19.61 + 5 \log h$ and $\phi^* = 0.020\ h^3\ \mathrm{Mpc}^{-3}$) is an acceptable representation of the luminosity function over the entire range of magnitudes ($M_{b_J} \leq -12.4 + 5 \log h$), our data strongly suggest a steepening of the luminosity function for $M_{b_J} \geq -17 + 5 \log h$. Such a steepening, well fitted by a power law with slope $\beta \sim 1.6$, is in agreement with what has been recently found by similar analyses for both field galaxies (Marzke et al. 1994) and galaxies in clusters (e.g., Driver & Phillipps 1996).

This steepening at the faint end of the luminosity function is almost completely due to galaxies with emission lines: in fact dividing galaxies into two samples, i.e., galaxies with and without emission lines, we find significant differences in their luminosity functions. In particular, galaxies with emission lines (which are $\sim 50\%$ of the total) show a steeper slope and a fainter M^*.

The normalization and the α and M^* parameters of our luminosity function are in excellent agreement with those of the AUTOFIB redshift survey (Ellis et al. 1996). Viceversa, our normalization is a factor ~ 2 higher than that found for both the APM (Loveday et al. 1992) and the Las Campanas (Lin et al. 1996) redshift surveys. Also the faint end slope of our luminosity function is significantly steeper than that found in these two surveys.

The galaxy number density for $M_{b_J} \leq -16 + 5 \log h$ is well determined ($\bar{n} = 0.08\ h^3\ \mathrm{Mpc}^{-3}$). Its estimate for $M_{b_J} \leq -12.4 + 5 \log h$ is more uncertain, ranging from $\bar{n} = 0.28\ h^3\ \mathrm{Mpc}^{-3}$, in the case of a fit with a single Schechter function, to $\bar{n} = 0.54\ h^3\ \mathrm{Mpc}^{-3}$, in the case of Schechter function and power law fit. The corresponding luminosity densities in these three cases are $\rho_{LUM} = (2.03, 2.23, 2.31) \times 10^8\ h\ \mathrm{L}_\odot\ \mathrm{Mpc}^{-3}$, respectively.

References

Driver, S.P., Phillipps, S., 1996, Astrophys.J. in press
Ellis, R.S., Colless, M., Broadhurst, T., Heyl, J., Glazebrook, K., 1996, Mon. Not. R. astron. Soc. 280, 235
Lin, H., Kirshner, R.P., Shectman, S.A., et al. 1996, Astrophys.J. 464, 60
Loveday, J., Peterson, B.A., Efstathiou, G., Maddox, S.J., 1992, Astrophys.J. 390, 338
Marzke, R.O., Huchra, J.P., Geller, M.J., 1994, Astrophys.J. 428, 43

INPUT CATALOGUE FOR THE 2DF QSO REDSHIFT SURVEY

R.J. SMITH[1,2], B.J. BOYLE[2], T. SHANKS[3], S.M. CROOM[3],
L. MILLER[4] AND M. READ[4]
[1] *Institute of Astronomy, Cambridge*
[2] *Anglo Australian Observatory*
[3] *Department of Physics, University of Durham*
[4] *Royal Observatory Edinburgh*

1. The 2dF QSO Redshift Survey

Observations that radio-quiet QSOs exist in average galaxy cluster environments (Smith *et al.* 1995 and references therein) demonstrate that QSOs can be used to derive important information on the structure of the Universe at the largest scales. Previous studies of QSO clustering have been frustrated by the lack of large QSO redshift surveys. Although QSO clustering is detected in the largest existing QSO catalogues (see Shanks & Boyle 1994), it is difficult to place strong limits on the cosmological evolution of QSO clustering or the level of clustering at large scales ($> 10h^{-1}$ Mpc) with current QSO catalogues.

Over the next 2 years the proposed 2dF survey will result in a 20–30 fold increase in the number of QSOs in a suitable, homogeneous catalogue with which to study large-scale structure; yielding approximately 25,000 QSO redshifts in the ranges; $0.3 < z < 2.2$ and $18.25 < B < 21.0$ over two $75° \times 5°$ strips on the sky.

2. The Input Catalogue

The input catalogue is based on candidates selected by their ultra-violet excess from APM scans of ~ 120 U and J UKST plates/films of 30 UKST fields. (The UKST J plates give a broad band B magnitude). In constructing this catalogue, it is of utmost importance that we take great care in the minimization of systematic photometric errors across the survey area. We find that there is a correlation between the local sky background, as determined by the APM, and the photometric error, so we have developed

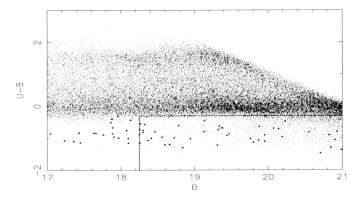

Figure 1. Colour-Magnitude plot showing all the stellar objects detected on one particular survey field (F864). Filled circles show the location on the plot occupied by all previously known QSOs in this region of sky (Veron-Cetty & Veron catalogue v.7 (1995) with some X-ray selected QSOs from Almaini (1996)). Dotted lines give an example set of selection criteria.

Total area of survey	: 740 deg^2	Number of candidates	: 46 077
Magnitude limits	: $18.25 \leq B_J \leq 21.0$	Expected completeness	: 90%
Colour selection	: $(U-B_J) \leq -0.36$	Number of expected QSOs	: 25 000

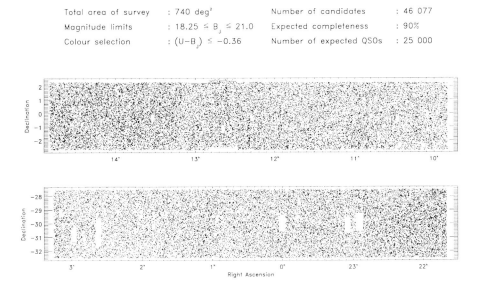

Figure 2. The completed UVX catalogue. The RA and dec of all the QSO candidates are shown in their two $75° \times 5°$ strips. Increasing contamination by Galactic sub-dwarfs at lower Galactic latitudes gives an obvious gradient ($< 50\%$) in number density along the survey strips. Completeness estimates are based on our success at recovering previously identified QSOs (see Figure 1) and on statistical consideration of the effect of our quoted errors.

an algorithm to make a magnitude dependent correction to all stellar objects' magnitudes, which is an extension of the process used by Maddox *et al.* (1990) in the creation of the APM galaxy survey. After this correction, we have directly confirmed that the systematic photometric errors in the B magnitudes (calibrated using CCD sequences in each UKST field) are less than 0.1 mag from field-to-field.

After correcting the J plates, we matched up the U data. Any small residual photometric variations in the magnitudes showed up as colour shifts over the field and were removed by applying a small field-dependent correction (< 0.05 mag) to the calibrated U magnitudes to ensure that the median $U - B$ colour of the stars remained constant over each UKST field.

As a first step towards analysis of this data, we have measured the two point correlation function, $w(\theta)$, for this survey. We detect a significant signal at small angular scales, adequately fit by a power-law of the form $w(\theta) = 0.002\theta^{-0.7}$. At larger scales, ($> 1$ degree), the amplitude of the angular correlation function is always less than 0.005.

This is the first large-area UVX survey where considerable attention has been devoted to securing as uniform a photometric catalogue as possible. We intend to include data from the UKST R plates in these fields in the near future to generate a truly wide-field, multi-colour stellar catalogue which will be made available on the WWW within the next year.
Our URL address is

http://www.aao.gov.au/local/www/rs/qso_surv.html

References

Almaini (1996) PhD Thesis, University of Durham
Maddox, Efstathiou and Sutherland (1990) Mon.Not.R.astron.Soc. 246, 433
Shanks and Boyle (1994) Mon.Not.R.astron.Soc. 271, 753
Smith, Boyle and Maddox (1995) Mon.Not.R.astron.Soc. 219, 537

REDSHIFT SURVEY OF 951 IRAS GALAXIES IN THE SOUTHERN MILKY WAY

N. VISVANATHAN[1] AND T. YAMADA[2]
[1] *Mount Stromlo and Siding Spring Observatories*
[2] *Astronomical Institute, Tohoku University*

1. Introduction

In this paper we present the results of a redshift survey of IRAS galaxies behind the whole southern Milky Way region $210° < l < 360°$ at $|b| < 15°$. The galaxy candidates in the southern Milky Way were selected from IRAS Point Source Catalogue (IPSC) by applying flux density and infrared colour criteria (Strauss *et al.* 1990; Rowan-Robinson *et al.* 1991). The selected sample is flux limited with $f_{60} > 0.6 Jy$. A visual search for galaxy-like objects have been carried out in UK-Schmidt Infrared and IIIaJ films in the selected IRAS positions. In spite of the large Galactic extinction in this region, 966 galaxy candidates have been identified (Yamada *et al.* 1993).

2. Redshift Measurements

Of the 966 galaxy candidates we have 951 confirmed galaxies in the region. Our sample of confirmed galaxies are very incomplete in the inner parts of the galaxy $|b| = 5°$ and 67 percent complete in the latitude zones, $5° < b < 15°$ and $15° < b < -5°$. A plot of the distribution of observed velocities of our sample of 951 galaxies in the present survey shows a broad maximum between 3000 to $6000 \, \mathrm{km\, s^{-1}}$ which can be taken to mean the survey is nearly complete up to $5000 \, \mathrm{km\, s^{-1}}$. Also a long tail is seen extending to $20,000 \, \mathrm{km\, s^{-1}}$ which is due to decreasing incompleteness of our sample at larger redshifts. It is interesting that the IRAS galaxy sample outside the Milky Way for $f_{60} > 0.6 Jy$ shows a peak around $5000 \, \mathrm{km\, s^{-1}}$ (Rowan-Robinson *et al.* 1991).

3. Discussion

Cone diagrams exhibiting the detailed distribution of all the identified IRAS galaxies with $v < 10,000 \,\mathrm{km\,s^{-1}}$ in the region $210° < l < 360°$ give the following picture. In negative latitudes near $l = 240°$ at $v = 2500 \,\mathrm{km\,s^{-1}}$, the Puppis cluster is identified. The best estimate of the position of the cluster is $l = 240°$ b $=-7°$, and its redshift is $2400 \,\mathrm{km\,s^{-1}}$. At l=245°, b $= -5°$ a new concentration of galaxies is visible at $7500 \,\mathrm{km\,s^{-1}}$. Also a void is seen at $l = 240°$ to $270°$, v=2500 to $3500 \,\mathrm{km\,s^{-1}}$. In positive latitudes at $270° < l < 360°$, $v < 5000 \,\mathrm{km\,s^{-1}}$ the extension of the Hydra, Centaurus (both Hi and Lo velocity components) and Antlia clusters are seen. A clump of galaxies is also seen in the positive latitudes near l =280°, v = $5500 \,\mathrm{km\,s^{-1}}$. Further the rich cluster A3627 (l =325°, b =−7°, v=$4500 \,\mathrm{km\,s^{-1}}$) studied in detail by Kraan-Kortewag et al. (1996) could be identified. The GA region $270° < l < 350°$ at $|b| < 15°$ is dominated by overdensity of galaxies representing the extension of Hydra-Centaurus complex in the positive latitudes and the Pavo-Indus complex in the negative latitudes. The broad over density of galaxies seen in our data corresponds with the distant concentration seen at $4500 \,\mathrm{km\,s^{-1}}$ in the Supergalactic Plane survey (Dressler 1991) that covers the same range in longitudes as us, but a larger range in latitude b ($-30°$ to $-10°$ and $+10°$ to $+45°$). We conclude that the overdensity seen in our survey connects the Centaurus-Hydra complex (l=302°, b=−22°) to the major concentration Pavo Supercluster (l=332°, b=−24°) through the Milky Way. Thus our study strengthens the conclusion of the previous surveys (Dressler 1991; Strauss et al. 1990; Visvanathan 1994; Fisher et al. 1995) that this whole region of increased density is responsible for the peculiar velocity field observed in the local region. Also the centroid of this extensive concentration could be situated at the Milky Way itself (l= 320°, b=0°) at v = $4500 \,\mathrm{km\,s^{-1}}$. Support for our conclusion comes from Kolatt et al. (1995) who finds a peak in mass-density in the Potent reconstruction of peculiar velocity data at l= 320°, b=0° at v=$4000 \,\mathrm{km\,s^{-1}}$.

References

Dressler, A., 1991, Astrophys.J.Suppl., 75, 241
Fisher,K.B. et al., 1995, Astrophys.J.Suppl., 100, 70
Fisher,K.B. et al., 1995, Astrophys.J.Suppl., 100, 70
Kolatt, T., Dekel, A. &Lahav, O., 1995, Mon.Not.R.astron.Soc., 275, 797
Kraan-Korteweg, R.C. et al., 1996, Nature, 379, 519
Rowan-Robinson, M., et al., 1991, Mon.Not.R.astron.Soc., 253, 485
Strauss, M.A., et al., 1990, Astrophys.J., 361, 49.
Visvanathan, N.,1994, in "Cosmic velocity fields," ed F.R. Bouchet and M. Lachiez-Rey, editions Frontiers, 125
Yamada, T., et al., 1993, Astrophys.J.Suppl., 89, 57

A DEEP 20 CM RADIO MOSAIC OF THE ESP GALAXY REDSHIFT SURVEY

I. PRANDONI[1], L. GREGORINI[1], P. PARMA[1], G. VETTOLANI[1], H.R. de RUITER[2] AND M.H. WIERINGA, R.D. EKERS[3]
[1] *IRA – CNR, Bologna*
[2] *OAB, Bologna*
[3] *ATNF*

1. The Optical Sample

In two strips of $22° \times 1°$ and $5° \times 1°$ near the SGP Vettolani *et al.* have made a deep redshift survey as an ESO Key Project (the ESO Slice Project galaxy redshift survey). All the galaxies down to $b_J \sim 19.4$ were observed with the OPTOPUS multi-fibre spectrograph on the 3.6 m telescope in La Silla, yielding 3348 redshifts.

The survey has a typical depth of $z = 0.1$. It fully samples the optical luminosity function down to $B = -15$ and various galaxy populations *e.g.*, spirals, ellipticals, dwarfs) are present.

Interestingly, emission lines (OII, Hβ, OIII) have been found in a large fraction of the galaxy spectra ($\sim 47\%$), suggesting strong evolution of the galaxy population in terms of enhanced star formation.

For further information on the ESP galaxy redshift survey see *e.g.*, Vettolani *et al.*, this volume, p. 346.

2. The Radio Survey

In the last two years we used the Australia Telescope Compact Array (ATCA) at 20 cm to image the entire area of the optical survey (27 deg^2). The ATCA supports a mosaic observing mode which allows efficient coverage of large areas of sky by interleaving short observations of a grid of pointings.

Since our optical sample is rather deep but narrow and 'normal' galaxies are typically low-power radio sources, deep radio observations were needed.

Observing times of the order of 1.2 hr/field allowed us to reach a 3σ radio limit of ~ 0.2 mJy, corresponding to a detection threshold of $P < 10^{21}$ W Hz^{-1} at $z < 0.1$.

The observing campaign (34 blocks of 12^h) started in November 1994 and has been completed in January 1996. Data reduction has also been completed.

We have obtained 16 big mosaiced radio maps, each covering 1.7 sq. degr. with spatial resolution of $16 \times 8''$. The noise level, after cleaning, is $\sim 70\,\mu$Jy and is fairly uniform (as needed for statistical studies) within each map and from map to map.

3. The Radio Properties of ESP Galaxies

On the entire region surveyed, we searched for radio emission associated with the redshift survey galaxies. We pushed the search down to a 3σ-threshold (which is allowed when sky positions are known). The searching box ($7.5 \times 7.5''$) was chosen so as to get the best 'identification to contamination ratio.'

Radio emission was found for 524 galaxies, corresponding to a detection rate of 16.4%. Spurious detections are expected to be less than 2% of the total sample of 3196 galaxies searched for, and incompleteness has been estimated to be $\sim 1\%$.

Typically radio detected ESP galaxies are associated to very faint, point-like radio sources ($\sim 86\%$ of them have $S_{peak} < 1$ mJy).

3.1. RADIO EMISSION VS. LINE ACTIVITY

The analysis of the correlations between optical and radio properties of ESP galaxies is now under way. As a first result, we found that a large fraction ($\sim 60\%$) of the radio detections is associated to galaxies showing one or more emission lines. This suggests that in normal galaxies radio emission is mostly induced by star formation, traced by the OII line (Kennicut 1983, Kennicut 1992). The same evidence comes from the cumulative distribution of galaxies with and without emission lines as a function of the radio to optical luminosity ratio, R (Condon 1980). For R values below ~ 100 the probability of being a radio source is higher for galaxies which show line activity than for galaxies which do not (see Figure 1).

4. A New Deep Catalogue of Radio Sources

We are producing a new catalogue consisting of all the radio sources present in the region surveyed above a 6σ-threshold. In a preliminary analysis of a 4 deg^2 area, we detected 360 radio sources above 0.4 mJy. A large fraction

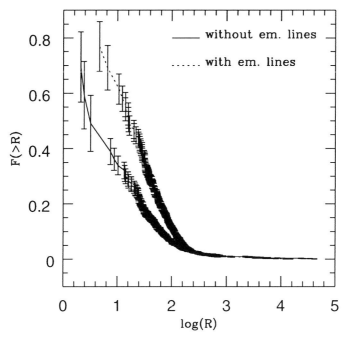

Figure 1. Cumulative distribution for radio detected ESP galaxies with (dot line) and without (solid line) emission lines as a function of the "radio excess" R.

of them ($\sim 40\%$) are sub-mJy objects. This leads us to expect a total number of ~ 2500 radio sources in the entire area observed (27 deg^2) and ~ 1000 sub-mJy sources. This catalogue will therefore be especially useful in studying the sub-mJy population which is still poorly understood (*e.g.*, Condon 1984).

References

Condon, J.J., 1980, Astrophys.J., 242, 894
Condon, J.J., 1984, Astrophys.J., 287, 461
Kennicut, R.C., 1983, Astron.Astrophys., 120, 219
Kennicut, R.C., 1992, Astrophys.J., 388, 310

RADIO EMISSION FROM HIGH REDSHIFT GALAXIES: VLA OBSERVATIONS OF THE HUBBLE DEEP FIELD

E.A. RICHARDS
University of Virginia & NRAO

1. Introduction

To study galaxy populations and their evolution at the highest possible redshifts, a small area of the sky, the Hubble Deep Field (HDF) was imaged to an unprecedented sensitivity of R = 29.5 (Williams *et al.* 1996). As a complement to the HST observations, we have used the VLA at 8 GHz to image an area 5′.4 in diameter (FWHM) centered on the HDF to an *rms* sensitivity of 2 μJy. With a radio resolution of about 3″, we have 33 sources above 9.5 μJy, seven in the 4 arcmin^2 HDF field of which six have clear optical IDs. There are an additional 12 IDs in the HST flanking fields. The optical counterparts of the radio sources are a mixture of ellipticals, spirals, and irregulars, consistent with earlier surveys of comparable depth (Windhorst *et al.* 1995). With a median redshift <z> $\sim$ 1, the radio galaxies we are sampling are somewhat more distant than the classical starbursting galaxies which dominate less sensitive radio surveys. Our HDF identifications are predominately with post-starburst galaxies, moderate power AGN, and blue irregulars (Fomalont *et al.* 1996).

2. Primeval Galaxies or Nascent AGN?

Six faint radio sources in our survey are identified with high redshift galaxy candidates ($z \geq 3$) as evidenced by their optical spectral energy distributions (SED) (Steidel *et al.* 1996). Two of these radio emitters have confirmed redshifts of $z = 2.845$ and $z = 3.158$ (Steidel *et al.*). (One of the former has a measured J,H, and K flux (Cowie *et al.* 1996), and a seven band photometric $z = 2.358 \pm 0.007$.) Thus we may be uncovering a new population of high redshift radio sources. At redshifts $z = 2.3$–3.5, and

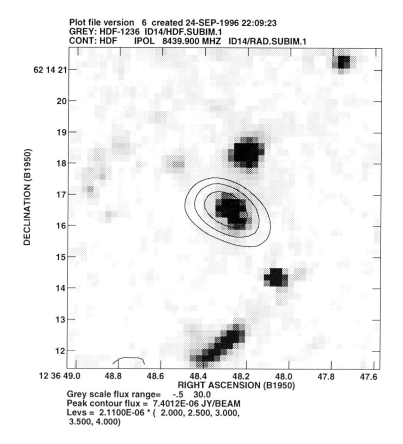

Figure 1. A J = 21.6 galaxy (Cowie *et al.*): the radio contours show the unresolved ($\theta < 3''$) radio source at 8 GHz. With a $z = 2.845$, the radio power is log P = 24.3 implying either a very high star formation rate or a powerful AGN. The radio/optical coincidence is better than $0.2''$ and the identification confidence is > 99%.

assuming a typical radio spectral index of -0.8, the power of these sources at 1 GHz is $10^{25.5}$ W/Hz, typical of strong FRI radio sources. Two different mechanisms may be responsible for this intense radio emission. The sources may be massive, star-forming galaxies, perhaps akin to the local ultraluminous IRAS galaxies. On the other hand, these galaxies may be relatively normal galaxies with an embedded AGN "monster."

References

Cowie, L. L. *et al.* 1996, Astron.J. submitted
Fomalont, E. B., *et al.* 1996, Astrophys.J.Lett. in press
Steidel, E. *et al.* 1996, Astrophys.J.Lett. 462, 17
Williams, R. *et al.* 1996, Astron.J. in press
Windhorst, R. A., *et al.* 1995, Nature 375, 471

EXPOSING NEW COMPONENTS OF THE X-RAY BACKGROUND WITH MULTI-WAVELENGTH SKY SURVEYS

E.C. MORAN
Lawrence Livermore National Laboratory

1. Introduction

The evolving class of objects responsible for the majority of the cosmic X-ray background (XRB) remains at large nearly three and a half decades after the discovery of the XRB. Surveys of sources selected on the basis of their X-ray properties *alone* provide an unbiased picture of the X-ray sky, but to date they have not been ideal for the discovery of rare types of X-ray sources at faint fluxes: large-area X-ray surveys have been restricted to bright sources, while deep X-ray surveys have been limited to very small patches of sky. X-ray selection coupled with another selection criterion, *e.g.*, a radio or infrared detection, complements "pure" X-ray surveys by (1) permitting the exploration of large areas of sky to faint flux limits for types of extragalactic X-ray sources not well represented in other surveys, and (2) assisting the location of the optical counterparts of these X-ray sources. Using this approach, I have searched for new components of the XRB among the faintest X-ray sources detected by the *Einstein Observatory*.

2. The Einstein Two-Sigma Catalog

For certain applications, the limiting signal-to-noise threshold for sources in an astronomical catalog may be considerably lower than commonly accepted values (*i.e.*, 4σ or 5σ), provided that one can determine the statistical reliability of the catalog. To facilitate the study of X-ray sources fainter than those contained in the *Einstein* Medium Sensitivity Survey (EMSS; Gioia *et al.* 1990), we have constructed a new catalog of 46,186 sources and fluctuations exceeding 2σ significance in 2520 high-latitude *Einstein* IPC images. We have employed various tests to validate our source-search algorithm for both high- and low-significance sources, and to identify and remove the small number of spurious sources induced by our detection procedure. Based on the known properties of the *Einstein* optics, the background characteristics of the IPC, and the measured X-ray $\log N - \log S$ relation, we have modeled the number of real sources expected in the catalog in order to evaluate its statistical properties below 4σ significance. Our model suggests that $\sim 28\%$, or $\sim 13,000$, of the sources in the Two-Sigma Catalog are real celestial X-ray sources. This is an increase of ~ 9100 over the number found in previous analyses of the same IPC images. Full details of the manufacture and evaluation of the Two-Sigma Catalog are described in Moran *et al.* (1996).

3. New Components of the Cosmic X-ray Background

The primary motivation for assembling large samples of faint X-ray sources is to search for possible new components of the X-ray background. As a means of selecting real celestial X-ray sources in the Two-Sigma Catalog for further study, we have applied astronomical catalogs at other wavelengths as filters. The cross-correlation of these catalogs with the Two-Sigma Catalog produces samples of hundreds of faint X-ray sources that are reliable at the 90% level. The specific filters we have used are catalogs from surveys of the radio and infrared sky.

Optical spectroscopy of 77 unidentified radio- and IR-selected Two-Sigma Catalog sources has turned up several surprises, illustrating the merits of selecting X-ray sources by a variety of methods. In addition to the types of objects one would expect to find in these samples, we have discovered high-redshift quasars (one at $z = 4.30$), which are absent in the EMSS, X-ray-luminous radio-loud elliptical galaxies with optical spectra devoid of emission lines, and infrared-bright AGNs with optical spectra dominated by starburst galaxy features (Moran *et al.* 1996). Follow-up X-ray observations are being carried out with *ROSAT* and *ASCA* to clarify the nature of these objects and to determine if any of them represents a previously unrecognized component of the X-ray background.

References

Gioia, I. M., *et al.* 1990, Astrophys.J.Supp. 72, 567.
Moran, E. C., Helfand, D. J., Becker, R. H., & White, R. L. 1996, Astrophys.J. 461, 127.

Part 7. Data Processing Techniques

STATISTICAL METHODOLOGY FOR LARGE ASTRONOMICAL SURVEYS

E.D. FEIGELSON[1] AND G.J. BABU[2]
[1] *Dept. of Astron & Astrophys, Pennsylvania State University*
[2] *Dept. of Statistics, Pennsylvania State University*

Abstract. Multiwavelength surveys present a variety of challenging statistical problems: raw data processing, source identification, source characterization and classification, and interrelations between multiwavelength properties. For these last two issues, we discuss the applicability of standard and new multivariate statistical techniques. Traditional methods such as ANOVA, principal components analysis, cluster analysis, and tests for multivariate linear hypotheses are underutilized in astronomy and can be very helpful. Newer statistical methods such as projection pursuit, multivariate splines, and visualization tools such as XGobi are briefly introduced. However, multivariate databases from astronomical surveys present significant challenges to the statistical community. These include treatments of heteroscedastic measurement errors, censoring and truncation due to flux limits, and parameter estimation for nonlinear astrophysical models.

1. Introduction

Between the 16th and 19th centuries, astronomy and statistics were closely allied fields. Many of the foundations of mathematical statistics were laid by astronomers such as Tycho Brahe, Galileo, Tobias Mayer and Adrien Legendre (Stigler 1986). But this relationship weakened during the late 19th century, as statistics turned to applications in the social sciences and industry, astronomy reaped benefits from mathematical physics. A byproduct of this shift is that most astronomers are trained by physicists and receive little or no formal education in statistics. Most astronomers are thus only vaguely aware of the tremendous advances in statistical theory and practice of the last few decades. Similarly, with the notable exception of galaxy clustering studies by Jerzy Neyman and Elizabeth Scott in the 1950–60s, statisticians became unaware of the tremendous developments in astronomy.

Mutual interest in astrostatistics has reemerged during the past decade. The comparison of astronomical data to astrophysical questions is becoming increasingly complex, outpacing the capabilities of traditional statistical methods. About 500 astronomical papers annually have 'statistics' or 'statistical' in their abstracts, yet they rarely refer to contemporary statistical texts or monographs for methodological guidance. Statistical procedures implemented in *Numerical Recipes* (Press *et al.* 1992) are used on a daily basis.

Recent cross-disciplinary efforts in astrostatistics have produced valuable resources. A number of conferences have been held in Europe (*e.g.*, Rolfe 1983; Jaschek & Murtagh 1990; Subba Rao 1997) and the U.S. (Feigelson & Babu 1992; Babu & Feigelson 1997), astrostatistical sessions at large meetings are being organized, an introductory monograph on astrostatistics has emerged (Babu & Feigelson 1996), and the Statistical Consulting Center for Astronomy is active (Feigelson *et al.* 1995; http://www.stat.psu.edu/scca). A monograph on multivariate data analysis, with FORTRAN codes and bibliography of astronomical applications, is very relevant to the issues discussed here (Murtagh & Heck 1987).

2. Statistics and Astronomical Surveys

Large astronomical surveys from new high-throughput detectors and observatories are powerful motivators for more effective statistical techniques. Observatories now frequently generate gigabytes of information every day, with terabyte-size raw databases which produce reduced catalogues of 10^6–10^9 objects. These catalogues, which may include up to dozens of observational properties of each object, often contain heterogeneous populations which must be isolated prior to detailed analysis. Although there are many types of astronomical surveys with many different goals, the statistical problems arising in their analysis can often be divided into three stages. We treat the first two stages very briefly here to concentrate on the final phase.

Reducing raw data into images The treatment of the raw data from the telescope or satellite observatory can be very complex, and has embedded within it many choices of statistical methods. These methods are typically described in internal technical memoranda which are rarely published or publically examined, and sometimes are invisible except for comments in source code. The IRAS Faint Source Survey Explanatory

Supplement (Moshir et al. 1992) offers a glimpse into this complex netherworld: a median filter is applied to reduce noise; outliers are detected to remove particle events; overlapping scans are combined and interpolated; fluxes are estimated with a trimmed mean; signal is extracted with a $S/N \geq 3.5$ criterion; distinct sources are devined by a complicated source merging procedure; sky positions are derived from recursive Kalman filtering and connected polynomial segmant fitting to satellite gyroscope time series data. The IRAS analysis benefits from robust statistcal procedures, such as the median and trimmed mean rather than the usual mean, which have been developed by statisticians over the past 20 years (e.g., Hoaglin et al. 1983). The problems addressed here are specific to each instrument and survey, and general advice has limited value.

Reducing images to catalogues The analysis of astronomical images can be very complicated. In sparsely occupied images from photon-counting detectors (as in X-ray and gamma-ray astronomy), efforts concentrate on detecting sources above an uninteresting background. Methods include maximum likelihood analysis based on the Poisson distribution, matched filtering and Voronoi tesselations. In fully occupied grey-scale images, a wide variety of image restoration methods have been applied to deconvolve point spread functions and reduce noise: least squares fitting; Lucy-Richardson method; maximum entropy and other Bayesian methods; neural networks, Fourier and wavelet filtering (e.g., Narayan & Nityananda 1986; Perley et al. 1989; Hanisch & White 1993; Starck & Murtagh 1994; Lahav et al. 1995). Many of these methods rest upon developments in statistical methodology.

Much work has also been directed to the automated analysis and classification of objects on images, particularly the discrimination of stars from galaxies on optical band photographic plates and CCD images. Each object is characterized by a number of properties (e.g., moments of its spatial distribution, surface brightness, total brightness, concentration, assymetry), which are then passed through a supervised classification procedure. Methods include multivariate clustering, Bayesian decision theory, neural networks, k-means partitioning, CART (Classification and Regression Trees) and oblique decision trees, mathematical morphology and related multi-resolution methods (Bijaoui et al. 1997; White 1997). Such procedures are crucial to the creation of the largest astronomical databases with 1–2 billion objects derived from digitization of all-sky photographic surveys.

The scientific product of multi-wavlength surveys is frequently a large table with rows representing individual stars, galaxies, sources or locations and columns representing observed or inferred properties. Often a single survey effort will produce multi-wavelength results, as in the four infrared bands of IRAS, the five photometric colors of the Sloan Digital Sky Survey, or spectral bands in the ROSAT All-Sky Survey. Analysis of such data is the domain of *multivariate analysis*. We therefore concentrate on multivariate statistical methodology in the following sections.

3. Fundamentals of Multivariate Analysis and Clustering

A multivariate analysis often begins with the computation of simple statistics of the sample: the mean and standard deviation of each variable; linear (Pearson's r) or rank (Spearman's ρ or Kendall's τ) correlation coefficients between pairs of variables. Statisticians often divide each value by the sample standard deviation for that variable (known as 'standardizing' or 'Studentizing' the sample), while astronomers often take a log transform or consider the ratio of two variables with the same units.

Study of pair-wise relationships between variables provides a valuable but fundamentally limited view of the data. A multivariate database should be viewed as a cloud of points (or vectors) in p-space which can have any form of structure, not just planar correlations parallel to the axes. The sample covariance patris S contains information for this more general approach, and lies at the root of many methods of multivariate analysis developed during the 1930-60s. The method most widely used in astronomy is *principal components analysis*. Here the 1st principal component is $e_1^T X$ where e_k is the eigenvector of S corresponding to the kth largest eigenvalue. This is equivalent to finding by the direction in p-space where the data are most elongated using least-squares to minimize the variance. The second component finds the elongation direction after the first component is removed, and so forth. Important applications in astronomy include the stellar spectral classification (Deeming 1964), eludication of Hubble's tuning-fork spiral galaxy classification system (Whitmore 1984), and characterization of relationships between emission lines, broad absorption lines and the continuum in quasar spectra (Francis *et al.* 1992).

In *canonical analysis*, the variables are divided into two preselected groups and the eigenvectors of the cross-sample covariance matrix $S_{11}^{-1/2} S_{12} S_{22}^{-1} S_{21} S_{11}^{-1/2}$ gives the principal linear relationships between the two sets of variables. This might be used to relate stellar metallicity variables with kinematic variables to study Galactic chemical evolution, or stellar magnetic activity indicators with bulk star properties to study dynamo theory.

A sample, collected from one or more multiwavelength surveys, often will not constitute a single type of astronomical object. Variance-covariance structure residing within the matrix S may thus reflect heterogeneity of the sample, rather than astrophysical processes within a homogeneous class. It is thus important to search for groupings in p-space using multivariate clustering or classification algorithms. Dozens of such methods have been proposed. Unfortunately, most are procedural algorithms without formal statistics (*i.e.*, no probabilistic measures of merit) and there is little mathematical guidance which produces 'better' clusters.

Hierarchical clustering procedures produces small clusters within larger clusters. One such procedure, 'percolation' or the 'friends-of-friends' algorithm is a ffavorite among astronomers. It is called *single linkage clustering* obtained by successively removing the longest branches of the unique *minimal spanning tree* connecting the n points in p-space. Single linkage produces long stringy clusters. This may be appropriate for galaxy clustering studies, but researchers in other fields usually prefer *average or complete linkage* algorithms which produce more compact clusters. The many varieties of hierarchical clustering arise because the scientist must chose the metric (*e.g.*, should the 'distance' between objects be Euclidean or squared?), weighting (*e.g.*, how is the average location of a cluster defined?), and criteria for merging clusters (*e.g.*, should the total variance or internal group variance be minimized?).

An alternative method with a more rigorous mathematical foundation is k-*means partitioning*. It finds the combination of k groups that minimizes intragroup variance. However, it is necessary to specify k in advance.

4. Methodological Challenges from Astronomical Surveys

Many astronomical surveys are not amenable to traditional multivariate analysis and classification, and present serious needs for methodological advances by statisticians. Four major difficulties are outlined here.

First, fluxes or other measured quantities are subject to **heteroscedastic measurement errors with known variances**. That is, each variable of each object has an associated measurement of the variable uncertainty, and these uncertainties can differ for each object. Surprisingly, statistical methodology is very poorly developed for such situations. For instance, there is no clustering algorithm that weights points by their known measurement errors. Only the LISREL model of the multivariate linear regression problem can begin to treat known heteroscedastic measurement errors (Jöreskog & Sörbom 1989).

Second, objects may be undetected at one or many wavebands, leading to upper limits or **censored data** in one or many variables. A mature field of statistics known as survival analysis, developed principally for biomedical and industrial reliability applications, has been developed for censored datasets. A suite of survival methods is now widely used in astronomy (Feigelson 1992). However, most survival statistics apply only to univariate problems; Cox regression, the principal multivariate technique, permits censoring only in the single dependent variable. A more general partial correlation coefficient based on Kendall's τ, which permits censoring in any or all variables, has recently been developed for astronomers (Akritas & Siebert 1996). But a full multivariate survival analysis is not yet available.

Third, astronomical surveys are virtually always suffer **truncation** in one or more variables due to sensitivity limits of the telescopes. This can create spurious structure in the variance-covariance matrix and makes the sample distribution a biased estimate of the underlying population. As with censoring, little statistical attention has been directed towards such datasets, except for linear regression problems in econometrics (Maddala 1983).

Fourth, following the traditions of celestial mechanicians of previous centuries, modern astronomers often seek to constrain **parameters of non-linear astrophysical models**. Multivariate methodology was largely developed to assist social sciences and industry where such modeling does not arise. While least-squares regression techniques can be extended from linear to non-linear functions (*e.g.*, the orthogonal distance regression package ODRPACK), such methods fail in the presence of heteroscedastic measurement errors, censoring and truncation. Often the model is so complex, particularly if survey selection effects are included within it, that the results are available only through Monte Carlo simulation. A possible approach to such parameter estimation problems is through half-space projections (Babu & Feigelson 1996).

While these issues have yet to be adequately addressed by statisticians, some recent methodological advances can have significant benefits to astronomers. First, a number of approaches have emerged to facilitate both linear and nonlinear modeling of multivariate datasets. **Projection pursuit** regression uses local linear fits and sigmoidal smoothers to model nonlinear behavior (Huber 1985; Friedman 1987). **Multivariate Adaptive Regression Splines** (MARS) and a variety of similar methods fit the data with multidimensional splines (Friedman 1991). These methods are based on reasonable, but not unique, proceedures for parsimoniously choosing the number of parameters that avoid overfitting the data.

Second, astronomers can greatly benefit from visualization tools that permit powerful exploration of complex multivariate datasets. **XGobi** provides a 2-dimensional 'grand tour' of the database by displaying various projections of the data, with flexible interactive choice of variables, color brushing and projection pursuit options. **ExplorN**, operating on Silicon Graphics computers, gives a d-dimensional grand tour, saturation brushing and parallel coordinate plots. **XNavigator** travels through the database along local principal components.

Finally, we note that this brief paper omits many topics in multivariate statistics with potential importance for astronomy. These include nonparametric methods, Bayesian approaches, outlier detection and robust methods, multicollinearity and ridge regression, goodness-of-fit measures, nonparametric density estimation, wavelet analysis, bootstrap resampling and cross-validation, mathamatical morphology, and many aspects of traditional multivariate analysis. The methodology for understanding multivariate databases is vast and constantly growing.

5. Astrostatistics References and Codes

Multivariate statistics are briefly reviewed in an astronomical context by Babu & Feigelson (1996), and are more thoroughly described (with FORTRAN codes) by Murtagh & Heck (1987). Many monographs presenting multivariate statistics are available, such as Johnson & Wichern (1992).

While commercial statistical packages are the most powerful tools for implementing statistical procedures, a cosiderable amount of software is in the public domain on the World Wide Web. An informative essay on statistical software by Wegman (1997) can be found at
> http://www.galaxy.gmu.edu/papers/astr1.html.

Information on commercial statistical software packages such as SAS, SPSS and S-PLUS is available at
> http://www.stat.cornell.edu/compsites.html.

Significant archives of on-line public domain statistical software reside at StatLib (http://lib.stat.cmu.edu) and the *Guide to Available Mathematical Software* (http://gams.nist.gov). StatLib provides many state-of-the-art codes useful to astronomers such as XGobi, ODRPACK, loess and MARS. Penn State operates the *Statistical Consulting Center for Astronomy* (http://www.stat.psu.edu/scca) for astronomers with statistical questions, and is initiating a site with links to statistical software on the Web (http://www.astro.psu.edu/statcodes).

Acknowledgements

This work was supported by NSF DMS 9626189, NASA NAGW-2120 and NAS 5-32669.

References

Akritas, M.G. and Siebert, J. (1996) Testing for partial association using Kendall's τ with censored astronomical data. Mon.Not.R.astron.Soc, in press.
Babu, G.J. and Feigelson, E.D. (1996) Astrostatistics. Chapman & Hall, London.
Babu, G.J. and Feigelson, E.D. (1997) Statistical Challenges in Modern Astronomy II. Springer-Verlag, New York.
Bijaoui. A., Rué, F and Savalle, R. (1997), in Statistical Challenges in Modern Astronomy II. Springer-Verlag, New York.

Deeming, T.J. (1964) Stellar spectral classification. I. Application of component analysis, Mon.Not.R.astron.Soc 127, 493.
Feigelson, E.D. (1992) Censoring in astronomical data due to nondetections (with discussion), in Statistical Challenges in Modern Astronomy, (E. D. Feigelson & G. J. Babu, eds.), Springer-Verlag, New York. p. 221.
Feigelson, E.D. and Babu, G.J. (1992) Statistical Challenges in Modern Astronomy. Springer-Verlag, New York.
Feigelson, E.D., Akritas, M., and Rosenberger, J. (1995) Statistical Consulting Center for Astronomy, in Astronomical Data Analysis Software and Systems IV (R.A. Shaw et al., eds.). Astron. Soc. Pacific, San Francisco.
Francis, P.J., Hewett, P.C., Foltz, C.B., and Chaffee, F.H., (1992) An objective classification scheme for QSO spectra, Astrophys.J 398, 476.
Friedman, J.H. (1987) Exploratory projection pursuit, J. Amer. Stat. Assn., 82, 239.
Friedman, J.H. (1991) Multivariate adaptive regression splines (with discussion), Annals of Statistics 19, 1.
Hanisch, R.J. and White, R.L. (eds.) (1993) The Restoration of HST Images and Spectra II, STScI: Baltimore.
Hoaglin, D.C., Mosteller, F. and Tukey, J.W. (eds.) (1983) Understanding Robust and Exploratory Data Analysis. Wiley, New York.
Huber, P.J. (1985) Projection Pursuit, Annals of Statistics 13, 435.
Jaschek, C., and Murtagh, F. (eds.) (1990) Errors, Bias and Uncertainties in Astronomy. Cambridge University Press, Cambridge.
Johnson, R.A. and Wichern, D.W. (1992) Applied Multivariate Statistical Analysis, 3rd edition, NJ: Prentice Hall.
Jöreskog, K. G., and Sörbom, D., 1989, LISREL 7 A Guide to the Program and Applications, SPSS Inc., 444 N. Michigan Ave., Chicago IL 60611.
Lahav, O. et al. (1995) Galaxies, human eyes and artificial neural networks, Science 267, 859.
Maddala, G.S. (1983) Limited-dependent and quantitative variables in econometrics. Cambridge University Press, Cambridge.
Moshir, M. et al. (1992) IRAS Faint Source Survye Explanatory Supplement Version 2. IPAC, Pasadena CA.
Murtagh, F., and Heck, A. (1987) Multivariate Data Analysis. Reidel, Dordrecht Neth.
Narayan, R. and Nityananda, R. (1986) Maximum entropy image restoration in astronomy, Ann. Rev. Astro. Astrophys. 24, 127.
Perley, R.A., Schwab, F. and Bridle, A.H. (1989) Synthesis Imaging in Radio Astronomy. Astron. Soc. Pacific, San Francisco.
Press, W.H. et al. (1992) Numerical Recipes: The Art of Scientific Computing, Cambridge University Press, Cambridge.
Rolfe, E.J. (ed.) (1983) Statistical Methods in Astronomy. ESA SP 201, European Space Agency Scientific & Technical Publications. Noordwijk Neth.
Starck, J.-L. and Murtagh, F. (1994) Image restoration with noise suppression using the wavelet transform, Astron.Astrophys. 288, 342.
Stigler, S.M. (1986) The History of Statistics: The Measurement of Uncertainty Before 1900. Harvard University Press, Cambridge.
Subba Rao, T. (ed.) (1997) Applications of Time Series Analysis in Astronomy and Meteorology. Chapman & Hall, London.
White, R.L. (1997) Object classification in astronomical images, in Statistical Challenges in Modern Astronomy II (G.J. Babu & E.D. Feigelson, eds.), Springer, New York.
Whitmore, B.C. (1984) An objective classification systems for spiral galaxies. I. The two dominant dimensions, Astrophys.J 278, 6.
Wegman, E.J., Carr, D.B., King, R.D., Miller, J.J., Poston, W.L., Solka, J.L. and Wallin, J. (1997) Statistical software, siftware and astronomy, in Statistical Challenges in Modern Astronomy II (G.J. Babu and E.D. Feigelson, eds.). Springer-Verlag, New York.

DIGITAL COLOUR MAPPING OF THE SKY FROM SUPERCOSMOS

H.T. MacGILLIVRAY
Royal Observatory Edinburgh, UK

1. Introduction

The major photographic surveys of the sky (undertaken with the ESO, Palomar and UK Schmidt telescopes) still hold enormous potential regarding the information content of the original plate material. The problem to date has been to extract that information in a meaningful and quantitative way, hence giving rise to the large number of digitisation programmes currently underway and which are reported elsewhere in this volume.

Until recently, such digitisation efforts have resulted solely in the production of **monochrome** images of the sky. Now, however, with the SuperCOSMOS scanning programme (aimed at systematically digitising these survey material) we have succeeded in combining the monochrome images from the separate wavebands in order to produce true **colour** digital images (an area which has for long been exclusively the domain of photography).

2. Method and Applications

The method is straightforward and directly similar to the photographic equivalent: digital images from blue, green and red plates are combined to produce the final colour-composited digital image (see Plate 1). Unfortunately, for much of the sky, 'green' plates are not available. However, sensible colours are still achieved using the blue and red data alone.

There are many advantages to having available a digital colour map of the sky. Among them: (i) the colours can be optimised and calibrated; (ii) the images are directly amenable to computer manipulation (pan, zoom, unsharp masking, contrast enhancement, *etc.*; (iv) the colours (in their own right) highlight interesting features worthy of further follow-up study; (iv) interesting objects (*e.g.*, quasars, brown dwarfs) are readily detected.

We plan to distribute the SuperCOSMOS colour maps on CD-ROM.

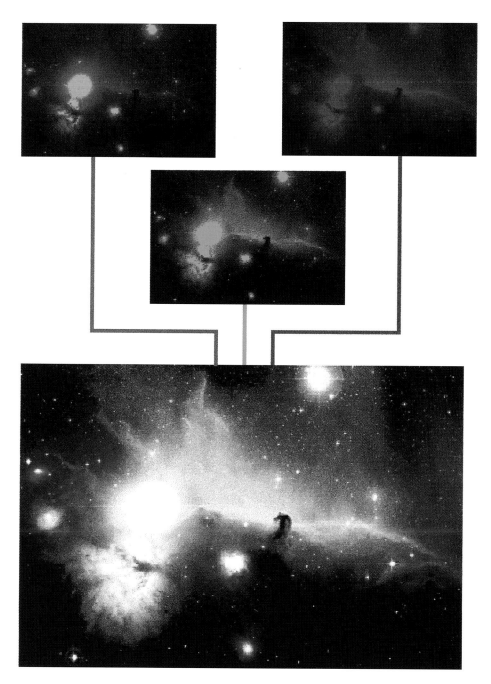

Figure 1. Illustrating the superposition of digital images from the blue, green and red plate data to produce the final colour digital image.

COLOUR EQUATIONS FOR TECHPAN FILMS

D.H. MORGAN[1] AND Q.A. PARKER[2]
[1]*Royal Observatory Edinburgh*
[2]*Anglo-Australian Observatory*

1. Introduction

The UK 1.2m Schmidt Telescope (UKST) at Siding Spring Observatory in Australia has been routinely using panchromatic Tech-Pan emulsion on film as a detector since 1992. Details of how the telescope was adapted to take film and how hypersensitization techniques have been modified for film have been described by Parker (1992). The film has a much finer grain than its equivalent spectroscopic emulsion on glass (IIIa-F) and consequently records images typically one magnitude fainter than IIIa-F when properly hypersensitized. The improved depth, resolution and low noise also permit better differentiation between stars and galaxies (see Parker *et al.* 1994). Consequently Tech-Pan film has become one of the most useful and commonly used emulsions at the UKST for a variety of passbands and may in time be used for a full or partial sky survey.

Although its panchromatic spectral sensitivity broadly matches IIIa-F, it is important to determine what, if any, colour terms are present in the passbands commonly used with Tech-Pan film and see whether they differ from those found for IIIa-F by Blair & Gilmore (1982).

2. Observations

Four short exposure (2–5 min) Tech-Pan films were taken with the UKST for this purpose in the R waveband using the OG590 filter. The field was F284 (centre $20^h\ 04^m$, $-45°$) which includes the Harvard Standard Region E8.

The films were measured using the SuperCOSMOS machine in Image Analysis Mode at the ROE (Miller *et al.* 1992). Only the central 3° were measured in order to avoid vignetting problems,

3. Analysis

The SuperCOSMOS instrumental magnitudes were calibrated using low-order polynomials fitted to 42 standard stars measured in the Cousins system (Menzies *et al.* 1989) and not thought to be variable. A few stars deviated from the calibaration fit by more than 2.5σ as a result of image contamination from nearby stars or minor plate defects. These were excluded from the calibration. The differences between the standard photoelectric magnitudes and the magnitudes calculated from the calibrated SuperCOSMOS data were then plotted against photoelectric colour and linear fits were constructed. The differences between the four measurements of the colour term were commensurate with the error of each fit.

The mean colour term obtained is:

$$R-R' = (-0.033 \pm 0.010) \times (R-I) : \quad -0.1 < (R-I) < +0.9$$

It is small and close to the value derived by Blair & Gilmore (1982) for IIIa-F on glass: $(+0.000 \pm 0.050) \times (R-I) : -0.0 < (R-I) < +0.9$. Therefore, replacing IIIa-F with Tech-Pan film will not introduce any significant photometric problems for R-band photometric programmes.

Acknowledgements

The authors would like to thank the staff of the UK Schmidt Telescope Unit (AAO) and of the SuperCOSMOS Unit (ROE) for their help in producing the films and measurements.

References

Blair M. and Gilmore G. (1982) Publ. Astron. soc. Pacific, 94, 742
Miller L., Cormack W., Paterson M., Beard S. and Lawrence L., (1995) In Digitised optical Sky Surveys, Astr Sp Sci Lib, 174, p. 133, eds H.T. MacGillivray & E.B. Thomson, Kluwer
Menzies J.W., Cousins A.W.J., Banfield R.M. and Laing J.D. (1989) SAAO Circ, No 13, p1
Parker Q.A. (1992) Kodak Tech Pan Estar-based emulsion, AAO Internal Report
Parker Q.A., Morgan D.H. and Phillipps S. (1993) In IAU Commission 9 Working Group on Wide-Field Imaging Newsletter, No 3, p60
Parker Q.A., Phillipps S., Morgan D.H., Malin D.F., Russell K.S., Hartley M. and Savage A. (1994) In Astronomy from Wide-Field Imaging, Proc. IAU Symp. No 161, p. 129, eds H.T. MacGillivray *et al.*, Kluwer

AN OBJECTIVE APPROACH TO SPECTRAL CLASSIFICATION

A.J. CONNOLLY AND A.S. SZALAY
Johns Hopkins University

1. Introduction

The next generation of spectroscopic surveys, both Galactic and extragalactic (*e.g.*, SDSS, 2dF), present the challenge of classifying spectra in an efficient and objective manner. The standard approach to this problem has been to visually classify spectra based on a number of spectral features (*e.g.*, the equivalent widths of emission lines). The size of new spectral surveys ($> 10^6$ galaxies) and the desire to compare the luminosity and environments of galaxies with their spectral properties make these techniques infeasible. We describe here an automated classification scheme that is being developed for the SDSS.

2. The Karhunen-Loève Transform

The application of Principal Component Analysis to multivariate astrophysical data has been described in detail in a number of publications (Efstathiou and Fall 1984, Connolly *et al.* 1995). For details of the Karhunen-Loève transform (KL; Karhunen 1947, Loève 1948) and its use in the analysis of spectroscopic data we refer the reader to Connolly *et al.* (1995). Here, we describe an extension to these standard techniques that applies to censored data (*i.e.*, data that contain spectral regions where the flux is unknown).

We can describe an observed spectrum, $f'_i(\lambda)$, in terms the true spectrum, $f_i(\lambda)$, and a mask, $m_i(\lambda)$. The mask is defined to be zero where the data are unknown and one in wavelength regions where the data are secure. If we decompose this spectrum onto a previously defined eigenbasis,

$e_j(\lambda)$, then, $f'_i(\lambda) = m_i(\lambda) \sum_{i=1}^{n} y_{ij} e_j(\lambda)$, where, y_{ij}, are the decomposition coefficients.

Clearly the observed spectrum and the eigensystem are not orthogonal over the masked spectral range. Therefore, projecting the observed spectrum onto this eigensystem will lead to biased coefficients. We, therefore, determine an error, E, that describes the difference between $f'_i(\lambda)$ and $\sum_{j=1}^{n} y_{ij} e_j(\lambda)$ over the wavelength region where the data are good (Everson and Sirovich 1995), $E_i = \int d\lambda \sum_i [f'_i(\lambda) - \sum_{j=1}^{n} y_{ij} e_j(\lambda)]^2$. By minimizing E with respect to y_{ij} we can define, $y_i = M_{ij}^{-1} g_j$, where $M_{ij} = \int_{\lambda \forall m(\lambda)=1} d\lambda e_i(\lambda) e_j(\lambda)$ and $g_j = \int_{\lambda \forall m(\lambda)=1} d\lambda f'_j(\lambda) e_j(\lambda)$.

In Figure 1 we show that these corrected coefficients can be used to optimally interpolate across the masked spectral regions. The typical rms deviation between the true and interpolated regions is $\sim 10\%$). By iteratively replacing the masked regions with the interpolated values the above analysis can be extended to the case where we wish to derive the eigensystem from a sample of galaxy spectra that cover different spectral ranges (Everson and Sirovich, 1995).

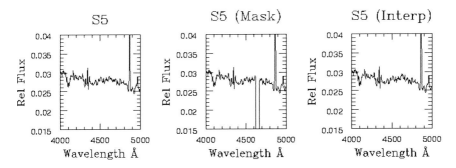

Figure 1. Spectra with missing data, *e.g.*, due to sky lines, can be optimally interpolated across using the derived eigensystem. In this figure (a) shows the original spectrum, (b) the spectrum with a 40 Å region masked out and (c) the interpolated spectrum.

3. A Spectral Classification Scheme

From the decomposition coefficients described above, we can classify galaxy spectra into different spectral types. Figure 6 in Connolly *et al.* (1995) shows this for the Kinney *et al.* (1996) spectral energy distributions. By plotting the mixing angles between the first three eigenspectra we find that galaxies from elliptical to starburst form a linear sequence *i.e.*, to first order, galaxies separate out into a simple monotonic spectral sequence.

To extend this analysis to grouping galaxies into subclasses along this spectral sequence (analogous to the sub groups in stellar classification) requires that we consider individual features within each spectrum (*e.g.*, lines or line complexes). This naturally results in additional dimensions and increased complexity to any classification scheme. While this is straightforward to undertake statistically the question remains how do you describe the classification of a galaxy spectrum in a simple and physical manner.

For the SDSS we choose to approach this problem by building on the work of Morgan and Mayall (1957). In their Yerkes Y system (and revisions thereof) they describe the spectral types of galaxies in terms of stellar classifications (*e.g.*, classes of galaxies go through A, AF, F, G, and K stellar types). In defining the subclasses they consider only a limited spectral range for each galaxy (3850–4100 Å). From the Kinney *et al.* (1996) eigenspectra we have shown that the statistical spectra derived from the KL analysis separate naturally into different stellar types (G, O and A). To subdivide further we apply the KL analysis to smaller spectral regions (*e.g.*, the 4000 Å break). In such a way a hierarchical classification of galaxy spectra can be built up.

4. Conclusions

(1) The Karhunen-Loève transform provides an elegant method for the derivation of eigenfunctions that describe multidimensional datasets. These techniques have been modified to account for data with imperfect spectral coverage, *e.g.*, due to sky lines or different rest-frame wavelength coverage. From this, an optimal interpolation scheme can be derived to account for the masked spectral regions.

(2) Decomposing spectra into whose eigenspectra their coefficients can be used as an objective spectral classification scheme. This natural classification can be described in terms of a one parameter problem and, in the mean, is correlated with independent morphological classifications.

(3) The resolution of any classification can be optimized to a particular range of spectral types by adopting an iterative scheme where we vary the normalizations of the input SEDs.

References

Connolly, A.J., Szalay, A.S., Bershady, M.A., Kinney, A.L. and Calzetti, D. 1995. A.J. 110, 1071.
Efstathiou, G. and Fall, M.S., 1984. Mon. Not. R. astron. Soc., 206 453
Everson, M. and Sirovich, L., 1995. J. Opt. Soc. Am. A, 12, 1657
Karhunen, H., 1947. Ann. Acad. Science Fenn, Ser. A.I. 37
Kinney, A.L., Calzetti, D., Bohlin, R.C., McQuade, K., and Storchi-Bergmann, T., 1994. Astrophys. J., 467, 38
Loève, M., 1948. Processus Stochastiques et Mouvement Brownien, Hermann, Paris, France

GSPC-II: A CATALOG OF PHOTOMETRIC CALIBRATORS FOR THE SECOND GENERATION GUIDE STAR CATALOG.

M. POSTMAN[1], B. BUCCIARELLI[2],
C. STURCH[1,3], T. BORGMAN[1], R. CASALEGNO[2],
J. DOGGETT[1] AND E. COSTA[4]
[1] *Space Telescope Science Institute*
[2] *Osservatorio Astronomico di Torino*
[3] *Computer Sciences Corporation, Astronomy Programs*
[4] *University of Chile*

1. Introduction

The Guide Star photometric Catalog (GSPC-I; Lasker *et al.* 1988) is an all-sky set of photoelectrically determined BV sequences created to provide photometric calibrators for the Guide Star Catalog (Paper-I: Lasker *et al.* 1990, Paper-II: Russell *et al.* 1990, Paper-III: Jenkner *et al.* 1990). Although the GSPC-I has been the basis of preliminary photometric calibrations for the Digitized Sky Survey (DSS; Doggett *et al.* 1995), its relatively bright cutoff at about 15th magnitude limits its capability to support calibration of sky surveys, *e.g.*, the new GSC-II (McLean *et al.* 1996, this volume, p. 431).

2. The GSPC-II Project

The goal of the GSPC-II project is to provide CCD stellar sequences with 5% photometric accuracy in the Johnson-Kron-Cousin B, V, and R passbands down to 18 to 20 mag, near the centers of all Schmidt survey plates for the calibration of the digitized scan data as part of the GSC-II project. Positional precision of the observations (*i.e.*, for object identification) should be consistent with GSC-I, typically 0.5″. For the north, there are 584 northern ($\delta \geq 6°$) sequences, centered on the 6°-grid of the original National Geographic Society—Palomar Observatory Sky Survey. For the south, there are 894 ($\delta \leq 0°$) sequences, centered on the 5°-grid of the UK SERC Southern Sky Survey and its equatorial extension. The SES and SERC-EJ plate centers to be used in the southern GSC-II are unchanged from GSC-I. The POSS-II plate centers, however, follow the convention used for the southern surveys and are centered on a 5°-grid. The locations of the GSPC-I sequences and their GSPC-II extensions are adequate for calibrating most of the POSS-II plates; supplemental observations are planned for the rest.

3. Program Status and Database Construction

Supported by a continuing collaborative international effort, the GSPC-II is nearing completion. The northern survey has benefited from the cooperation of several observatories—Mont Megantic, Wise, KPNO, Lowell, McDonald, Mt. Hopkins and Mt. Laguna—and it is, at present, over 90 percent complete. In the southern hemisphere the program has been carried on thanks to long term observing status at CTIO and ESO observatories. To date about 70 percent of the southern sequences have been obtained. At the present rate of four observing runs per year, it is expected that the observations will be completed at the beginning of 1998. An intermediate catalog of reduced data, containing multiple object photometry, has been compiled. Inter-observatory comparisons as well as monitoring of common and calibration fields allow one to check the photometric quality of GSPC-II sequences, at the same time ensuring all-sky data homogeneity.

The Distributed Information Retrieval from Astronomical files (DIRA) database has been selected for storing and accessing GSPC-II photometry. OATo staff are maintaining the system for this purpose and are currently populating the database with all usable observations acquired by both ST ScI and OATo staff. Error analysis and status reporting for the entire project will be supported with appropriate database utilities.

References

Doggett, J., et al. 1995, "Astronomical Data Analysis Software and Systems V", Astron.Soc.Pacific Conference Series, 101, 159, Jacoby and Barnes (eds)
Lasker, B. M., et al. 1988 Astrophys.J. Suppl. 68, 1
Lasker, B. M., et al. 1990, Astron.J. 99, 2019
Jenkner, H., et al. 1990, Astron.J. 99, 2081
McLean, B., et al. 1997, this volume, 431
Russell, J. L., et al. 1990, Astron.J. 99, 2059

TECHNIQUES FOR SCHMIDT PLATE REDUCTIONS WITH APPLICATION TO GSC1.2

J. E. MORRISON AND S. RÖSER
Astromisches Rechen-Institut, Heidelberg, Germany

1. Introduction

Over the last few years there has been considerable progress in developing successful algorithms for obtaining astrometric quality positions from Schmidt plates which compensate for deficiencies of the polynomial approach; the *sub-plate* method (Taff 1989), the *mask* method (Taff et al. 1990), the *collocation* method (Bucciarelli et al. 1993) the *filter* method (Röser et al. 1995) and a filter weighting according to the *method of infinitely overlapping circles* (Morrison et al. 1996b.) However, none of these nor any other studies have investigated the magnitude dependence of the position estimates outside the magnitude range of standard reference catalogs. Often Schmidt plates cover the magnitude range from 6^m to 19^m (for the GSC it is 6^m to 15^m). Presently available reference catalogs, however, have a limiting magnitude of $V \approx 10^m$. Therefore, no magnitude dependent term for fainter stars can be reliably found by reducing the measurements based only on comparisons with reference stars.

2. Method of Removing Systematic Errors

What is needed is a dense all-sky reference catalog covering the same magnitude range of the Schmidt plates with sufficiently accurate positions at the epoch of the plate material. The Astrographic Catalog (AC) is an interesting prospect, it contains 10 millions measurements of roughly 4 million stars with a $0.3''$ rms error per star position, and a limiting magnitude of 12^m (but in many cases it is as faint as 13.5^m). However, the AC has a mean epoch of 1903 and contains no proper motions; therefore there are roughly 80 years difference between the epochs of the two catalogs and neither catalog contains proper motion information. We have developed a method of reduction where this difference in the epochs is *inconsequential*.

The AC was used in two steps; to remove the systematic errors which are a function of magnitude and radial distance from the plate center and to remove those that are only a function of location on the plate. In both steps the average systematic errors were found by 'stacking' all the AC plates onto the GSC plate-based coordinate system. For each GSC/AC match the differences in position were binned and averaged in the GSC plate system. The grid pattern for the two steps was different, but in both cases the large number of AC stars and the high degree of overlap of the AC plates (50% in α and 50% in δ) resulted in each bin containing an enormous number of matches (at least, tens of thousands). Therefore, we can assume that after the averaging the random errors in the AC and GSC positions cancel out, leaving only the signal of the average systematic distortions in the GSC plates. Note also that the systematic errors on the AC plates cancel out because the the AC and GSC plate centers are uncorrelated.

Part of the effects of the unaccounted proper motion are also taken care of by the averaging process. The components of proper motion can be separated into the peculiar motions of the stars plus the induced proper motion arising from the galactic rotation and solar motion. By assuming the peculiar motion of the stars to be random, after the averaging process, the effects caused by these motions cancel out leaving only those from the galactic and solar motion. We will show later, in our method of utilizing the AC, these physical effects either cancel or subtract out. As a consequence the final positions at epoch are unaffected by any physical motions of the stars in the Galaxy.

Numerous tests we have performed on the magnitude effect have proven that the overwhelming part of it is radial. Therefore, concerning this effect, we are only interested in determining for each GSC/AC match the difference (GSC−AC) in radial distance from the GSC plate center. For a set of specified magnitude ranges this difference is found for all the matches and then binned and averaged in thin rings (2.7′) centered on the center of the GSC plate-based coordinate system. In this process the proper motions induced by galactic and solar motion cancel out. The magnitude effect is then corrected by spline fits to the radial difference between the GSC and AC positions as a function of distance from the plate center and magnitude.

After the above correction was applied to the GSC positions, the removal of the position-only dependent systematics was accomplished by constructing a vector mask of residuals (*i.e.*, a grid of points in the GSC plate-based system containing the binned and averaged GSC−AC positional residuals). This mask represents the average systematic distortions plus the constant from the remaining proper motions. From previous studies using the PPM, we know that the mask of distortions is nearly zero in the central region. The average residual vector in this region is simply the constant caused by the galactic and solar motion over the 80 years. Subtracting this value from all the grid points, leaves the mask representing the mean positional errors on the Schmidt plates. The resulting mask can then be applied to GSC positions based on their location on the plate.

In conclusion, using a combination of the filter and mask method we have removed the mean systematic deformations which have plagued the GSC Schmidt plates (and other similar Schmidt plates). Also using a rather unconventional approach, we have developed a method where a nonuniform, inhomogeneous, imprecise, single early epoch reference catalog (the AC) can be a *powerful* tool for removing the mean of magnitude dependent systematics.

References

Bucciarelli, B, Lattanzi, L. G., and Taff, L. G. 1993, Astrophys.J., 103, 1689
Lasker, B. M., *et al.* 1990 Astron.J., 99, 2019
Morrison, J. E., *et al.* 1996a, Astron.J., 111, 1405
Morrison, J. E., Smart, R. L., Taff, L. G., 1996b, submitted to Mon.Not.R.astron.Soc.
Röser, S., Bastian, U., and Kuzmin, A. V. 1995, I.A.U. Colloquium 148, Astron.Soc.Pacific Conference Series, Vol 84, ed. J. M. Chapman *et al.*
Taff, L. G. 1989, Astron.J., 98, 1912
Taff, L. G., *et al.* 1990, Astrophys.J., 353, L45
Taff, L. G., Lattanzi, M. G., and Bucciarelli, B. 1990, Astrophys.J., 358, 359

WADING THROUGH THE QUAGMIRE OF SCHMIDT-PLATE COORDINATE SYSTEMATICS

C.-L. LU[1], I. PLATAIS[1], T.M. GIRARD[1],
V. KOZHURINA-PLATAIS[1], W.F. VAN ALTENA[1],
C.E. LÓPEZ[2] AND D.G. MONET[3]
[1] *Yale University Observatory, New Haven, CT 06511, USA*
[2] *Félix Aguilar Observatory, San Juan, Argentina*
[3] *US Naval Observatory, Flagstaff, AZ 86002, USA*

1. Magnitude-dependent Errors

We attempted to quantify the magnitude-dependent systematics in a sample of Schmidt plates by comparison to positions from the Yale/San Juan Southern Proper Motion program which offers star positions and absolute proper motions down to $B = 18$ with a mean density of about 50 stars per square degree and a positional accuracy of 0.1″ (Platais et al. 1995).

The Schmidt-plate coordinates used in this study are those from the USNO_A0.9 all-sky catalog (Monet 1996). In total for 53 SPM fields we extracted 46 corresponding Schmidt survey fields (30 UKST and 16 Palomar) with more than 1000 stars in common. First, the variations in differences "SPM–USNO_A0.9" with position were removed using the infinite overlapping circle method (Taff et al. 1990). These variations undoubtedly were inherited from the Guide Star Catalog which served as the reference catalog. Second, the post-geometric-correction residuals plotted as a function of magnitude clearly indicate a trend with magnitude. Morrison et al. (1996) reported the existence of fixed-pattern, magnitude-dependent systematics in the Guide Star Catalog positions. In addition to that, our comparisons unambiguously show a plate-to-plate *variable* magnitude equation.

2. Conclusions

Magnitude-dependent systematics, which vary from plate to plate, are present in Schmidt-plate based coodinates at a level indicated in Figure

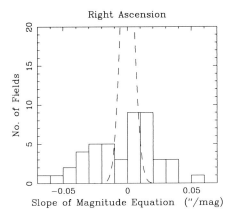 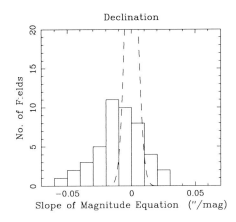

Figure 1. The distribution of fitted magnitude-equation slopes for all 46 Schmidt fields. The dashed curve shows the expected distribution in the case of *no* magnitude equation.

1. Assuming the geometrical distortions are corrected using a reference catalog at $B \approx 11$, the magnitude-dependent offsets will be largest at the faint end. Similarly, the absolute proper motions are expected to suffer from magnitude equation, although this time for the bright stars. If ignored, the magnitude equation has obvious implications for faint positional reference catalogs and for proposed galactic structure studies.

References

Monet, D. G. 1996. DDA meeting (April 15–17) held in Washington, DC, abstract
Morrison, J. E., Röser, S., Lasker, B. M., Smart, R. L. and Taff, L. G., 1996. Astron. J., 111, 1405.
Platais, I., Girard, T. M., van Altena, W. F., Ma, W.-Z., Lindegren, L., Crifo, F. and Jahreiß, H., 1995. Astron. Astrophys., 304, 141.
Taff, L. G., Bucciarelli, B. and Lattanzi, M. G., 1990. Astrophys. J., 361, 667.

PROGRESS IN WIDE FIELD CCD ASTROMETRY

N. ZACHARIAS
USRA/USNO Washington DC, USA

1. Astrometric Performances of 5 Telescopes

Instrumental parameters and astrometric results from five telescopes are summarized in Table 1. The KPNO 0.9m has field-corrector optics. The CTIO 0.9m is a classical Cassegrain. The 4-meter telescopes both have doublet field-correctors. The USNO 0.2m (8-inch) astrograph has a 5-element lens which is designed for a 9° flat field of view for photographic plates. A bandpass of 570–650 nm is used at the USNO 0.2m while most frames with the other telescopes have been taken through a Gunn r (600–710 nm) filter. Stellar images on the CCD frames have been fitted with a 2-D circular symmetric Gaussian profile, giving the centroiding error σ_{fit}. Plots of σ_{fit} vs. instrumental magnitude look similar in shape for all telesopes. The asymptotic fit precision, σ_{afp}, is the limit in σ_{fit} achieved for bright stars, given in milli pixels (mpx) in Table 1. This centering error is overestimated by an amount depending on the deviation of the real image profile from the model function (Winter 1997).

A measure of the repeatability of the observations, σ_{ff}, has been obtained from frame to frame comparisons of centrally overlapping frames using a linear transformation model. The error σ_{atm} due to the turbulence in the atmosphere accounts for about 50% of σ_{ff} for $\approx$ 200 sec exposure times (Zacharias 1996).

For the reflecting telescopes in this investigation, a significant part of the general field distortion pattern is the third order optical distortion (D3) term. D3, σ_{D3} and the center of distortion have been determined by x, y-transformations of frames overlapping by about 50% in area (Zacharias *et al.* 1995). A significant offset of the center of the D3 term with respect to the center of the CCD has been found for each of the 0.9m telescopes, which varies from one observing run to another. The small and constant

TABLE 1. Telescope parameters and astrometric results

telescope	pixel μm	scale $''/px$	FOV arcmin	σ_{afp} mpx	σ_{ff} mpx	σ_{ff} mas	D3 px/px^3	σ_{D3} px/px^3
CTIO 4m	24	0.43	15 × 15	12	13	6	−1.53e-9	0.02e-9
KPNO 4m	24	0.47	16 × 16	13	15	7	−2.07e-9	0.01e-9
CTIO .9m	24	0.40	13 × 13	16	15	6	−0.45e-9	0.03e-9
KPNO .9m	24	0.68	23 × 23	12	11	7	−0.49e-9	0.02e-9
USNO .2m	9	0.90	23 × 15	12	11	10	2.2e-13	<0.2e-13

TABLE 2. Characteristics of a new astrometric survey

CCD detector	4k × 4k	KODAK
readout noise	15	e^- / pixel
field of view	60 × 60	arcmin
exposure time	120	seconds, guided
observing throughput	15	frames/hour
req. observing time	≈ 3500	hours/hemisphere, 2-fold
	≤ 2	year at a good site
estimated catalog accuracy	20	mas, R = 6...13.5 mag
	30	mas, R = 15.0 mag
	70	mas, R = 16.0 mag
average	2000	stars / frame
total	40	million stars / hemisphere
long exposure access	≥ 100	RORF sources/hemisphere

D3 term for the USNO astrograph has been determined from external plate solutions using a 5 degree field.

The mean FWHM, mean image elongation and σ_{afp} are important parameters for assessing the astrometric quality of the CCD observations. The USNO astrograph is more diffraction than seeing limited. Thus larger than average FWHMs indicate a focus setting problem while larger than average mean image elongations indicate a guiding problem. For the KPNO and CTIO 0.9m telescopes the mean image elongation is strongly correlated with focus setting because of astigmatism present at the field edges, particularly at the CTIO 0.9m. A large FWHM with these and the 4-meter telescopes is an indication for poor seeing.

2. High Precision Astrometric Catalog

Planning has begun at the USNO for a global, high precision, astrometric sky survey using the 0.2m astrograph equipped with a CCD camera (Table 2) (Gauss *et al.* 1996). Current studies (Zacharias & Rafferty 1996, Zacharias 1997) show that the projected accuracy, which includes an estimate of the systematic errors, is achievable.

3. Astrometric Calibration Fields

Following is a recommendation for standard fields to be used for astrometric calibrations at current epochs, selected from the radio-optical reference frame (RORF) list. Most fields are close to the galactic plane (small b). Coordinates of the field centers are for J2000. A status flag (f) indicates particularly good (g) or poor (p) observational coverage as of today.

```
north                  equator                 south
  h m s    d ' "  b f    h m s    d ' "  b f    h m s    d ' "  b f
010246  +582411   5 p  033931  -014636  43    064814  -304420  14
023752  +284809  29    050113  -015914  25    092752  -203451  21
064632  +445117  18 g  074554  -004418  12    111827  -463415  13
095457  +174331  48    090910  +012136  31    142756  -420619  17
183250  +283338  17 g  165833  +051516  27    170053  -261052  10 p
211529  +293338  13    210139  +034131  27    191110  -200655  13 p
220315  +314538  19 g  225718  +024318  49
```

References

Gauss,F.S., Zacharias,N., Rafferty,T.J., Germain,M.E., Holdenried,E.R., Pohlman,J. and Zacharias,M.I. 1996, Bull.Am.Astron.Soc. in preparation
Winter,L. 1997, diss. Univ.of Hamburg, in preparation
Zacharias,N. de Vegt,C. Winter,L. Johnston,K. 1995, Astron.J., 110, 3093
Zacharias,N., Rafferty,T.J. 1995, Bull.Am.Astron.Soc,, 27, 1302
Zacharias,N., 1996, in press Publ.Astron.Soc.Pacific, December issue
Zacharias,N., 1997, in prep. for Astron.J.

LINKING THE RADIO AND OPTICAL FRAMES WITH MERLIN

S.T. GARRINGTON [1], R.J. DAVIS[1],
L.V. MORRISON[2] AND R.W. ARGYLE[2]
[1] *Nuffield Radio Astronomy Laboratories, Jodrell Bank, UK*
[2] *Royal Greenwich Observatory, UK*

Abstract. MERLIN positions of 12 radio stars are used to link the provisional Hipparcos reference frame to the International Celestial Reference Frame. The accuracy of the link using these radio stars is 2.3 milliarcseconds. Further observations are planned to check the accuracy of the link in the future.

1. Introduction

The challenge for astrometrists working at wavebands other than radio is to link their reference frames to the International Celestial Reference Frame (ICRF) which, though very accurate, is defined by VLBI positions of several hundred quasars. These are usually very weak or even absent at other wavelengths.

The high-precision Hipparcos astrometric catalogue is the obvious choice for the optical counterpart of the ICRF. This catalogue has positions, proper motions and parallaxes of 120,000 stars with a mean accuracy of 1.5 mas at 1991.25.

There are several methods of establishing the link and these have been reviewed by Lindegren and Kovalevsky (1995). Potentially, one of the most accurate methods is to compare the optical positions and proper motions of radio stars in the Hipparcos catalogue with their radio counterparts which are measured with respect to the ICRF. The difficulty is that apart from a few stars, the sources are weak and very variable at radio wavelengths, with peak intensities of less than 10 mJy. Notwithstanding, a team led by J.-F. Lestrade began an observational campaign in 1984 to measure the differential positions of radio stars using VLBI phase-referencing from adjacent ICRF sources. By this method, the radio positions and proper motions of 11 radio stars in the northern hemisphere have been determined and the link of Hipparcos to the ICRF established with an accuracy of 0.6 mas at the epoch 1991.25 and the residual rotation reduced to the level of 0.3 mas yr^{-1} (Kovalevsky and Lindegren, 1996).

In order to obtain an independent check on this VLBI solution for the link, and with a view to extending the number of stars used, an astrometric programme using MERLIN was established.

We report here on the astrometry of 13 radio stars, nine of which are not in the VLBI programme. The stars observed are given in Table 1.

2. Radio Star Candidates

The targets in this investigation were chosen for several reasons:
- A small radio emission cross-section (~1–2 mas) *i.e.*, single stars or close binaries.
- Known radio emitters at the several mJy level.
- Preferably not in the VLBI list of Lestrade *et al.* (1994).

TABLE 1. Post-fit residuals, ICRF-MERLIN

Star	HIC	$\Delta\alpha\cos\delta$ (mas)	$\Delta\delta$ (mas)	Star	HIC	$\Delta\alpha\cos\delta$ (mas)	$\Delta\delta$ (mas)
LSI 61° 303	12,469	−3.1	−0.3	σ^2 CrB	79,607	0.0	+1.7
Algol[1]	14,756	−3.7	+2.4	29 Dra[2]	85,852	−	−
HD22403[3]	16,879	+4.9	−10.3	BY Dra	91,009	−1.4	−1.8
EI Eri[4]	19,431	−9.7	−14.8	FF Aqr	108,644	−3.8	+0.7
DM UMa[5]	53,425	+14.9	+1.3	λ And	116,584	−1.6	+4.8
FK Com	65,915	+6.6	−8.8	II Peg	117,915	−0.9	−0.5
HR5110	66,257	−1.3	−2.3	II Peg	117,915	+4.3	−4.9

[1] The line of nodes in Pan *et al.* (1993) is wrong by 180°.
[2] 29 Dra = DR Dra; rejected because of relatively high χ^2 value notified by Hipparcos team. (It is a 2.5 year period spectroscopic binary).
[3] HD22403; lowest signal to noise ratio.
[4] EI Eri = HD26337; resolution in declination is worst near the equator, although FF Aqr produces a small residual.
[5] DM UMa; calibrator is a double source, 33 mas across.

3. Discussion

The MERLIN solution for the (x, y, z) axes of the frame can be compared with that for the finally adopted version of the Hipparcos catalogue. These offsets in milliarcseconds, in the sense MERLIN *minus* Hipparcos, are:

$$\Delta\epsilon_x = +1.5 \pm 2.3; \qquad \Delta\epsilon_y = -0.6 \pm 2.0; \qquad \Delta\epsilon_z = +0.5 \pm 2.2.$$

The final orientation of the published Hipparcos catalogue is very close to the VLBI determination. Therefore, the offsets above effectively show how close the MERLIN solution is to that of the VLBI. Only three stars are common to the VLBI and MERLIN datasets (LSI 61° 303, HR5110, Algol); so the two solutions are virtually independent and their close agreement lends confidence to the stability of the link, which could have been distorted by significant offsets between the optical and radio emission in some of the binary stars.

4. Conclusions

The accuracy of the link has clearly demonstrated the potential of MERLIN for differential astrometry over several degrees. In the best cases MERLIN may achieve sub-mas accuracy, but even for weak radio stars of only a few mJy and separations of several degrees, MERLIN can achieve accuracies of a few mas.

The VLBI solution for the rotation of the Hipparcos frame relative to the ICRF has an uncertainty of $\pm 0.3\,\mathrm{mas\,yr^{-1}}$. The maintenance of the Hipparcos frame will therefore depend on regular checks on its alignment to the ICRF. By the year 2000 the misalignment could reach $\sim$3 mas. By doubling the number of radio stars to $\sim$20, MERLIN will be capable of measuring such an offset.

Acknowledgements

Hans Schrijver provided the pre-publication Hipparcos information on possible astrometric binaries in the MERLIN list of radio stars, and Lennart Lindegren provided the astrometric information on the Algol triple system. MERLIN is a national facility operated by the University of Manchester on behalf of PPARC.

References

Kovalevsky, J and Lindegren, L. 1996 Astron.Astrophys. (in prep).
Lestrade, J.-F., et al. 1994 IAU Symp.166, pp 119–126.
Lindegren, L. and Kovalevsky, J. 1995 Astron.Astrophys. 304, 189.
Pan, X., Shao, M. and Colavita, M.M., 1993, Astrophys.J.Lett. 413, L129.

Part 8. Catalogues

OVERVIEW OF THE TYCHO CATALOGUE

E. HØG[1], C. FABRICIUS[1], V.V. MAKAROV[1],
D. EGRET[2], J.L. HALBWACHS[2], G. BÄSSGEN[3],
V. GROßMANN[3], K. WAGNER[3], A. WICENEC[3],
U. BASTIAN[4] AND P. SCHWEKENDIEK[4]
[1] *Copenhagen University Observatory, Denmark*
[2] *Observatoire de Strasbourg, France*
[3] *Astronomisches Institut der Universität Tübingen, Germany*
[4] *Astronomisches Rechen-Institut, Germany*

1. Overview of the Tycho Catalogue

The final Tycho Catalogue (ESA 1997b) has been derived from 37 months of observations with the star mapper of the astrometric satellite Hipparcos. The Hipparcos Catalogue (ESA 1997a) with about 120,000 stars is the result of the main Hipparcos mission and has, *e.g.*, been described by Kovalevsky *et al.* (1995). Both catalogues will be published in 1997.

The Tycho Catalogue provides astrometry (positions, parallaxes and proper motions) and two-colour photometry (in B_T and V_T) for more than one million stars brighter than $V_T = 11.5$ mag. The median precision (standard error) is 25 mas in position and 0.10 mag in the $B_T - V_T$ colour index. These values apply at the median magnitude $V_T = 10.5$ mag for stars of median colour index $B_T - V_T \simeq 0.7$ mag. The Tycho Catalogue contains 1,052,031 entries (stars) observed by Tycho, supplemented by 6301 entries from the Hipparcos Catalogue that were not observed by Tycho. The Tycho Catalogue contains roughly 40,000 stars brighter than $V_T = 9$ mag which are not contained in the Hipparcos Catalogue. For these stars the median precision is 7 mas in position, parallax and annual proper motion and 0.019 mag in $B_T - V_T$. Double stars with separations larger than 2 arcsec and with moderate magnitude difference are usually resolved.

The Tycho Catalogue, and its photometric annex, referred to as the Tycho Epoch Photometry Annex, is strictly an observational catalogue. It contains data derived exclusively from the Hipparcos satellite's star mapper observations, with the exception of certain cross-identifications.

V_T	<6.0	6–7.0	7–8.0	8–9.0	9–10.0	10–11.0	>11.0	All	<9.0
Median V_T	5.38	6.63	7.62	8.62	9.61	10.58	11.19	10.47	8.33
N (TYC)	4553	9550	27,750	78,029	211,107	515,029	205,934	1,052,031	119,882
N (not HIP)	4	55	3485	36,511	182,773	506,720	205,275	934,901	40,055
Median standard errors in astrometry (mas):									
Position	1.8	2.6	4.0	6.7	12.9	27.2	39.2	24.6	5.6
Parallax	2.5	3.6	5.3	8.6	16.4	34.3	49.6	31.2	7.2
P.M./yr	2.3	3.3	5.0	8.3	16.0	33.5	48.6	30.2	7.0
Median standard errors in astrometry (mas):									
B_T	0.003	0.006	0.010	0.018	0.036	0.084	0.128	0.074	0.014
V_T	0.003	0.005	0.008	0.014	0.027	0.064	0.122	0.057	0.012
$B_T - V_T$	0.005	0.008	0.014	0.024	0.049	0.117	0.200	0.104	0.019
$B - V$	0.004	0.007	0.012	0.020	0.041	0.098	0.171	0.087	0.017

TABLE 1. This table gives the number of stars in TYC and the number of TYC stars not included in HIP, along with the corresponding median standard errors for stars within the given intervals of V_T magnitude (the column 'All' also including entries for which V_T is not available). Systematic errors in astrometry are less than 1 mas and 1 mas/yr, although the external standard errors (the true accuracies) may be 50 per cent larger than the quoted standard errors for faint stars. In photometry, systematic errors may reach the level of the quoted standard errors for faint stars. The photometry for about 20,000 stars is considered to be uncertain, for example when the standard errors are larger than 0.3 mag.

The reduced data comprise two parts. The main catalogue (the Tycho Catalogue, or TYC) contains the astrometric and summary photometric data for each star. The Tycho Epoch Photometry Annex contains summary photometric data for all stars, along with 'epoch photometry' (photometry at each epoch of observation) for a subset of stars observed with sufficiently high signal-to-noise ratio. In structure, the Tycho Catalogue and the Tycho Epoch Photometry Annex resemble the corresponding machine-readable parts of the Hipparcos Catalogue (HIP) and the associated Hipparcos Epoch Photometry Annex.

Solar system objects observed as part of the Tycho experiment are contained within a general annex of solar system observations by the Hipparcos satellite.

1.1. COMPLETENESS OF THE TYCHO CATALOGUE

Some Hipparcos Catalogue stars were not observable by the star mapper. The dynamic range of the star mapper detector resulted in non-linearity at the brightest magnitudes (Sirius was not observable). The faintest Hipparcos Catalogue stars fell below the detection threshold of the star mapper detectors. Stars in very dense clusters and other dense fields could not be observed by Tycho, thus leaving the resulting Tycho Catalogue incomplete in such regions.

All 6301 single star entries and double and multiple star components contained within the Hipparcos Catalogue but not observed by Tycho have nevertheless been included in the Tycho Catalogue for completeness, and assigned a corresponding TYC number. In these cases a truncated astrometric and photometric descriptor taken from the Hipparcos Catalogue has been included in the Tycho Catalogue. The position of the entry taken from the Hipparcos Catalogue is also included in order to assist cross-identification between the catalogues.

2. Tycho Photometry

Tycho photometry was obtained in two colour bands, B_T and V_T, closely corresponding to B and V in the Johnson UBV system. The Tycho colour index, written $B_T - V_T$ or $(B - V)_T$, is not explicitly given, but may be simply derived from the difference of the published magnitudes. Approximate values of the Johnson V magnitude and colour index $B - V$ are also provided, derived by a suitable transformation. Because it is a strict observable, and unaffected by the uncertainties inherent in such a transformation, the Tycho colour index rather than the derived Johnson colour index is recommended for use whenever appropriate.

A simple linear transformation from the Tycho B_T, V_T magnitudes to B, V magnitudes in the Johnson photometric system is:

$$V \simeq V_T - 0.090\,(B_T - V_T)$$
$$B - V \simeq 0.850\,(B_T - V_T)$$

In the interval $-0.2 < B_T - V_T < 1.8$ mag the systematic errors of this simple transformation do not exceed 0.015 mag for V and 0.05 mag for $B - V$ for unreddened main-sequence stars.

3. Reference Stars

A distinct flag indicates whether the Tycho Catalogue entry is considered as a 'recommended' astrometric reference star. This classification of the entry requires, e.g., non-duplicity and good astrometric quality, resulting in 886,000 stars being flagged as astrometric reference stars.

Most of the Tycho Catalogue magnitudes have a sufficient accuracy for calibration of magnitudes derived from photographic survey plates in colour bands near to B or V. A selection of non-double and non-variable stars having a standard error < 0.1 mag results in about 520,000 Tycho photometric reference stars suitable for such a purpose.

4. Concluding Remarks

The Tycho Catalogue constitutes an astrometric reference system with twice as many stars as PPM which is the best system up to now with high star density, cf. Table 2 by Høg (1995). The 30 mas standard error of positions for TYC at the epoch 1990 is ten times smaller than that of PPM and the systematic errors are 100 times smaller. The accuracy is maintained for one or two decades from 1990 if PPM proper motions are used or, even better, if proper motions are derived for all Tycho stars by means of positions from the Astrographic Catalogue, reduced to the Hipparcos reference system, as proposed by Röser & Høg (1993) and discussed by Röser, Schilbach & Hirte (1995).

As a photometric reference catalogue in two colour bands the TYC is about an order of magnitude larger though less accurate than the collection of all ground-based photometric data with UBV photometry for about 100, 000 stars by Mermilliod & Mermilliod (1994).

References

ESA, 1997a, The Hipparcos Catalogue, ESA SP-1200
ESA, 1997b, The Tycho Catalogue, ESA SP-1200
Høg E. 1995, in: E. Høg and P.K. Seidelmann (eds.), Astronomical and Astrophysical Objectives of Sub-Milliarcsecond Optical Astrometry, 317, Kluwer Academic Publ.
Kovalevsky J., Lindegren L., Froeschle M., van Leeuwen F., Perryman M.A.C., Falin J.L., Mignard F., Penston M.J., Petersen C.S., Bernacca P.L., Bucciarelli B., Donati F., Hering R., Høg E., Lattanzi M.G., van der Marel H., Schrijver H., Walter H.G. 1995, Astron.Astrophys. 304, 34
Mermilliod J.-C. & Mermilliod M, 1994, Catalogue of Mean UBV Data on Stars, Springer Verlag, New York, Berlin, Heidelberg, 1994.
Röser, S., Høg, E., 1993, 'Tycho Reference Catalogue: A Catalogue of Positions and Proper Motions of one Million Stars.' In: Workshop on Databases for Galactic Structure. Ed.: A.G. Davis Philip, B. Hauck and A.R. Upgren. Van Vleck Observatory Contr. No.13, 137. L. Davis Press, Schenectady, N.Y.
Röser S., Schilbach E. & Hirte S., 1995, ESA SP-379, 143

MAIN PROPERTIES OF THE HIPPARCOS CATALOGUE

F. MIGNARD
OCA/CERGA, Grasse, France

Abstract. The Hipparcos mission has come to a close with the end of the data processing on August 1996, only three years after the last observations were performed. About 118,000 stellar objects were observed and subsequently solved for their position, parallax and proper motion together with repeated determinations of their magnitude in the Hipparcos band. This paper outlines the main properties of the solution and explains how and when this information will be made available to the astronomical community.

1. Introduction

The ESA astrometric mission Hipparcos was the first space experiment fully dedicated to astrometry. The observation program came to its end in August 1993 after 37 months of effective observation which was followed by three more years of data analysis. The primary result is an astrometric catalogue of nearly 118,000 entries nearly evenly distributed over the sky with an astrometric accuracy of 1 mas or better for the brightest stars. This set of stars formed a pre-defined observing programme selected after a wide consultation of the astronomical community from their scientific interest and the possibility to allocate sufficient observational time to each of them. This task of preparing the stellar program was deemed as so challenging that it was entrusted by ESA to a dedicated Consortium, the Input Catalogue Consortium, with the goal of sorting out the 200 scientific proposals, amounting to more than 200,000 different stars, to obtain a working catalogue containing as many high priority objects as possible. The same group was also to coordinate the necessary new ground based observations so that the accuracy of the Input Catalogue meets the Hipparcos specifications.

2. The Hipparcos Mission

The main scientific goals were those of astrometry :
- To provide a non-rotating celestial reference frame to which the motions of objects in the solar system or stars in the Galaxy could be referred
- To provide basic observational data for the studies of stellar properties within the Galaxy, such as distances, luminosities, masses and velocity vector.

The satellite was launched, by Ariane, in August 1989, and carried out 37 months of high quality observations before the mission was terminated following serious problems with the guidance system and the failure of the on-board computer in 1993. The satellite and the operational principles have been described in the literature and the interested reader will find technical information in the three volumes of the ESA SP-1111 report and in a special issue of Astronomy and Astrophysics dedicated to Hipparcos (Kovalevsky *et al.*, Lindegren *et al.* 1992).

On a quantitative basis the initial objectives of the mission were to determine the five astrometric parameters of each program stars to a precision of 2 mas for the core of the catalogue at about 8.5 magnitude, along with the astrometry to 10-20 mas and two-color photometry for an additional 400,000 stars (the Tycho experiment). An optimized scanning law allowed to cover the whole sky more or less regularly in such a way that every star has been observed at 25 to 70 epochs according to its ecliptic latitude. Regarding the photometry the number of individual observations per star ranges between 70 to 300, with again the ecliptic latitude as being the main factor of variation.

The Hipparcos Catalogue whose properties are described below, results from the merging of two largely independent solutions produced by scientists of the FAST (*Fundamental Astrometry by Space Technic*) and NDAC (*Northern Data Analysis Consortium*) consortia under the respective leadership of J. Kovalevsky (France) and L. Lindegren (Sweden). The treatment represents a major effort of more than 40 people extending over about 15 years between the design of the algorithms, the simulations and the processing of real data.

3. Summary of the Astrometric Results

A brief summary of the initial scientific goals is given in Table 1. The figures refer to the accuracy based on phase A and B assessments and on simulations carried out during the mission planning. Compared to the best available fundamental catalogues, this represented an improvement in accuracy of a factor 20 and nearly 100 in term of number of objects.

The overall properties of the final catalogue are given in Table 2. The effective observation period is 37 months, slightly shorter than the time base, because of the occurrence of several interruptions in the observing program caused by hardware problems.

The Hipparcos epoch has been so chosen as to be close to the mean observation epoch for all the stars. The true mean epoch for each star may be different from the catalogue epoch by at most six months earlier or later. It was considered as most convenient to generate a catalogue with a single epoch, although this does not optimize the correlations between position and proper motions. The true mean epoch can be computed by searching for the time offset required to minimize the correlation between the right ascension and proper motion in right ascension, or that between the declination and proper motion in declination. These two epochs may differ by few weeks.

TABLE 1. The initial goals of the mission

Number of stars	120,000
Limiting magnitude	12.5
Positional accuracy	2 mas
Accuracy of parallax	2 mas
Accuracy of proper motions	2 mas/yr
Mission duration	2.5 yr

TABLE 2. Overall properties of the catalogue

Measurement period	1989.85–1993.21
Catalogue epoch	1991.25
Reference system	ICRS
Mean-Sky density	≈ 3 stars/deg^2
Number of astrometric solutions	117,955
Systematic errors	< 0.1 mas

In the Hipparcos program there was no direct observation of extragalactic object and the reference frames of the Hipparcos solutions were very consistent but had no reason to be rotation-free. Several radio stars were in the observing program and were also included in a VLBI program (Lestrade et al. 1994). A final rotation has allowed to align the final Hipparcos solution to the ICRS reference system. Therefore the Hipparcos Catalogue can been seen as the best available optical realization of the extragalactic reference frame.

4. The Astrometric Results

The main results of the absolute astrometry are summarized in Table 3 and in Figures 1–2. The major advance brought by Hipparcos is obviously the direct measurement of the trigonometric parallax for all the program stars, although for the most distant stars the parallax may have only a statistical, rather than individual, significance. The relative accuracy is more important in this respect than the fact the parallax may be known within 1 or 2 mas. With more than 20,000 stars measured with $\sigma(\pi)/\pi$ better than 10 percent, including several thousands binary stars, one can say that this is the start of a new era for the luminosity calibration and the mass determination. The number of doubles stars with good relative and absolute astrometry, including distances, is also an outstanding output of the mission.

The analysis of the Hipparcos parallaxes against a set of known very distant stars or against the best photometric parallaxes indicates that there is no systematic effect larger than 0.1–0.2 mas in the Hipparcos value.

TABLE 3. Summary of the Astrometric solution

	$\alpha \cos \delta$	δ
Median precision of positions (Hp <9) (mas)	0.77	0.64
Median precision of proper motions (Hp <9)(mas/yr)	0.88	0.74
Median precision of parallaxes (Hp <9)(mas)	0.97	
Number of stars with $\sigma(\pi)/\pi < 0.1$	20,800	
Number of stars with $\sigma(\pi)/\pi < 0.2$	50,000	
Number of double star systems	12,195	
Range of separations (arcsec)	0.1–25	
Range of magnitude difference	0.0–3.5	
Median accuracy of the separation (mas)	11	
Nuumber of suspected astrometric binaries	2,900	

The median accuracy in Table 3 is typical for the bulk of the catalogue for stars of magnitude 8 to 8.5. The variation of the mean accuracy of

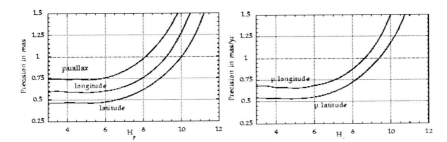

Figure 1. Precision of the astrometric parameter in the final Hipparcos solution as a function of the magnitude Hp.

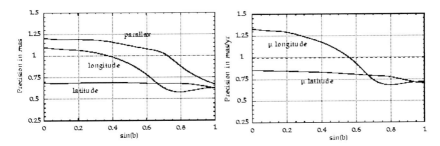

Figure 2. Precision of the astrometric parameter in the final Hipparcos solution as a function of the ecliptic latitude

the five astrometric parameters as a function of the magnitude is shown in Figure 1. At the faint end the photon noise was the limiting factor and the diagram gives the expected increase in the standard error with the magnitude. For stars brighter than 8 mag, the accuracy does not improve as expected, as the limiting factor is chiefly instrumental and weakly sensitive to the magnitude. The difference between the accuracy of longitude and latitude comes as no surprise and is a mere consequence of the geometry of the scanning. Besides the brightness, one must add that the accuracy depends strongly on the ecliptic latitude, as illustrated in Figure 2 for the same parameters. There is virtually no effect in latitude, but a very conspicuous one in longitude, with about a factor two between the standard error for ecliptic stars compared to polar stars. The minimum at about $b \approx 47$ degrees follows again from the scanning law, with the axis of rotation of the satellite kept at 43 degrees from the direction of the Sun.

5. The Results of the Photometric Treatment

Although the Hipparcos mission was primarily planned and optimized to produce good astrometric measurements, it was soon realized that the signal conveyed much information on the star brightness, provided a good monitoring of the instrument sensitivity was undertaken.

At each grid crossing a magnitude was determined for all programme stars in a well defined photometric system, kept constant during the mission. Altogether the missions ends up with 13×10^6 such measurements spread over 118,200 stars, that is to say an average of 110 magnitude determinations for each star, with an accuracy of 0.01–0.02 mag for a typical Hipparcos star of magnitude 8.5. This constitutes the most comprehensive photometric survey so far, combining both precision and homogeneity of the photometric system all over the sky. Summary figures are given in Table 4.

TABLE 4. Summary of the photometric solution

Number of entries with solution	118, 204
Photometric band	Hp
λ_{eff}	550 nm
Photometric accuracy for one observation ($Hp = 8.5$)	0.012 mag
Average number of observations per star	110
Total number of photometric observations	13×10^6
Median photometric accuracy ($Hp < 9$)	0.0015 mag
Number of variables detected	6, 000
Number of suspected variables	5, 700

The number of variables detected or suspected depends on various statistical thresholds set in the variability analysis. Although this was done with great care and based on several statistical tests, it may happen that further analyses based on the Hipparcos data and additional ground based data could end with different figures.

The actual distribution of the standard errors of the accumulated photometry (the standard error of the median of the ≈ 110 measurements of a star) is shown in Figure 3. All the constant bright stars are on the left of the diagram with $\sigma(Hp) < 0.001$ mag. The rightmost column must be associated with the variable stars, since the scatter among the individual observations is large as a result of the variability. The left diagram gives the accuracy at the transit level, that is to say for the individual observations.

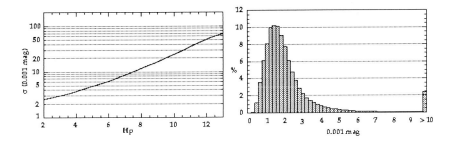

Figure 3. Precision of the photometric solution in the Hipparcos Catalogue. Left : Variation of the precision of the photometric accuracy as a function of the magnitude. Right : The distribution of the standard errors in the accumulated photometry

6. Conclusion

The Hipparcos final product will comprise 16 printed volumes, a set of CD-ROMs with the Hipparcos and Tycho Catalogues in ASCII files, and a second set including all the files in binary form along with a dedicated software permitting interrogation, sample constructions on any of the fields or their logical combination, and graphic display. The availability to the general astronomical community in due for June 1997. Advanced data release was granted to PIs and early proposers respectively in August 1996 and January 1997 for approved scientific programs. The first scientific results based on the exploitation of the Hipparcos data, will be presented at a meeting to be held in May 1997 in Venice.

Acknowledgements

This short overview reflects under the name of a single author the full dedication of many scientists, engineers and technicians who have devoted most of their professional activity to the project during the last fifteen years. Their name and affiliation will appear in the Hipparcos documentation. They are here collectively thanked for their invaluable contribution.

References

Kovalevsky J. *et al.* 1992, The FAST Data Analysis Consortium : Overview of the reduction software, Astron.Astrophys., 258, 326.
Lestrade, J. F., Jones, D.L. *et al.* 1995, Astron.Astrophys., 304, 182.
Lindegren L. *et al.* 1992, The NDAC Data Analysis Consortium : Overview of the reduction methods, Astron.Astrophys., 258, 18.

THE ASTROGRAPHIC CATALOGUE

A Gold Mine for Proper Motions

D.H.P. JONES
Royal Greenwich Observatory, Cambridge, UK

1. Introduction

The most reliable method of cross-wavelength identification is the coincidence of accurately determined positions, both referred to the same reference frame. But objects move with time and the more accurate their positions, the more important their proper motions are.

2. The Astrographic Catalogue

A useful sky survey should cover the whole sky to a uniform flux limit at a given wavelength. The need for such surveys was realized in the 19th century *e.g., Bonner Durchmusterung*. The earliest whole sky survey of accurate positions was the Astrographic Catalogue and *Carte du Ciel* planned at a conference in Paris in 1887. Eighteen observatories agreed to collaborate; each would observe a separate band of declination. The aims were

1. To use a common design of telescope with a plate scale of $60''.0 \text{mm}^{-1}$. and a field of $2°$.
2. To produce a catalogue of all stars brighter than 11, accurate to $0''.3$ rms in the measured coordinates and $0''.5$ in Right Ascension and Declination. Exposures of 6 min, 3 min and 20 sec were taken on each plate.
3. To produce charts reaching to magnitude 14. With the emulsions then available it was believed that a 40 min exposure would be needed; most observatories took three exposures in the form of an equilateral triangle.

The number of observatories eventually rose to 22. Although Greenwich published its catalogue by 1906, the last zone was not published until 1962. The catalogue has never been widely used because of the inconvenience and inaccuracy of converting the rectangular coordinates to Right Ascension and Declination and inhomogeneities between the different zones in limiting magnitude, accuracy and presentation. The published catalogue contains four million stars.

3. Recent Work

There has been an upsurge of interest in the Astrographic Catalogue in recent years; (e.g., Röser & Høg 1993; Urban et al. 1996a, 1996b; Kuzmin, this volume, p. 409). While there are substantial differences between zones, all have limiting magnitudes closer to 12 photographic than the 11 expected. The accuracy differs markedly between different zones, from $0''.19$ to $0''.42$ and there is evidence that the difference arises from the methods of measurement which differ from zone to zone. It appears that the plates are essentially homogeneous between zones in accuracy but not in limiting magnitude. If this is confirmed then re-measurement with a modern automatic machine should improve all the zones to the level of the best.

At the 1994 IAU, Commission 24 (de Vegt & Morrison 1995) set up a new working group to consider the re-measurement of the Astrographic Catalogue.

TABLE 1. Working Group on Re-measuring Astrographic Plates

Members	Status	Members	Status
B Bucciarelli		S Röser	
TE Corbin		R Smart	(Consultant)
CC Dahn		SE Urban	(Consultant)
DHP Jones	(Chairman)	Chr de Vegt	

As a pilot study, a plate borrowed from the Vatican Observatory has been re-measured on a conventional PDS. Table 2 shows the goodness of fit when the measures are reduced with the ACRS catalogue and six degrees of freedom. The corrected values have been found by subtracting the ACRS errors quadratically. The goodness of fit is better than $0''.19$ compared to $0''.42$ for the printed catalogue.

A better estimate of the accuracy of the measures can be found from the difference of the 6 and 3 minute exposures. The Vatican catalogue is based on the 3 min exposure. The r.m.s. errors of one measurement, listed in Table 3, show a steady increase with machine magnitude. One micron is believed to be the irreducible measuring error and corresponds to $0''.06$. If the limiting magnitude is set at 13.5 then an accuracy of $0''.18$ should result, in accordance with other recent work (Hiesgen, this volume, p. 415;

TABLE 2. Goodness of fit (r.m.s.) of Vatican Plate

Exposure	α	δ	α (corr)	δ (corr)
6 min	0″.28	0″.23	0″.11	0″.15
3 min	0″.26	0″.26	0″.08	0″.19
ACRS	0″.25	0″.18		

TABLE 3. Accuracy as a function of magnitude

Magnitude	limits	Number	σ_x (μ)	σ_y (μ)
8	9	22	0.9	1.3
9	10	46	0.9	1.3
10	11	86	1.1	1.0
11	12	139	1.2	1.3
12	13	229	2.1	2.0
13	14	224	4.2	3.8

Geffert *et al.* 1996). Continuing problems are caused by the réseau lines which were photographed on to each plate; and the difficulty of defining the centre of a triangle of images which are resolved for faint stars but blended for bright.

References

de Vegt, Chr. and Morrison, L.V. 1995 WGM 3 International Catalogue Projects, Highlights of Astronomy, 10, pp. 683–695

Geffert, M., Bonnefond, P., Maintz, G. and Guibert, J. 1996 The astrometric accuracy of "Carte du Ciel" plates and proper motions in the field of the open cluster NGC 1647 Astron. Astrophys. Suppl.118, 277

Röser,S. and Høg,E. 1993 Tycho Reference Frame Catalogue (TRC): A Catalogue of Positions and Proper Motions of one Million Stars Workshop on Databases for Galactic Structure,pp. 137–143

Urban, S.E., and Corbin, T.E. 1996a New reductions of the Astrographic Catalogue; Conventional plate adjustment of the Cape Zone Astron.Astrophys.305, 989

Urban, S.E., Martin, J.C., Jackson, E.S. and Corbin, T. E. 1996b New Reductions of the Astrographic Catalogue; Plate adjustments of the Algiers, Oxford I and II, and Vatican Zones Astron. Astrophys. Suppl. 118, 163

COMPLETION OF THE STERNBERG ASTRONOMICAL INSTITUTE ASTROGRAPHIC CATALOGUE PROJECT

V. NESTEROV[1], A. GULYAEV[1], K. KUIMOV[1], A. KUZMIN[1],
V. SEMENTSOV[1], U. BASTIAN[2] AND S. RÖSER[2]
[1] *Sternberg Astronomical Institute, Russia*
[2] *Astronomisches Rechen-Institut, Germany*

1. Introduction

The first astronomical photographic survey, the *Carte du Ciél* was initiated in 1887 by a group of French astronomers. The observational campaign was started in 1891, while the last of more than 22,000 total plates were photographed in 1950; most of observations (more than 90%) were performed prior to 1920. Detailed description of the *Carte du Ciél* development can be found elsewhere (Kolchinsky 1989, Eichhorn 1974, Debarbat *et al.* 1987).

The main outcome of the project was the *Astrographic Catalogue*(AC) (or, more correctly, *Carte du Ciél* Astrographic Catalogues to emphasize zonal arrangement). It comprises close to 8.5 million measurements of rectangular coordinates and brightness estimates. The majority of stars were photographed on two plates; therefore, the total number of catalogue stars is circa 4.5 million.

Due to the early epoch (around 1905) and high positional accuracy ($0.25''$ may be expected after performing global block adjustment and tying AC to *HIPPARCOS* reference frame), the AC constitutes an excellent first epoch for massive derivation of proper motions. Given a modern epoch with at least the same level accuracy of positions, resulting proper motions will have unprecedented quality of 3–4 mas/year.

In 1987, within the framework of the preparation of the input catalogue for the *LOMONOSOV* space astrometry mission, the Sternberg Astronomical Institute (SAI) AC project was initiated. The main stages of the project are described in the following sections.

2. The Astrographic Catalogue: Machine-readable Version

AC data are published in 254 volumes, nearly 90% of which were present in SAI library. Other volumes were kindly supplied by colleagues from Pulkovo and Kazan Observatories (Russia), Golosiivo and Odessa Observatories (Ukrain), Tartu University Observatory (Estonia) and Astronomisches Rechen-Institut (Heidelberg, Germany).

Keypunching of AC data was started in 1987 and took nearly 4 years. Total amount of work may be estimated as 50 manyears. The first preliminary machine-readable version subject to processing appeared in 1990 and the final version for both hemispheres was completed January 1994.

Verification of the keypunched AC data set included both manual and automated procedures, which assured that data comply to formats and records sequencing used in published volumes, all the fields are present and all the exceptions (field incomplete or uncertain) are marked with special flags. All the misprints reported in published erratum lists were accounted for.

Machine-readable version of the AC includes all the published measurements of 19 completed zones of Astrographic Catalogue plus 406 published plates of the original unfinished Potsdam section. Total number of plates is 22,652; total number of measurements is 8,633,975. Number of plates, measurements and observation epochs statistics are given in Table 1.

3. Astrometric Calibration of the AC Plates Material

Astrometric calibration of AC plates should provide reliable and consistent transformation of the measured Cartesian coordinates into positions referred to a modern reference system (*e.g*, the present standard—ICRS). This could in principle be achieved by either standard procedure of individual plates reduction (conventional plate adjustment, CPA) or by rigorous block-adjustment ("plate-overlap technique" (Eichhorn 1960).

Conventional plate adjustment requires a reference catalogue of sufficient density and accuracy (both systematic and random) to derive accurate estimates of the parameters of plate-to-sphere transformation (or, in astrometric terminology, plate constants). Prerequisites are defined by:

- size of AC plate: $2.1 \times 2.1°$,
- number of expected plate parameters: observatory specific systematic errors may lead to as many as 18 unknowns per plate and
- standard error of measurement: 100 to 300 mas per coordinate.

Modern reference catalogues that could be used for the AC CPA purposes are: Astrographic Catalogue Reference Stars (ACRS, Corbin and Urban 1991), Positions and Proper Motions (PPM, Röser and Bastian 1991,

TABLE 1. AC zones: Final assignment

Zone	Declination limits	Plates taken	Measurements	Epoch: Earliest, Latest and Median
Melbourne	−90° to −65°	1149	392,615	1892 1940 1897.4
Sydney	−64° to −52°	1400	744,034	1891 1948 1907.3
Cape	−51° to −41°	1512	901,244	1897 1911 1902.2
Perth	−40° to −32°	1376	604,365	1902 1924 1911.0
Cordoba	−31° to −24°	1360	467,404	1903 1915 1911.9
Hyderabad South	−23° to −17°	1260	521,867	1901 1928 1918.5
Tacubaya	−16° to −10°	1260	516,646	1900 1938 1904.0
San-Fernando	−09° to −03°	1260	346,142	1891 1917 1896.7
Algiers	−02° to +04°	1260	330,459	1891 1911 1903.7
Toulouse	+05° to +11°	1260	433,087	1893 1935 1909.5
Bordeaux	+11° to +17°	1260	355,071	1893 1925 1904.9
Paris	+18° to +24°	1261	436,494	1891 1927 1895.2
Oxford	+25° to +33°	1500	631,816	1892 1936 1907.6
Uccles	+34° to +35°	320	158,660	1939 1950 1944.6
Hyderabad North	+36° to +39°	592	242,550	1928 1937 1930.6
Helsingfors	+40° to +46°	1008	284,661	1892 1909 1894.6
Catania	+47° to +54°	1009	320,631	1894 1931 1902.8
Vatican	+55° to +64°	1046	479,976	1891 1926 1908.3
Greenwich	+65° to +90°	1153	322,238	1892 1905 1896.5
Potsdam	+32° to +39°	460	144,015	1893 1900 1895.6
Totals	−90° to +90°	22,652	8,633,975	1891 1950 1904.4

Bastian *et al.* 1992) and the final *HIPPARCOS* catalogue (upon completion, Kovalevsky *et al.* 1995).

Analysis of the properties of these catalogues under the constraints listed above immediately shows that *NONE* of three catalogues can serve as a reference catalogue for AC CPA purposes, mainly due to their low accuracy at AC epoch and insufficient density. Therefore, conventional plate adjustment could not be used as an approach to high-precision astrometric calibration of the *Astrographic Catalogue*.

Rigorous block-adjustment (also called global block adjustment, GBA) does not in principle require a reference catalogue at the time of solving equations of observations for unknowns (plates constants and/or stars positions)—it can proceed in any arbitrary coordinate system, while comparatively small (but highly accurate) reference catalogue could be introduced at final stage to define necessary rotation of an arbitrarily chosen system to any desired one.

Global block adjustment starts with a preliminary step—definition of plate models which describe telescope and measuring machine specific systematic errors. Parameters of such a model later become unknowns in a global system of equations to be solved.

Being unable to define plate models for AC zones by common means, we should go back to conventional plate adjustment, though this time with a completely different purpose: to define zone-specific reduction models on the basis of available reference catalogues. Obviously, this approach is limited to plate-scale systematics which will in turn be statistically significant with respect to the chosen reference catalogue.

Summing up, we use conventional plate adjustment of AC plates material with ACRS as a reference catalogue in order to define observatory specific plate-scale systematic errors. By means of CPA we also construct a *Provisional AC Catalogue*, to be used as a list of initial positions of stars within global block adjustment procedure. Meanwhile, we believe the provisional positions to be of quite good quality—accuracy of catalog positions is between 0.25″ and 0.45″ depending on the section—and therefore valuable on its own.

The original approach to the global block-adjustment of the photographic sky survey as developed by Eichhorn and was based on the assumption of the absence of epoch difference between overlapping plates. This is obviously not true in the AC case—internal (within section) overlaps expose significant epoch variations, and the situation is much worse with the cross-section overlaps.

The best approach would be to introduce high-quality modern-epoch positions of the substantial number of AC stars into global block-adjustment. Modern-epoch positions will facilitate elimination of proper motions from GBA equations and, on the other hand, will provide the direct link of the AC positions to modern reference frame. Application of this approach may be expected within the Tycho Reference Catalogue project (TRC), *cf.* (Röser and Høg 1993), which is aimed at derivation of the high-quality proper motions for about 1,000,000 Tycho stars using the AC as early epoch.

4. Photometric Calibration of the Astrographic Catalogue

The Astrographic Catalogues also provide brightness estimates along with measured coordinates of stars. These estimates, while being extremely non-uniform in quality and reduction schemes applied, are still highly important. Photometric data reduced into a modern system will be of interest in general and specifically within the framework of construction of the astrometric catalogue based on Astrographic Catalogues. Moreover, in order to retain consistency, coma and magnitude effects, if present, should be studied in terms of magnitudes derived from original AC brightness estimates.

Unfortunately, available photometric standards, *e.g.*, (Hauck *et al.* 1990) do not allow to access individual variations of photometric systems of AC plates and therefore are poorly suited for the AC photometry calibration task. The only reasonable approach may be realized with the forthcoming TYCHO catalogue which will cover nearly 25% of AC stars. It will provide high-precision photometry necessary for individual AC brightness estimates calibration.

5. Applications of the AC: Present and Prospect

The Astrographic Catalogue is first and foremost a reservoir of proper motions, and therefore most of its applications deal with proper motions. We will note few projects of the kind:

- Positions and Proper Motions Catalogue (completed), where Astrographic Catalogue observations were used to obtain proper motions 2–3 times more precise than that of preceding reference catalogues,
- Tycho Reference Catalogue (proceeding) aimed at even more precise proper motions of more than 1 million stars.

On the other hand, AC should not be overlooked as a deep enough photographic survey of high reliability. For instance, due to manual objects selection and measuring the AC is nearly free of artifacts. This consideration enabled a number of projects aimed at identification of different types of objects that present special interest on the basis of AC positions and proper motions. Examples of applications of the latter type are:

- Accurate astrometry for HD/HDE/HDEC stars (Nesterov *et al.* 1995),
- Catalog of positions and proper motions of variable stars (Gulyaev and Ashimbaeva 1996).

Acknowledgements

We consider it necessary to thank our colleagues at Astrometry Division of Sternberg Astronomical Institute who participated in the AC project: Dr. N.N. Kabaeva, E.A. Turova, Dr. G.V. Romanova, O.D. Solovyeva, Dr. A.A. Volchkov, E.V. Dolganova, E.Yu. Gureeva and N.M. Evstigneeva. The project was supported for nearly 10 years by Sternberg Astronomical Institute; additional support of the Russian Academy of Sciences and Deutsche Forschungsgemeinschaft is gratefully acknowledged.

References

Bastian U., Röser S., Yagudin L., Nesterov V. *et al.* 1993. PPM Star Catalogue. Positions and Proper Motions of 197,179 Stars South of $-2.5°$ declination for Equinox and Epoch J2000.0. V. III, IV. Heidelberg: Spektrum Akademischer Verlag.

Corbin T.E. and Urban S.E. 1991. Astrographic Catalogue Reference Stars, Washington, U.S. Naval Observatory.

Eichhorn H. 1960. Über die Reduktion von Photographischen Sternpositionen und Eigenberwegungen, Astron. Nachr., 285, 233-237.

Eichhorn H. 1974. Astronomy of Star Positions, New York: F. Ungar Publ.

Debarbat S., Eddy J.A., Eichhorn H.K., Upgren A.R. eds., Proc. IAU Symp. 133 "Mapping the Sky," Paris, 1987.

Gulyaev A.P., Nesterov V.V. eds. 1992, Four Million Stars Catalogue, Moscow University Press (in Russian).

Gulyaev A.P., Ashimbaeva N.T. 1996, Astrometric Catalogue of Variable Stars, submitted to AZh (in Russian).

Hauck B., Nitschelm C., Mermilliod M., Mermilliod J.C. 1990, The General Catalogue of Photometric Data, Astron. Astrophys. Suppl., 85, 989.

Kolchinsky I.G. 1989, On the 100th anniversary of the International Cooperative Projects "Carte du Ciel" and "Astrographic Catalogue," in: History of Astronomy, Nauka, Moscow, 100 (in Russian).

Kovalevsky J., Lindegern L., Froeschle M. *et al.* 1995, Construction of the intermediate Hipparcos astrometric catalogue, Astron. Astrophys., 304, 1.

Nesterov V.V., Kislyk V.S., Potter H.I. 1991, An Astrometric Catalogue of Four Million Stars, Proc. IAU Symp. 141 "Inertial coordinate systems on the sky," ed. Lieske J.H. and Abalakin V.K., Leningrad,

Nesterov V., Kuzmin A., Ashimbaeva N., Röser S., Bastian U. 1995, The Henry Draper Extension Charts: A catalogue of accurate positions, proper motions, magnitudes and spectral types of 86,933 stars, Astron. Astrophys. Suppl., 110, 367.

Röser S., Bastian U. 1991, PPM Star Catalogue. Positions and Proper Motions of 181,731 Stars north of $-2.5°$ declination for Equinox and Epoch J2000.0. V. I, II. Heidelberg: Spektrum Akademischer Verlag.

Röser S., Høg E. 1993, Tycho Reference Catalogue, Proc. Workshop "Databases for Galactic Structure," L. Davis Press, New York.

SALVAGING AN ASTROMETRIC TREASURE

The European Network for the use of the Carte du Ciel

M. HIESGEN[1], P. BROSCHE[2], A. Ortiz GIL[2], J. COLIN[3],
A. FRESNEAU[1], M. GEFFERT[2], H. T. MACGILLIVRAY[5],
S. HECHT[4], E. KALLENBACH[4], M. ODENKIRCHEN[2],
C. SCHÄFFEL[4] AND H.-J. TUCHOLKE[2]
[1] *Observatoire Astronomique de Strasbourg, France*
[2] *Sternwarte der Universität Bonn, Germany*
[3] *Observatoire de Bordeaux, France*
[4] *Institut für Mikrosystemtechnik, Mechatronik und Mechanik der Technischen Universität Ilmenau*
[5] *Royal Observatory, Edinburgh*

1. Introduction

"Salvaging an astrometric treasure" is an european effort to retrieve the astrometric and photometric information contained in the early Carte du Ciel plates. Five institutes from three european countries are involved. The project was supported by the European Community.

2. The Plates of the Carte du Ciel

The "Carte du Ciel" (CdC) and "Astrographic Catalogue" (AC) projects were the very first photographic all-sky surveys. They started after the 1887 "Congrés Astrophotographique International" held in Paris. The observations were made from 1893 until 1930. While the AC plates were measured completely, the CdC plates were not up to now. They were taken to make paper charts of all stars down to B=14^m. About 7000 plates covering almost 75% of the entire sky were taken for the Carte. These plates are spread all over the world in 15 observatories.

3. A New Measuring Machine

To access these plates, a transportable measurement machine which can be sent around to digitize the plates at their storage locations was developed.

The desired positional accuracy of this machine is $< 1\,\mu$m, the maximum time to digitize a $160 \times 160\,mm^2$ plate is 30 min. Engineering studies at the Technical University of Ilmenau (Germany) have led to the construction of a prototype scanner approaching these goals. This machine uses air bearings and achieves movement of the plate holder by an integrated electrodynamic x-y-$\delta\phi$ drive without contact between object holder and base (Kallenbach 1995). The measurements with the final version will be done by a large CCD with effective size of 10–15 μm per pixel. This machine will cost less than $100 k.

4. First Results

To develop the reduction process and to analyse the quality of the CdC, we measured nearly 100 plates of the Paris zone on different available measuring machines. We can show that stars up to B=15.5^m can be detected on the plates. This means that about 20 million stars can be used to derive proper motions with an accuracy of 2 mas/y when using present epoch surveys (Lasker et. al. 1996). Previuos work using the AC have shown that this accuracy is possible.

In most cases, three exposures were taken on each plate with a small shift between them. These triple images show a triangular geometry of up to 300 μm size. The presence of several exposures and its peculiar geometry leads to the merging of the three exposures when the star is relatively bright ($\sim$ B$=8^m$). In addition, the plates were provided with a rectangular grid to make the finding of stars on the charts easier.

Emulsion defects, dirt and dust are also present on many of the plates. These factors make it very difficult to determine accurate astrometric and photometric data of merged stellar images and those affected by réseau lines. To detect the triple objects, we developed a new algorithm using a local-maximum-finder. The identification of the 3 exposures has to be done automatically without knowing the triangle geometry and size. The results of these processes can be seen in Figure 1.

We used a model consisting on three added bi-dimensional Gaussian curves including a saturation parameter to be fitted to these triple images. A principal component analysis of the 22 fit parameters has shown that at least 7 dimensions are necessary to represent the manifold of the parameters. In the majority of cases, we obtained an internal positional accuracy of around $0.12''$. Internal magnitudes can be calibrated with an accuracy of 0.12^m. This is independent of the elongation of the images caused by optical aberrations.

Some pilot projects which have used a limited number of plates have shown the value of the CdC for galactic astrometry. Geffert et al. (1996)

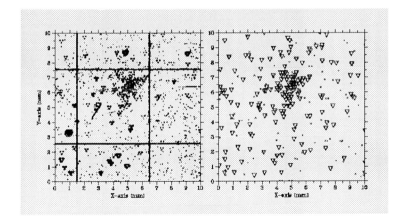

Figure 1. Part of a 1911 Carte du Ciel plate: Contour-plot of raw pixel data (left) and all thereon detected objects (right). Crosses show the single objects, triangles the detected triple images.

have used CdC plates for a proper motion study of the open cluster NGC 1647, others used these plates for astrometry of single objects (Dick *et. al.* 1993) or globular clusters (Hiesgen *et al.* 1996b). There are some more groups working with CdC plates on single targets (Lattanzi *et. al.* 1994, Jones 1996).

The most important reason of digitizing the whole CdC is be to provide equitorial coordinates and B-magnitudes around epoch 1910 for stars down to $B = 15.5^m$.

References

Brosche, P., Geffert, M., 1988, Proceedings of Mapping the Sky, IAU Symp. 133, p. 403, S. Debarbat *et al.* (eds)
Dick, W. R., Tucholke, H.-J., Brosche, P., Galas, R., Geffert, M., Guibert, J., 1993 Astron.Astrophys. 279, 267
Fresneau, A., 1990, Astron.J. 100, 1223
Geffert, M., Bonnefond, P., Maintz, G., Guibert, J., 1996 Astron.Astrophys.Supp. 118, 277
Hiesgen, M., Brosche, P., Ortiz Gil, A., 1996a Astr. Ges. Abstr. Ser. 12, 240
Hiesgen, M., Geffert, M., Maintz, G., 1996b Astron.Astrophys., in preparation
Jones, D., 1996, this volume, 406
Kallenbach, E., König, K., Saffert, E., Schäffel, Chr., Eccarius, M., 1995 "3rd Conference on Mechatronics and Robotics" October 1995, Paderborn, Germany
Lasker, B., Sturch, C., Doggett, J., Laidler, V., Wolfe, D., Loomis, C., 1996 this volume
Lattanzi, M. G., Massone, G., Munari, U., 1991, Astron.J. 102, 177
Ortiz Gil, A., Brosche, P., Hiesgen, M., 1995 Astr. Ges. Abstr. Ser. 11, 92

A NEW HIGH DENSITY, HIGH PRECISION ASTROMETRIC CATALOG

M.I. ZACHARIAS[1,2], G.L. WYCOFF[2], G.G. DOUGLASS[2],
T.E. CORBIN[2] AND N. ZACHARIAS[1,2]
[1] *Universities Space Research Association*
[2] *U.S.Naval Observatory, Washington, DC, USA*

1. Introduction

The Version 1.0 of the U.S. Naval Observatory Twin Astrographic Catalog (TAC) contains positions of 705,679 stars (Zacharias *et al* 1996). The TAC 1.0 covers over 90% of the sky from $+90°$ to $-18°$ declination; 260 out of 5180 plates are not yet measured. Plates were taken in both the yellow (508–578 nm) and in the blue (410–486 nm) bandpass between 1978 and 1986 at the U.S. Naval Observatory (USNO) in Washington, DC.

2. Catalog Properties

Preliminary photometry from these plate measures was obtained by using the Hipparcos Input Catalogue (HIC); 697,140 stars have a V and 675,800 stars a B magnitude. The limiting magnitudes are about V=11.5 (Figure 1) and B=12.0. The formal standard error of a mean magnitude is 0.11 mag or better for 50% of the stars.

The International Reference Stars (IRS) catalog (FK5/J2000 system) was used as the reference star catalog for the astrometric reductions. Magnitude-dependent systematic errors were found and preliminarily corrected (Zacharias and Douglass 1995). The average precision of a catalog position is 90 mas per coordinate at the mean epoch of observation (Figure 2).

Proper motions are being obtained by combining the TAC 1.0 results with new reductions of the Astrographic Catalogue (AC, Urban *et al.* 1996), as the latter work progresses. The precision of these proper motions varies between 2 and 8 mas/yr (Figure 3) depending on the quality and epochs of the AC plates.

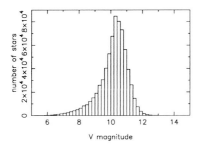

Figure 1. Distribution of V magnitudes of 697,140 TAC 1.0 stars from the yellow lens plates

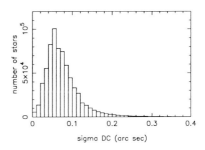

Figure 2. Mean positional errors in δ (mean epochs of observation) for 692,133 stars with at least 2 observations

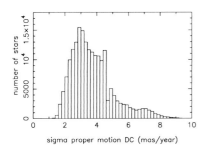

Figure 3. Estimated errors of declination proper motions for 195,622 stars of 3 AC test zones

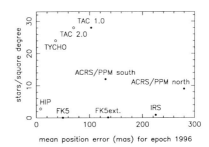

Figure 4. Star density vs. mean position error (epoch 1996) of astrometric catalogs (full = existing, open = future data).

The TAC 1.0 is almost 3 times more precise than the PPM or ACRS in the northern hemisphere at current epochs (Figure 4) and the star density is about 3 times higher. The TAC 1.0 has a higher star density than the Tycho catalog and provides independent, high precision positions for a large fraction of the Tycho stars at an epoch about 10 years earlier than the Tycho mean epoch. The AC, TAC and Tycho data combined will give high quality proper motions for over half a million stars in 1997. The catalog and details are at

aries.usno.navy.mil/ad/tac.html.

References

Urban, S.E., Martin, J.C., Jackson E.S., and Corbin, T.E. 1996 New Reductions of the Astrographic Catalogue:Plate adjustments of the Algiers, Oxford I and II, and Vatican Zones, Astron.Astrophys., 118, 163

Zacharias, M.I., and Douglass, G.G. 1995 Preliminary Results of the U.S. Naval Observatory Twin Astrographic Catalog (TAC) Plate Reductions, Bull.Am.Astron.Soc. 27, 857

Zacharias, N., Zacharias, M.I., Douglass, G.G., and Wycoff, G.L. 1996 The Twin Astrographic Catalog (TAC) Version 1.0, Astron.J., 112, 2336.

CONTENTS, TEST RESULTS, AND DATA AVAILABILITY FOR GSC 1.2

S. RÖSER[1], J. MORRISON[1], B. BUCCIARELLI[2],
B. LASKER[3] AND B. McLEAN[3]
[1] *Astronomisches Rechen-Institut, Heidelberg, Germany*
[2] *Osservatorio Astronomico di Torino, Italy*
[3] *Space Telescope Science Institute*

1. Introduction

A collaboration between STScI and ARI has produced a new astrometric reduction of the Guide Star Catalog (GSC, Lasker *et al.* 1990). This new version, GSC 1.2, has dramatically reduced the systematic errors present in GSC 1.1. The positions in GSC 1.1 are affected by plate-based systematic distortions which are largest at the plate edges (1.0″, north; 1.2″, south) (Taff *et al.* 1990a). These positions also suffer from systematic errors which are a function of magnitude and radial distance from the plate center (Morrison *et al.* 1996). This effect is small for radii under 2.7° from the plate center, then rapidly increases producing an average offset of the faint stars (15^m) versus the reference stars (10^m) of 0.9″ at the plate edges.

2. New Reduction: GSC1.2

Using the filter method (Röser *et al.* 1995) GSC 1.1 was placed onto the PPM system. The mean of the plate-based distortions were removed with a mask technique (Taff *et al.* 1990b) using the Astrographic Catalogue (AC) as the reference material. The AC was also used to remove the radial magnitude dependent systematic errors. (Morrison and Röser, this volume, p. 381).

Table 1 shows the comparison between the GSC1.2, PPM and Carlsberg Meridian Catalogue (CMC). Note, these values are influenced by the rms error of the PPM and the unknown proper motions in the CMC. Figure 1 shows the rms differences in both coordinates derived from different (overlapping) plates, averaged over all overlapping plate pairs for the −30° zone. The improvement with respect to GSC 1.1 is dramatic.

TABLE 1. RMS Results for GSC 1.2

Version	PPM ($V \geq 8$)		CMC ($V \geq 8$)	
	ra cos(dec)	dec	ra cos(dec)	dec
GSC 1.1	0.65″	0.53″	0.57″	0.54″
GSC 1.2	0.31″	0.31″	0.40″	0.40″

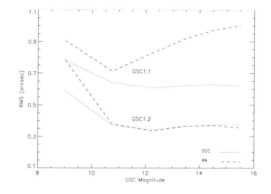

Figure 1. RMS-differences of GSC coordinates measured on overlapping plates

This new reduction has, on the average, eliminated the major systematics errors found in GSC 1.1. The astrometric data of GSC 1.2 are available on the web. The overall accuracy of the positions in GSC 1.2 is better than 0.3″. A few caveats on GSC 1.2 are: (1) it is on the PPM not the HIPPARCOS system, (2) it is not compatible with ST ScI Digitized Sky Surveys, and (3) at present *it must not be used for HST observation planning*.

References

Lasker, B. M., *et al.* 1990 Astron.J. 99, 2019
Morrison, J.E., *et al.* 1996, Astron.J. 111, 1405
Röser, S., Bastian, U., and Kuzmin, A. V. 1995, IAU Colloquium 148, ed. J. M. Chapman *et al.*
Taff, L. G., *et al.* 1990a, Astrophys.J., 353, L45
Taff, L. G., Lattanzi, M. G., and Bucciarelli, B. 1990b, Astrophys.J. 358, 359

THE 488,006,860 SOURCES IN THE USNO-A1.0 CATALOG

B. CANZIAN
USRA/U. S. Naval Observatory, Flagstaff

Abstract. The USNO-A1.0 catalog was generated from the U. S. Naval Observatory's digitization of the Palomar Observatory Sky Survey I O and E survey plates for fields with central declination $\delta \geq -30°$, and from the European Southern Observatory R and Science Research Council J survey plates for fields with central $\delta \leq -35°$ using the Precision Measuring Machine (PMM) located at the Flagstaff Station. It lists positions (α and δ in J2000) and brightnesses (red and blue magnitudes on the parent plate system) for all objects. A flag indicates if the entry also exists in GSC1.1.

1. Object Distribution

The PMM detected and measured objects at and beyond the nominal visual limiting magnitudes of $O = 21$, $E = 20$, $J = 22$, and $F = 21$. USNO-A1.0 was constructed from the correlation of the blue and red detection lists, and the only acceptance requirement was that there be matching detections with a 1 arcsec radius aperture. The object classification information was ignored: USNO-A1.0 is a catalog of astrometric standards.

The plot of the sky shown in Figure 1 was generated from the 488,006,860 sources contained in the USNO-A1.0 catalog. The sky is shown in Galactic coordinates (Galactic Center at center, north up, longitude increasing to the right) and the shading is proportional to the logarithm of the number of catalog sources per square degree. The increase in brightness in the vicinity of the south equatorial pole arises from the approximate factor of two increase in the number of PMM detections found on Southern survey plates as compared to Northern survey plates. The Large and Small Magellanic Clouds are obvious and dust in nearby clouds obscures parts of the Galactic plane. Dark spots mark the brightest stars in the sky, where detections are suppressed owing to flare on the plates.

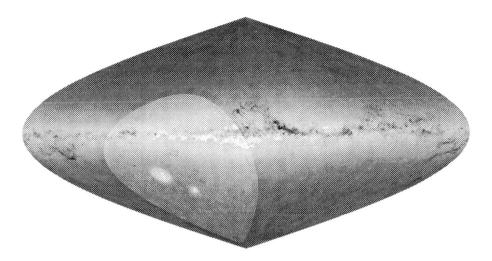

Figure 1. This is a representation of all the objects in the USNO-A1.0 catalog. The grayscale is coded to the logarithm of star density with black equal to 2.8 (about 700 deg^{-2}) and white equal to 5.1 (about 130,000 deg^{-2}).

2. Calibration

The USNO-A1.0 astrometry has been reduced with respect to GSC1.1, and so retains all the systematic errors of that catalog. Nonetheless, the USNO-A1.0 catalog is useful as a deep, all-sky reference catalog. The mean dispersion of the astrometric error in USNO-A1.0 is about 0.4 arcsec.

The bright photometry of USNO-A1.0 has been calibrated using the Tycho input catalog. The faint photometry has been calibrated using CCD photometry of 185 fields in common with the USNO parallax program. Systematic differences from plate to plate in the photometric slopes and zero points were removed by relaxing all plates to a common system using their overlap regions. The dispersion of the photometric error for faint (13–21st magnitude) stars is about 0.3 mag.

Acknowledgements

The USNO PMM has been funded by the U. S. Navy and Air Force, and USNO-A1.0 incorporates intellectual property rights held by Palomar Observatory, California Institute of Technology, the National Geographic Society, European Southern Observatory, the Particle Physics and Astronomy Research Council, the Anglo Australian Observatory, and the Space Telescope Science Institute.

CATALOGING OF THE DIGITIZED POSS-II: INITIAL SCIENTIFIC RESULTS

S.G. DJORGOVSKI[1], R.R. De CARVALHO[1,2], R.R. GAL[1],
M.A. PAHRE[1], R. SCARAMELLA[3] AND G. LONGO[4]
[1]*Palomar Observatory, Caltech, Pasadena, CA 91125, USA*
[2]*Observatorio Nacional, CNPq, 20921 Rio de Janeiro, Brasil*
[3]*Osservatorio Astr. di Roma, I-00040 Monteporzio, Italy*
[4]*Osservatorio Astr. di Capodimonte, I-80131 Napoli, Italy*

1. The Survey

The Second Palomar Sky Survey (POSS-II) is now nearing completion. It will cover the entire northern sky with 894 fields (6.5° square) at 5° spacings, with no gaps in the coverage. Plates are taken in three bands: IIIa-J + GG395, $\lambda_{eff} \sim 480$ nm; IIIa-F + RG610, $\lambda_{eff} \sim 650$ nm; and IV-N + RG9, $\lambda_{eff} \sim 850$ nm. Typical limiting magnitudes reached are $B_J \sim 22.5$, $R_F \sim 20.8$, and $I_N \sim 19.5$, *i.e.*, $\sim 1^m - 1.5^m$ deeper than the POSS-I. The image quality is improved relative to the POSS-I, and is comparable to the southern photographic sky surveys. For more details, see Reid *et al.* (1987), and Reid & Djorgovski (1993).

The plates are being digitized at STScI, using modified PDS scanners (see the papers by McLean, Lasker, *et al.* in this volume). Plates are scanned with 15-micron (1.0 arcsec) pixels, in rasters of 23,040 square, giving ~ 1 GB/plate, or ~ 3 TB of pixel data total for the entire digital survey (DPOSS). Preliminary astrometric solutions are good to ~ 0.5 arcsec, and will get better soon. Completion of the scanning should follow closely the completion of the plate taking, hopefully by mid-1998. The plates are also digitized independently at USNOFS by Monet *et al.*

There is a major ongoing effort at Caltech to process and calibrate the scans, while at the same time catalog and classify all objects detected down to the survey limit. We are using SKICAT, a novel software system developed for this purpose (Weir *et al.* 1993ab, 1994, 1995abc; Djorgovski *et al.* 1994; Fayyad *et al.* 1996). SKICAT incorporates some standard astronomical image processing packages, commercial Sybase DBMS, as well as a number of artificial intelligence (AI) and machine learning based modules.

As of late 1996, some 15% of the entire survey area has been processed, but this work will soon speed up considerably. We have started a collaborative effort between the Caltech DPOSS group and the Observatories of Rome and Naples (Italy) to complete the DPOSS processing in a timely manner (project CRONA). Data processing pipelines are now being set up at both Italian sites. Observatorio Nacional in Rio de Janeiro (Brasil) will join the consortium in 1997.

The resulting Palomar-Norris Sky Catalog (PNSC) will contain all objects down to an equivalent limiting magnitude of $B_J \sim 22^m$, with star-galaxy classification accurate to 90% or better, down to $B_J \sim 21^m$. The PNSC is expected to contain > 50 million galaxies, and > 2 billion stars, including $\sim 10^5$ quasars. We note that the size of the DPOSS data set, in terms of the bits, numbers of sources, and resolution elements, is $\approx 1,000 \times$ the entire IRAS data set, and is $\approx 0.1 \times$ the anticipated SDSS data set.

We will publish the catalogs as soon as the validation tests are complete, and the funding allows it, via computer networks and other suitable media. The anticipated project completion timescale is ~ 3 years.

2. The Data

A particular strength of SKICAT is the star-galaxy classification, which uses artificial induction decision tree techniques. By using these methods, and using superior CCD data to train the AI object classifiers, we are able to achieve classification accuracy of 90% or better down to $\sim 1^m$ above the plate detection limit; traditional techniques achieve comparable accuracy typically only $\sim 2^m$ above it. This effectively triples the number of usable objects for most scientific applications of these data, since in most cases one wants either stellar objects or galaxies.

Future technical developments include an improved treatment of very bright and/or extended objects, optimization of the object measurement module for crowded regions (*e.g.*, low Galactic latitudes), better structuring of the catalog database for efficient access and manipulation, and testing and implementation of novel methods for data exploration, including unsupervised classifiers and clustering analysis algorithms, *etc.* Some initial results have been presented by de Carvalho *et al.* (1995).

An extensive CCD calibration effort is now underway at the Palomar 60-inch telescope, and we expect it to expand to other sites soon. The data are calibrated in the Gunn *gri* system. We obtain 2 CCD images per sky survey field, sometimes more. Usually these CCD images are used both for magnitude zero-point calibrations, and for training of automated star-galaxy classifiers. In addition to the CCD calibrations, we use heavily smoothed sky measurements from the plate scans themselves (after the object removal) to "flatfield" away the telescope vignetting effects and the individual plate emulsion sensitivity variations.

As a result, we have demonstrated an unprecedented photometric stability and accuracy for this type of photographic plate material (Weir *et al.* 1995a). We have performed tests using both CCD sequences and plate overlaps, and find that our magnitude zero-points are stable to within a few percent, across the plates, between adjacent plates, and across the individual plates. Typical r.m.s. in the magnitude zero-points between different plates is 0.015^m–0.045^m in the r band, slightly worse in the g band, perhaps due to the larger color terms in the J/g calibration. Keeping the systematic magnitude zero-point errors below 10% is essential for many scientific applications of these data.

This may be best illustrated in the internal consistency of galaxy counts from sets of adjacent POSS-II plates. We compared the counts published by Picard (1991), who used COSMOS machine scans processed in a traditional way, with the counts from Weir *et al.* (1995a), who used DPOSS scans processed with SKICAT. While Picard has seen large plate-to-plate variations (a factor of 2, or more) in number counts at a given magnitude, Weir finds excellent agreement, to within the Poissonian errors, and reaches a magnitude deeper; yet both used the same kind of plate material.

Median random magnitude errors for stellar objects in all three bands start around 0.05^m at the bright end, and increase to $\sim 0.25^m$ at $g_{lim} \approx 22^m$, $\sim 0.20^m$ at $r_{lim} \approx 21.5^m$, and $\sim 0.25^m$ at $i_{lim} \approx 20^m$. For galaxies, these errors are typically higher by about 50% at a given magnitude.

3. Some Initial Scientific Applications

This large new database should be a fertile ground for numerous scientific investigations for years to come. The nature of the data dictates its uses: these images are not very deep by modern standards, but they do cover uniformly a very large solid angle. In addition to the obvious applications such as large-scale optical identifications of sources from other wavelengths (*e.g.*, radio, X-ray, IR), there are two kinds of studies which can be pursued very effectively with data sets of this size:

(1) Statistical astronomy studies, where the sheer large numbers of detected sources tighten the statistical errors and allow for more model parameters to be constrained meaningfully by the data.

(2) Searches for rare types of objects. For example, at intermediate Galactic latitudes, about one in a million stellar objects down to $r \approx 19.5^m$ is a quasar at $z > 4$, although we can find such quasars very efficiently.

We have already started a number of scientific projects using DPOSS, which not also serve as scientific verification tests of the data, but have helped us catch some errors and improve and control the data quality.

Galaxy counts and colors in 3 bands from DPOSS can serve as a baseline for deeper galaxy counts and a consistency check for galaxy evolution models. Our initial results (Weir *et al.* 1995a) show a good agreement with simple models of weak galaxy evolution (*e.g.*, Koo *et al.* 1995) at low redshifts, $z \sim 0.1$–0.3. We are now expanding this work to a much larger area, to average over the local large-scale structure variations. Our galaxy catalogs have been used as input for redshift surveys down to $\sim 21^m$, *e.g.*, in the Palomar-Norris survey (Small *et al.* 1997). Several other groups also plan to use them for their own redshift surveys.

Galaxy n-point correlation functions and power spectra of galaxy clustering provide useful constraints of the CDM and other scenarios of large scale structure. Our preliminary results from a limited area near the NGP (Brainerd *et al.* 1995; and in preparation) indicate that there is less power at large scales than was found by the APM group (Maddox *et al.* 1989) in their southern survey. We suspect that field-to-field magnitude zero-point calibration errors and errors in star-galaxy separation in the APM data may account for this discrepancy. High quality, uniform calibrations are absolutely essential for this task. We are now also starting to explore the correlations of our galaxy counts with H I, IRAS, and DIRBE maps in order to better quantify the foreground Galactic extinction. We expect to generate extinction maps superior to those now commonly used.

We are now starting a project to generate an objectively defined, statistically well defined catalog of rich clusters of galaxies. We estimate that eventually we will have a catalog of as many as 20,000 rich clusters of galaxies at high Galactic latitudes in the northern sky. Their median redshift is estimated to be $\langle z \rangle \sim 0.2$, and perhaps reaching as high as $\langle z \rangle \sim 0.5$.

There are many cosmological uses for rich clusters of galaxies. They provide useful constraints for theories of large-scale structure formation and evolution, and represent valuable (possibly coeval) samples of galaxies to study their evolution in dense environments. Studies of the cluster two-point correlation function are a powerful probe of large-scale structure, and the scenarios of its formation. Correlations between optically and X-ray selected clusters are also of considerable scientific interest. Most of the studies to date have been limited by the statistical quality of the available cluster samples. For instance, the subjective nature of the Abell catalog has been widely recognized as its major limitation. Still, many far-reaching cosmological conclusions have been drawn from it. There is thus a real need to generate well-defined, objective catalogs of galaxy clusters and groups, with well understood selection criteria and completeness.

We use only objects classified as galaxies in DPOSS catalogs, down to $r = 19.6^m$, where the accuracy of object classifications is $> 90\%$. We then use colors for selection of the candidate cluster galaxies: early-type galaxies should better delineate high-density regions. Next we apply the adaptive kernel method to create the surface density maps. Its major advantage is that it uses a two-step process which smooths well the low density regions, and at the same time leaves the high density peaks nearly unaffected. Finally we evaluate the statistical significance of the density peaks using a bootstrap technique. Typically we set our threshold at a $4.5\,\sigma$ level, where we successfully recover all of the known Abell clusters of richness class 0 and higher, and also find a large number of new cluster candidates which were apparently missed by Abell. We have also started spectroscopic follow-up of our cluster candidates at the Palomar 200-inch telescope.

Another ongoing project is a survey for luminous quasars at $z > 4$. Quasars at $z > 4$ are valuable probes of the early universe, galaxy formation, and the physics and evolution of the intergalactic medium at large redshifts. The continuum drop across the Lyα line gives these objects a distinctive color signature: extremely red in $(g-r)$, yet blue in $(r-i)$, thus standing away from the stellar sequence in the color space. Traditionally, the major contaminant in this type of work are red galaxies. Our superior star-galaxy classification leads to a manageable number of color-selected candidates, and an efficient spectroscopic follow-up. As of late 1996, over 25 new $z > 4$ quasars have been discovered. Images are available to other astronomers for their studies as soon as the data are reduced.

Our initial results (Kennefick *et al.* 1995ab) are the best estimates to date of the bright end of the quasar luminosity function (QLF) at $z > 4$, and are in excellent agreement with the fainter QLF evaluated by Schmidt *et al.* (1995). We have thus confirmed the decline in the comoving number density of bright quasars at $z > 4$. There are also some intriguing hints of possible primordial large-scale structure as marked by these quasars. However, much more data is needed to check this result.

We have also stared optical identifications of thousands of VLA FIRST radio sources (Becker *et al.* 1995). Our preliminary results indicate that there are ~ 400 compact radio source IDs per DPOSS field, and we expect a comparable number of resolved source IDs. Among the first 10 red stellar-like IDs we have observed spectroscopically in May 1996, we discovered a quasar at $z = 4.36$, VF 141045+340909. We estimate that a few tens of $z > 4$ quasars will be found in the course of this work. Eventually, we expect to have $> 10^5$ IDs for the VLA FIRST sources, plus many more from other surveys.

In the area of statistical gravitational lensing studies, we have explored the possibility of microlensing of quasars, by looking for a possible excess of foreground galaxies near lines of sight to apparently bright, high-z QSOs from flux-limited samples (Barton *et al.*, in prep.). We find, at most, a modest excess, roughly as expected from theory, in contrast to some previous claims which used similar data (*e.g.*, Webster *et al.* 1988). We are also planning to use our galaxy counts to explore the possible lensing magnification of background AGN by foreground large scale structure, as proposed, *e.g.*, by Bartelmann & Schneider (1994).

Other extragalactic projects now planned include a catalog of $\sim 10^5$ brightest galaxies in the northern sky, with a quantitative surface photometry and morphological information, automated searches for low surface brightness galaxies, an archival search for supernovæ from plate overlaps, derivation of photometric redshift estimators for galaxies, automated optical identifications of IR and X-ray sources, and so on.

Galactic astronomy should not be neglected. Star counts as a function of magnitudes, colors, position, and eventually proper motions as well, fitted over the entire northern sky at once, would provide unprecedented discrimination between different Galactic structure models, and constraints on their parameters. With $\sim 2 \times 10^9$ stars, such studies would present a major advance over similar efforts done in the past.

We can also search for stars with unusual colors or variability. We have started a search for stars at the bottom of the main sequence and field brown dwarf candidates, using colors: anything with $(r-i) > 2.5$ should be interesting. At high Galactic latitudes, about one star in a million is that red, down to the conservative limit used so far ($r < 19.5^m$). Such a survey can be made much more powerful with the addition of IR data.

The same techniques we use to search for galaxy clusters can then be applied to our star catalogs, in an objective and automated search for sparse globulars in the Galactic halo, tidal disruption tails of former clusters, and possibly even new dwarf spheroidals in the Local Group (recall the Sextans dwarf, found using similar data by Irwin *et al.* 1990).

4. Concluding Remarks

These, and other studies now started or planned, should produce many interesting and useful new results in the years to come. Availability of large data sets such as DPOSS over the Net or through other suitable mechanisms would also enable astronomers and their students anywhere, even if they are far from the major research centers or without an access to large telescopes, to do some first-rate observational science. This new abundance of good data may profoundly change the sociology of astronomy.

Nor should we discount serendipity: With a data set as large as DPOSS, there is even an exciting possibility of discovering some heretofore unknown types of objects or phenomena, whose rarity would have made them escape the astronomers' notice so far.

This is a foretaste of things to come: with DPOSS, GSC-II, and surveys to follow (*e.g.*, SDSS, 2MASS, *etc.*), we are changing the very concept of an astronomical catalog, into a living, permanently evolving data set, which must come along with adequate tools for its exploration. We will need to learn new skills, develop new data mining and exploration tools (including AI and machine learning techniques), new data structuring paradigms and standards, and above all, learn to ask new kinds of astronomical questions.

We are grateful to our many collaborators, including Nick Weir, Joe Roden, Julia Kennefick, Tereasa Brainerd, Usama Fayyad, Jeremy Darling, Vandana Desai, Emil Kartalov, Paul Stolorz, Alex Gray, Daniel Stern, and Isobel Hook; to Neill Reid, Jean Mueller, and others in the POSS-II survey team; to Barry Lasker, Brian McLean, and others in the digitization team at STScI; and to Massimo Capaccioli, Roberto Buonanno, and other CRONA-ies. This work was supported at Caltech by the NSF PYI award AST-9157412, grants from NASA, the Bressler Foundation, and Palomar Observatory. This paper is the CRONA Contribution No. 1.

References

Bartelmann, M., & Schneider, P. 1994, Astron.Astrophys., 284, 1
Becker, R., White, R., & Helfand, D. 1995, Astrophys.J., 450, 559
Brainerd, T., de Carvalho, R., & Djorgovski, S. 1995, Bull.Am.Astron.Soc., 27, 1364
de Carvalho, *et al.* 1995, Astron. Soc. Pacific Conf. Series, 77, 272
Djorgovski, S., Weir, N. & Fayyad, U. 1994, Astron. Soc. Pacific Conf. Series, 61, 195
Fayyad, U., Djorgovski, S.G. & Weir, N. 1996, in Advances in Knowledge Discovery and Data Mining, eds. U. Fayyad *et al.*, (Boston: AAAI/MIT Press), 471
Irwin, M., *et al.* 1990, Mon.Not.R.astron.Soc., 244, 16P
Kennefick, *et al.* 1995a, Astron.J., 110, 78
Kennefick, J.D., Djorgovski, S.G. & de Carvalho, R. 1995b, Astron.J., 110, 2553
Koo, D., Gronwall, C., & Bruzual, G. 1995, Astrophys.J., 440, L1
Maddox, S., *et al.* 1989, Mon.Not.R.astron.Soc., 242, 43P
Reid, I.N., *et al.* 1987, Publ.Astron.Soc.Pacific, 103, 661
Reid, I.N., & Djorgovski, S. 1993, Astron. Soc. Pacific Conf. Series, 43, 125
Schmidt, M., Schneider, D. & Gunn, J. 1995, Astron.J., 110, 68
Small, T., Sargent, W., & Hamilton, D. 1997, Astrophys.J.Supp., in press
Webster, R., Hewett, P., Harding, M., & Wegner, G. 1988, Nature, 336, 358
Weir, N., *et al.* 1993a, Astron. Soc. Pacific Conference Series, 43, 135
Weir, N., *et al.* 1993b, Astron. Soc. Pacific Conf. Series, 52, 39
Weir, N., *et al.* 1994, IAU Symp. 161, 205 eds. H. MacGillivray *et al.*
Weir, N., Djorgovski, S. & Fayyad, U. 1995a, Astron.J., 110, 1
Weir, N., Fayyad, U. & Djorgovski, S. 1995b, Astron.J., 109, 2401
Weir, N., *et al.* 1995c, Publ.Astron.Soc.Pacific, 107, 1243

THE SECOND GUIDE STAR CATALOGUE

B. McLEAN[1], G. HAWKINS[1], A. SPAGNA[2], M. LATTANZI[2],
B. LASKER[1], H. JENKNER[1,3] AND R. WHITE[1]
[1] *Space Telescope Science Institute*
[2] *Osservatorio Astronomico di Torino*
[3] *European Space Agency*

1. Introduction

Although the HST GSC–I (Paper-I: Lasker *et al.* 1990, Paper-II: Russell *et al.* 1990, Paper-III: Jenkner *et al.* 1990) has been used with great success operationally, it was always known that it was possible to improve the scientific and operational usefulness by an increase in scope to include multi-color and multi-epoch data. Once the GSC–II concept was established, it was evident that, even beyond the original motivations in HST operations, it would address a number of other astronomical needs such as increasing demands for fainter catalogues to support remote or queue scheduling capabilities and adaptive optics on the next generation of large-aperture, *new-technology* telescopes. In addition, the all sky nature of the GSC–II makes it a natural data source for research in galactic structure.

2. The GSC-II Project

Once the broad applications of the GSC–II were recognized and its resource requirements were considered with care, it was clear that the project needed to be expanded beyond its original ST ScI context. This led to the formation of a GSC–II consortium (STScI, the Italian Council for Research in Astronomy and the Astrophysics Division of the European Space Agency), with the formal goal of producing the GSC–II as an all-sky catalogue of positions, proper motions, magnitudes and colours, complete to 18th magnitude.

3. GSC-II Overview

The GSC–II will be constructed using the digitized sky survey plates available at STScI. The POSS-II (J and F plates), ER, and SES surveys have finer grain emulsions, hence better resolution, and a deeper limiting magnitude than the original POSS surveys while the first epoch plates can be used to provide a baseline to determine the stellar proper motions.

The Guide Star Photometric Catalog (GSPC–I; Lasker *et al.* 1988) is being supplemented by CCD observations to provide a set of standard stars to at least 18th magnitude over the entire sky for each Schmidt plate. The future availability of the HIPPARCOS and TYCHO catalogues will provide a dense grid of very accurate astrometric standards to tie the plates to the inertial reference frame and may allow the mapping of field variations in the sensitivity of the plate emulsions. The addition of *texture* values for each object and the ranking of all object parameters has proven to be an effective technique for training a decision tree classifier and providing more reliable object classifications.

The improvement in the projected precision of GSC–II compared to the original GSC based on the current prototype software is shown in Table 1.

TABLE 1. Properties of GSC-I and GSC-II

Property	GSC 1.1	GSC 2.0
Number Objects	1.8×10^7	2.0×10^9
Epochs	1	2–3
Passbands	1	>2
Limiting Magnitude	15	>18
Relative/Absolute Position Error	0.4″/ 1.0″	0.15″/ 0.5″
Proper Motion Error	n/a	<4.0 mas yr^{-1}
Magnitude Error	0.4 mag	0.1–0.2 mag
Classification Accuracy	99% stellar	>95% unbiased

References

Jenkner, H., *et al.* 1990 Astron. J. 99, 2081
Lasker, B. M., *et al.* 1988 Astrophys. J. Suppl. 68, 1
Lasker, B. M., *et al.* 1990 Astron. J. 99, 2019
Lasker, B. M. 1994, Proceedings of Astronomy from Wide Field Imaging, IAU Symp. 161, p. 235, MacGillivray *et al.* (eds)
Russell, J. L., *et al.* 1990 Astron. J. 99, 2059

THE ROSAT ALL-SKY SURVEY BRIGHT SOURCE CATALOG

W. VOGES, B. ASCHENBACH, Th. BOLLER,
H. BRÄUNINGER, U. BRIEL, W. BURKERT, K. DENNERL,
J. ENGLHAUSER, R. GRUBER, F. HABERL, G. HARTNER,
G. HASINGER, M. KÜRSTER, E. PFEFFERMANN,
W. PIETSCH, P. PREDEHL, C. ROSSO, J.H.M.M. SCHMITT,
J. TRÜMPER AND U. ZIMMERMANN
Max-Planck-Institut für extraterrestrische Physik, Germany

1. Screening Process

In order to ensure the quality of the source catalogue derived from the SASS processing an automatic as well as a visual screening procedure was applied to 1378 survey fields. Most (94%) of the 18,811 sources were confirmed by this screening process. The rest is flagged for various reasons. Broad band images are available for a subset of the flagged sources. Details of the screening process can be found at

www.rosat.mpe-garching.mpg.de/survey/rass-bsc/doc.html.

2. Catalogue Access

Lists of the catalogue sources—selected by certain properties—may be retrieved via the **ROSAT Source Browser** at:

www.rosat.mpe-garching.mpg.de/survey/rass-bsc/scr_by.html.
In certain cases, the list provided contains also links to source images. The catalogue can be retrieved via the WWW or anonymous ftp at

(ftp.rosat.mpe-garching.mpg.de; cd archive/survey/rass-bsc).
More information on the ROSAT All-Sky Survey Bright Source Catalogue can be found at:

www.rosat.mpe-garching.mpg.de/survey/rass-bsc.html
A list containing correlations with other catalogues and identifications is under preparation and will be available via the World Wide Web. A detailed description of the generation and the scientific content of the ROSAT BSC will be published in Astronomy & Astrophysics.

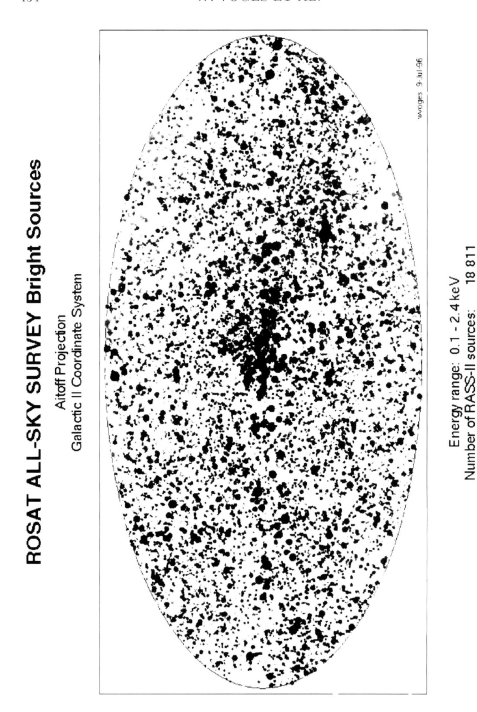

Figure 1. The ROSAT All-Sky Survey Bright Source Catalogue sources. The originally colour coded hardness ratio is translated into a grey scale (light grey: soft and black: hard X-rays). The size of the symbols corresponds to the source count rate.

Part 9. Multi-Wavelength Cross Identification

CROSS WAVELENGTH COMPARISON OF IMAGES AND CATALOGS

J.G. BARTLETT AND D. EGRET
Centre de Données astronomiques de Strasbourg, France

1. Introduction

We would like to start this discussion by attempting to make a useful, working distinction between *catalogs* and *databases* of astronomical objects. It seems to us that such a distinction could be made based upon the mode of access to the information: A catalog may be considered as a *list* of objects, almost invariably ordered by coordinates; a database, on the other hand, may be distinguished by its ability to extract a set of objects based on user-given criteria, such as all objects within a certain sky region with magnitudes brighter than m in the blue. Catalogs form the basis of the database, which adds the means of multi-criteria access to the information. In concrete terms, one usually thinks of SIMBAD, NED and LEDA as databases, while an ftp site containing electronic lists of objects may be thought of as a catalog storage warehouse.

The next comment we would like to make is related to the above definition and concerns the sizes of astronomical catalogs. In Table 1, we list the numbers of objects expected or currently existing in various catalogs. The list is by no means exhaustive, but only meant to give an idea of the number of catalogued objects as a function of spectral domain. It is clear that there is a large dichotomy between the Optical/Near-infrared (NIR) and the rest of the spectrum: for the former, one expects on the order of 10^8–10^9 sources, while in no other part of the spectrum is there more than a *few* 10^6 sources. In terms of numbers and storage volume, *the Optical/NIR dominates*, and this has important consequences for both the access to and the science performed with astronomical catalogs.

What is the origin of this dominance? As an answer to this question, we will take the opportunity to defend a little the oft-attacked photographic

TABLE 1. Number of Cataloged Objects by Spectral Domain

Optical/NIR		Other	
GSC-II:	10^9	FIRST:	10^6
USNO A1:	5×10^8	IRAS:	10^5
SuperCOSMOS/APM:	5×10^8	ROSAT:	10^5
DENIS/2MASS:	10^9		
	$\sim 10^8 - 10^9$		$< 10^6$

plate. The advantages of CCDs should not obscure the fact that the photographic plate has proven itself as a magnificent detector, combining a large field-of-view with high spatial resolution; *and* it is its own storage medium. Remember that a $6° \times 6°$ Schmidt plate with 1 arcsec resolution represents 4.7×10^8 pixels taken in a single exposure! No other detector to date can cover the focal plane with more efficiency. Although modern detectors are beginning to achieve the same sky coverage and resolution as photographic plates by scanning CCD arrays across the sky (such as DENIS, 2MASS and the Sloan Digital Sky Survey), it seems to us arguable that part of the reason for the dominance of the Optical/NIR rests with the one-hundred year legacy of photographic plates.

It must be said that, apart from the serious drawbacks of nonlinearity and calibration difficulties, photographic plates present a problem of access: the information is not readily manipulated, as is a computerized image. This has changed with the advent of plate scanning machines, which produce the desired electronic quantification. However, this digitization program suffers from the need of adequate storage media for the terabytes of data that a full sky Optical/NIR survey produces. This problem highlights the performance of a photographic plate as a storage medium, which, in terms of the ratio (viability)/(volume) over long periods of time, is as yet unmatched.

In fact, the key challenge facing astronomical archiving at present is this question of storage and access to such large quantities of data. As an example, the "classical," if such a term is now permitted, databases such as SIMBAD, NED or LEDA contain at most around a few million objects; multi-criteria access to a much larger number of objects seems to demand novel techniques, and forms a large part of the development effort of the Sloan Survey (Szalay & Brunner, this volume, p. 455). An interesting question in this context is the relation between efficient storage and science. Designing an archive structure for quick access requires some

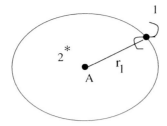

Figure 1. Geometry for the cross-identification of source A with source 1.

knowledge of the correlations and clustering of the data in the space defined by the measured parameters. This amounts to a statistical analysis of the properties of the survey objects.

Let us finish this section by returning to the implications of the Optical/NIR dominance. One important consequence of this dominance is that it would seem to guarantee that most cross-identification efforts of newly detected objects in other wavebands will rely on Optical/NIR catalogs. The questions we would pose are: "Is it useful to have a centralized *database* of the ensemble of Optical/NIR objects?"; and, if so, "Who has the resources to construct it?" These questions concern the construction of a true multi-criteria *database*, the fundamental elements of which, as we have seen, require something beyond that presently used by NED/SIMBAD/LEDA.

2. Cross Identifications: Nature of the Problem

In this section, we would like to discuss some statistical aspects of the general problem of identifying objects from two different catalogs, a procedure known as the cross-identification of sources. The two catalogs will be called *Letters* and *Numbers*.

2.1. AS A TEST OF HYPOTHESES

Consider the situation depicted in Figure 1, where we wish to cross-identify a source, A, from catalog *Letters* with objects from the catalog *Numbers*, two of which are shown within the error ellipse of source A (sources 1 and 2). Key to the problem are accurate positions in the two catalogs, because spatial proximity serves as a primary criterion. After selecting objects for consideration based on coordinates, one may think to apply supplementary criteria; for example, suppose that A were an IRAS source with galaxy colors, and that source 1 is an optical galaxy, while source 2 is a star. This additional information would favor an identification with the optical galaxy.

A quantitative approach must rely on a measure of the acceptability of a given cross-identification. If we ignore, for simplicity, any supplemental information, we may define the *likelihood* that source 1 is the same as source A as the *probability* that the two cataloged positions would be separated by distance r_1 *if they represented the same physical object*. Assume that the position of source 1 (and 2) is much more precise than that of source A; and, further, that the error in position A is described by a two-dimensional Gaussian. Then,

$$L_1 \equiv \bar{N}(<m_2) e^{-\bar{N}(<m_2)} \frac{1}{2\pi |M|^{1/2}} e^{-\frac{1}{2}\vec{r}_1^T \cdot M^{-1} \cdot \vec{r}_1}, \qquad (1)$$

where M is the covariance matrix and $|M|$ is its determinant. The Poisson term expresses the probability of finding one source from *Numbers* by chance alignment within r_1 at magnitudes brighter than m_2, when the expected number is $\bar{N}(<m_2)$. We denote this likelihood by the subscript 1, which will hereafter refer to the *hypothesis* $H1$ that sources 1 and A are the same. Notice that this procedure requires a clear quantitative statement of the positional errors (*i.e.*, a Gaussian distribution). The alternate hypothesis, that sources 1 and A are separated by distance r_1 by chance alignment, may be assigned the likelihood

$$L_0 \equiv \bar{N}(<m_2) e^{-\bar{N}(<m_2)} n(m_1). \qquad (2)$$

Here, $n(m_1)$ denotes the density of objects in the reference catalog (*Numbers*) at the magnitude of source 1, m_1. The subscript 0 will denote the hypothesis $H0$ that the sources, 1 and A, are not the same physical object.

The decision to cross-identify sources 1 and A now amounts to the statistical rejection of the null hypothesis, $H0$. For this purpose, we may employ the *likelihood ratio test* (Meyer 1975), which focuses on the quantity $\lambda = L_1/L_0$. We must now construct the probability distributions of λ under the assumptions $H0$ ($p_{H0}(\lambda)$) and $H1$ ($p_{H1}(\lambda)$). This step *requires* a set, sometimes called the *training set*, of previously known, sure-fire cross-identifications. We emphasize the importance of this training set and the fact that the need for such *a priori* information is unavoidable in any quantified approach. Suppose, then, that we have found the two distributions and that they resemble the curves shown in Figure 2.

We identify source A with source 1 with *confidence* α if the observed value $\lambda_{obs} > \lambda_*$, where λ_* is defined by $P_{H0}(\lambda > \lambda_*) = \alpha$ (note that capital P refers to the cumulative distributions). The confidence α represents the probability of a type I statistical error, or the chance that the cross-identification is wrong. The *power*, β, of our cross-identification is defined by $\beta = P_{H1}(\lambda > \lambda_*)$; the quantity $1 - \beta$ represents the probability of a type II statistical error and embodies the concept of the *reliability* of a conclusion that the sources are *not* the same. A wonderful example of this

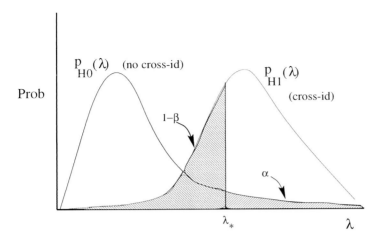

Figure 2. Likelihood Ratio Test: The distributions $p_{H0}(\lambda)$ and $p_{H1}(\lambda)$ are shown and labeled (see text). The shaded area to the right of λ_* represents α and that to the left represents $1-\beta$.

kind of approach is given by Lonsdale *et al.* (1996), and we are grateful to C. Lonsdale for helpful discussion on this topic.

2.2. AS A MAXIMIZATION OF LIKELIHOOD

In the case just described, the cross-identification was made without regard to the outcome of attempts to cross-identify other sources. This approach might be readily applied when catalog *Letters* contains many fewer objects than catalog *Numbers* (a situation we may describe by saying that "catalog *Numbers* is dense within catalog *Letters*"). Consider another case in which the two catalogs each have about the same source density, as schematically represented in Figure 3. In pursuing the cross-identification of source A with source 1, it now seems particularly urgent to incorporate the possibility that source 1 could also be identified with source B. One way of proceeding is to calculate the probability, P_i, $i = 1, 7$, of each of the seven possible outcomes of the cross-identification of the two catalogs: $(A, 1)(B, 2)$, $(A, 1)(B, 0)$, $(A, 2)(B, 1)$, $(A, 2)(B, 0)$, $(A, 0)(B, 1)$, $(A, 0)(B, 2)$, $(A, 0)(B, 0)$, where a 0 means that the source in *Letters* has no counterpart in catalog *Numbers*. The case i which has the largest probability represents the *most likely* cross-identification of the two catalogs, and therefore the proper choice. This approach is global in that it seeks the most satisfactory solution for all the sources considered at once. Note that this method is well adapted to treat systematic problems, such as a coordinate offset between the two catalogs; in fact, it incorporates what we

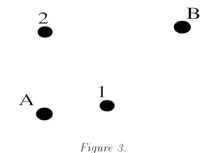

Figure 3.

mentally do to align two sky charts. The drawback of this approach, as formulated, is the large computing power required to consider all possible cross-identifications.

3. Two Software Systems

In this section, we present the characteristics of two software systems designed to aid the astronomer in making cross-identifications. These packages are interactive, and therefore not suitable for performing automatic cross-identifications of large catalogs. Their philosophy is, rather, to replace by interactive software the tasks of going to the library to obtain and combine copies of images (e.g., sky survey films), tables and catalogs for a specific region of the sky. The second package, **Aladin**, is particularly well adapted for the training/optimization of automated routines (recall the importance of the training set in Section 2), the cross-identifications of small, user catalogs and the resolution trouble cases, which is useful for astronomical database quality control and for "mopping-up" difficult objects flagged by an automatic routine.

3.1. SKYVIEW: HTTP://SKYVIEW.GSFC.NASA.GOV/SKYVIEW.HTML

This package is provided by the NASA Goddard Space Flight Center (McGlynn et al. 1996). **SkyView** exists in true interactive form or as a Web tool. On the Web, the user fills out a form requesting images and catalogs of interest; **SkyView** then returns a composite image overlaying the requested images and the positions of the cataloged objects in the region. Available images cover the entire electromagnetic spectrum, from Gamma rays (EGRET) to the radio, and may be superimposed by adjusting a color table or by plotting the contours of one image on another. A large range of catalogs are on-line. In addition to plotting the positions of the selected catalog objects, **SkyView** returns a table containing the stored information on each object. Access to SIMBAD as a name resolver permits the user to select a region of sky by giving the name of a well known object instead of sky coordinates.

3.2. ALADIN: HTTP://CDSWEB.U-STRASBG.FR/CDS.HTML

Aladin is an interactive sky atlas under development at the *Centre de Données astronomiques de Strasbourg* (CDS) (Bonnarel *et al.* 1996). Aladin provides unified access to the CDS archives by overlaying the positions of cataloged objects on digitized images of Schmidt plates. To obtain the information, the **Aladin** client queries the CDS SIMBAD, catalog ($\sim$ 1500 catalogs) and image archive servers. Additionally, the client accepts both user defined catalogs, permitting the astronomer to visualize his sources in conjunction with documented information, and user given image files, provided they conform to the World Coordinate System FITS extension. The image archive at CDS consists of the Space Telescope Science Institute's *Digital Sky Survey*, first and second epochs (as this becomes available), and higher resolution scans (0.7 arcsecs/pixel) of the Southern Galactic Plane, Ecliptic Poles and the Magellanic Clouds; these latter images are provided by the MAMA facility at the Paris Observatory and by SuperCOSMOS of the Royal Observatory, Edinburgh. All of the images include astrometric *and* photometric calibrations.

Beyond its capability of visualization, **Aladin** furnishes an active approach to cross-identification: The system includes image analysis tools for the detection, extraction and classification of objects found on the image. For example, the user may plot the positions of his sources and then request that all objects on the image found within a specified aperture, centered on his positions, be extracted and associated with the corresponding sources in a table, which may then be written to disk. In this way, he directly creates a new catalog of his sources cross-identified with objects from a pre-selected catalog.

Acknowledgements

We are very grateful for the helpful discussions that we enjoyed with F. Bonnarel, C. Lonsdale and S. Mei.

References

Bonnarel *et al.* 1996, this volume, 469
Lonsdale *et al.* 1996, this volume, 450
McGlynn *et al.* 1996, this volume, 465
Meyer, S.L. 1975, "Data Analysis for Scientists and Engineers," John Wiley & Sons (New York).

THE HAMBURG IDENTIFICATION PROGRAM OF ROSAT ALL-SKY SURVEY SOURCES

N. BADE[1], L. CORDIS[1], D. ENGELS[1],
D. REIMERS[1] AND W. VOGES[2]
[1] *Hamburger Sternwarte, Hamburg, Germany*
[2] *MPI für Extraterrestrische Physik, Garching, Germany*

1. Introduction

The ROSAT All Sky Survey (RASS), performed between July 1990 and February 1991, provided about 60,000 X-ray sources in a soft X-ray band (0.1–2.4 keV) with a flux limit of approximately $5 \times 10^{-13}\,\mathrm{ergs\,cm^{-2}\,s^{-1}}$ for exposure times of 400 sec (a typical value in low ecliptic latitudes). A wealth of information can be extracted from the RASS source content and many objects deserve (or have already deserved) extensive follow-up studies. However, this possibility is limited by the fact that the nature of most of the RASS sources is unknown *i.e.*, they are unclassified. A correlation with the SIMBAD data base yields only identifications for about one third of the RASS sources.

2. Optical Identification Program

Since 1991 we use objective prism and direct plates taken by the Hamburg Schmidt telescope located on Calar Alto (Spain) to classify the different object classes in the RASS. The plates were taken in order to conduct an optically based search for AGN. The KODAK IIIa-J emulsion was used which gives high sensitivity from 5400 Å down to the atmospheric cut off at 3400 Å. The unusual continuum in comparison to stars in this wavelength range and strong emission lines enable the classification of AGN. With a typical ratio of $\log(f_X/f_B) = 0.55$ (Bade et al., 1995) AGN down to $6 \times 10^{-13}\,\mathrm{ergs\,cm^{-2}\,s^{-1}}$ at the plate limit of $B \approx 18.5$ are reachable. Because this value is near the RASS limit, principally all typical AGN can be identified on our prism plates. But this value also implies that AGN with higher $\log(f_X/f_B)$ are below the plate limit.

Follow-up observations have shown that the success rate of the AGN classifications is above 95%. Besides AGN, several galactic X-ray sources (cataclysmic variables, M dwarfs, white dwarfs) can also be classified with a high degree of certainty.

To date, we classified 336 fields of $5.5 \times 5.5 \deg^2$ covering a total area of about $8000 \deg^2$. The work is based on the Hamburg Quasar Survey (Hagen et al., 1995) which has achieved a full sky coverage with objective prism plates for the northern hemisphere and galactic latitude $|b| > 20°$ this year. The HQS comprises 567 fields and it is planned to extend the identification process to all fields except those with extremely high star density or Galactic absorption $N_H > 10^{21}$ cm^{-2} (leaving then ~ 500 fields).

3. Identification Results

The area surveyed so far contains 10,800 different RASS sources which can be divided into the object classes in Table 1.

TABLE 1. Identification of *ROSAT* Sources

Number Sources	% of Sample	Object Type
3376	31.3%	AGN candidates
2407	22.3%	Bright stars
435	4.0%	M dwarfs
37	0.3%	White dwarfs
703	6.5%	Normal galaxies/Clusters
635	5.9%	Empty fields
3207	29.7%	Unidentified sources

Our identification project is embedded in the Hamburg Quasar Survey (HQS) and the AGN content of the RASS is our main scientific object. Follow-up observations are necessary, and in order to maximize the scientific output two approaches are possible and have been undertaken:

Due to the coverage of large parts of the extragalactic sky, the Hamburg identification program is especially suited for the selection of peculiar object classes which are well discernible on objective prism plates. One example are X-ray loud QSO at $z > 2$, showing the prominent Ly-α emission line in the spectra. Objective prism plates with KODAK IIIa-J emulsion allow the recognition of emission line redshifts up to $z = 3.2$. Most of the X-ray loud QSOs with $z > 2$ are radio loud and have been selected from radio–X-ray correlations. Their radio silent counterparts are very rare and can only be identified by optical means.

An important task for identification programs is the compilation of complete flux limited samples. In order to minimize the influence of Galactic absorption on the source selection our group is working on a count rate limited AGN sample in the hard ROSAT band (0.5–2.0 keV) covering 4000 deg^2 and containing several hundred AGN. This sample allows the investigation of the local luminosity function and the study of relations between X-ray and optical properties in a flux limited sample.

References

Bade N., Fink H., Engels D., Voges W., Hagen H.-J., Wisotzki L., Reimers D. 1995 Astron.Astrophys.110, 469

Hagen H.-J., Groote D., Engels D., Reimers D. 1995 Astron.Astrophys. 111, 195

CROSS-CORRELATION OF LARGE SCALE SURVEYS: RADIO-LOUD OBJECTS IN THE ROSAT ALL SKY SURVEY

W. BRINKMANN, W. YUAN, J. SIEBERT
MPI für Extraterrestrische Physik, Garching, FRG

1. Introduction

There are basically two different approaches to study the physical conditions and the energy transfer processes operating in astronomical objects:
- the detailed observations of a few prominent objects over a wide wavelength range.
- the study of the broad band properties of suitably choosen samples

Both approaches have some limitations when conclusions are drawn concerning the general (class) properties of different types of objects.
- Detailed observations of single sources give a maximum of information about that particular object. This is, however, in general 'well known'; *i.e.*, near by, bright, well studied, and in many cases NOT AT ALL representative for its class.
- Samples of objects are usually rather small, selected from easily accessible sky regions, and biased towards the 'available instrument.'

We argue that a maximum of information about the typical qualities of various object classes can be gained by combining data obtained from large, sensitive, unbiased sky surveys.

We demonstrate this with some examples of radio-loud AGN, obtained from the correlation of the ROSAT All-Sky Survey with large scale radio surveys, like the 5 GHz Green Bank Survey of the northern sky, providing a very large sample of more than 2000 radio-loud X-ray sources.

2. X-ray versus Radio Properties

Flux-ratio diagrams show a clear distinction between radio selected BL Lacs, X-ray selected BL Lacs, and quasars.

AGN's from the 87 GB-ROSAT sample (re-observed with the VLA) fall in the gap between the above 'extreme' type of objects (Brinkmann et al. 1996a). They lie near the position of radio-loud quasars shifted diagonally towards higher values of f_x/f_r and f_o/f_r. This appears to be related to the fact that the average radio fluxes are generally about one order of magnitude lower than those of the previously known quasars.

Thus, the class of 'radio-loud' AGN covers a huge range in radio fluxes and this range is not bimodal but rather continuous. This could imply that the previously found gap in the flux ratio diagram was merely caused by selection effects as well and that radio-loud quasars (and, perhaps, BL Lacs) show a nearly linear relation in their flux ratios.

There have been claims that the X-ray spectral index of radio-loud quasars correlates with the radio spectral index and, eventually, with the core dominance and the radio loudness. The large number (> 600) of ROSAT detected RL QSOs seem to confirm these correlations. However:

- it remains unclear which **physical quality** is causing this correlation
- the behavior seems not to be continuous, but class dependent and bimodal.

The X-ray loudness $\alpha_{ox} = -0.384 \log (l_{2\ keV}/l_{2500A})$ has been used frequently in the past for the discussion of the relative fraction of X-ray to optical emission in an evolving quasar source population. There is no evolution with redshift but α_{ox} seems to depend on the quasars's optical luminosity. A regression analysis yields $\alpha_{ox} \sim 0.1 \times \log l_o$ with a probability level for a correlation of $P_r < 10^{-8}$. However, at least a major fraction of the claimed correlation is caused by selection effects! In a simulation, assuming a constant α_{ox} with a certain dispersion as well as upper and lower luminosity limits for the X-ray data, the test sources occupy the same phase space region as the real data.

The boundaries of the distribution, causing the apparent correlation, became only visible from the large number of objects in the sample (Brinkmann et al. 1996b).

References

Brinkmann, W., et al. (1996a) Radio loud AGN in the ROSAT Survey, Astron.Astrophys. in press

Brinkmann, W., Yuan, W. and Siebert, J. (1996b) Broad band energy distribution of ROSAT detected quasars, Astron.Astrophys. in press

IDENTIFICATION OF A COMPLETE SAMPLE OF NORTHERN ROSAT ALL-SKY SURVEY X-RAY SOURCES

J. KRAUTTER[1], I. THIERING[1], F.-J. ZICKGRAF[1],
I. APPENZELLER[1], R. KNEER[1], W. VOGES[2],
A. SERRANO[3] AND R. MUJICA[3]
[1] *Landessternwarte Königstuhl, Heidelberg, Germany*
[2] *MPI für Extrat. Physik, Garching, Germany*
[3] *Instituto Nacional de Astrofisica, Optica y Electronica, Puebla, Mexico*

Abstract. We present results of the optical identification of a spatially complete, flux limited sample of about 700 ROSAT All-Sky X-ray sources contained in 6 study areas north of $\delta = -9°$ with $\mid b^{II} \mid > 20°$ (including one region near the North Galactic pole (NGP), another one near the North Ecliptic pole (NEP)). Countrate limits are $0.01\,\mathrm{cts\,s^{-1}}$ near the NEP and $0.03\,\mathrm{cts\,s^{-1}}$ for the other areas. The optical observations were performed at the 2.15-m telescope of the Guillermo Haro Observatory, Mexico, using the Landessternwarte Faint Object Spectrograph Camera which allows to carry out direct CCD imaging and multi-object spectroscopy. The limiting magnitude is about 19^m for spectroscopy and about 23^m for B and R direct imaging. Our analysis shows a dependency of the ratio of 'extragalactic' (*e.g.*, AGN, cluster of galaxies) to 'stellar' (*e.g.*, coronal emitters, active binaries) counterparts on N_H. In the area near the NGP (low N_H) 'extragalactic' counterparts dominate, while in the area with the highest N_H 'stellar' counterparts dominate.

THE OPTID DATABASE: DEEP OPTICAL IDENTIFICATIONS TO THE IRAS FAINT SOURCE SURVEY

C. LONSDALE[1], T. CONROW[1], T. EVANS[1], L. FULLMER[1],
M. MOSHIR[1], T. CHESTER[1], D. YENTIS[2],
R. WOLSTENCROFT[3], H. MacGILLIVRAY[3] AND D. EGRET[4]
[1] *Infrared Processing and Analysis Center/CalTech, Pasadena, USA*
[2] *Naval Research Laboratory, Washington DC, USA*
[3] *Royal Observatory Edinburgh, Edinburgh, Scotland*
[4] *CDS, Strasbourg, France*

1. Introduction

We use a new, robust, method to estimate the identification probabilities of optical matches from digitized plate catalogs (COSMOS/UKST Catalog of the Southern Sky Version 2, Yentis *et al.* 1992; The Guide Star Catalog Version 1.1, Lasker *et al.* 1990; The Tycho Input Catalog, Egret *et al.* 1992; The APM Northern Sky Catalogue, Irwin, Maddox and McMahon 1994) to sources in the IRAS Faint Source Survey (FSS; Moshir *et al.* 1992), including both the Catalog (FSC) and the Reject File (FSR), utilizing a new random matching procedure with the advantages that it: (1) eliminates systematic uncertainties due to many problems, such as uncertainty in $N(m)$; variations across the optical plate of magnitudes, plate limits or $N(m)$; misclassification of stars and galaxies; the assumption of Gaussian error ellipses, *etc.* and (2) properly calibrates the identification probabilities. We find that at high SNRs and high galactic latitudes essentially all IR star-colored sources have an optical identification with $P_{id} > 99\%$. At high SNRs and high galactic latitudes, $\sim 90\%$ of all IR galaxy-colored sources have an optical identification with $P_{id} > 90\%$.

This project was undertaken to provide a star/galaxy classification for all FSS sources and an optical (blue) magnitude, allowing users to make more sophisticated searches through the FSS for sources of a given star or galaxy type, optical/IR color range, and optical magnitude range, and to search for unusual objects such as high redshift galaxies, brown dwarfs and IR-bright QSOs. OPTID can also be used to significantly improve IRAS positions and to help discriminate true from false low signal-to-noise IR sources. To access OPTID telnet to xcatscan.ipac.caltech.edu.

2. Method

We assume there is only one correct optical match to each IR source, but we do not assume that it is either the closest or the brightest. Our method can easily be generalized for the case of multiple true matches (*e.g.*, an interacting galaxy pair). We allow the possibility that the match is fainter than the optical catalog limit.

For the *ith* candidate optical identification within the search area (chosen to be 4σ to optimize trade-offs between completeness and reliability): the Likelihood Ratio is $LR_i = \frac{Qe^{-r_i^2/2}}{2\pi\sigma_{maj}\sigma_{min}N(<m_i)}$, where Q is the probability true identification exists in the optical catalog, r is the source separation, σ is the positional uncertainty ellipse, and $N(< m_i)$ gives the integral background counts brighter than magnitude m_i. At each position we calculate LR for each candidate optical match to each IR source, but use it only as a relative weight for each match, because its absolute value is very sensitive to match magnitude. The Reliability of each association with a given LR is defined as the number of true associations divided by the number of true plus number of random associations for a given subclass of sources (binned by star/galaxy type, galactic latitude and SNR): $R_i(LR_i) = \frac{N_{true}(LR_i)}{N_{true}(LR_i)+N_{false}(LR_i)}$. Finally, we compute the Identification Probability P_{id} for each match by taking into account the number of matches found for each source and their individual Reliabilities, including the possibility that no match exists on the optical plate: $P_{id_i} = \frac{QR_i\Pi_{j=1}^n[(1-R_j)]/(1-R_i)}{S}$, and $P_{no-id} = \frac{(1-Q)\Pi_{i=1}^n(1-R_i)}{S}$, where $S = \sum_{i=1}^n[num(P_{id_i})] + num(P_{no-id})$.

References

Egret, D., Didelon, P., McLean, B.J., Russell, J.L., and Turon, C. 1992, Astron. Astrophys., 258, 217.
Irwin, M., Maddox., S. and McMahon, R.G. 1994. RGO Newsletter.
Lasker, B.M., Sturch, C.R., McLean, B.J., Russell, J.L., Jenkner, H., and Shara, M.M. 1990. Astron. J., 99, 2019.
Moshir, M., *et al.* 1992. Explanatory Supplement to the IRAS Faint Source Survey, Version 2 JPL D-10015 8/92 (Pasadena: JPL)
Yentis, D.J., Cruddace, R.G., Gursky, H., Stuart, B.V., Wallin, J.F., MacGillivray, H.T., and Collins, C.A. 1992. in proceedings of "Digitized Optical Sky Surveys," ed. H.T. MacGillivray and E.B. Thomson; Kluwer Academic Publishers 1992, page 67

Part 10. Databases

EXPLORING TERABYTE ARCHIVES IN ASTRONOMY

A.S. SZALAY AND R.J. BRUNNER
Dept. of Physics and Astronomy, The Johns Hopkins University

1. Introduction

Astronomy is about to undergo a major paradigm shift, with data sets becoming larger, and more homogeneous, for the first time designed in the top-down fashion. In a few years it may be much easier to "dial-up" a part of the sky, when we need a rapid observation than wait for several months to access a (sometimes quite small) telescope. With several projects in multiple wavelengths under way, like the SDSS, 2MASS, GSC-2, POSS2, ROSAT, FIRST and DENIS projects, each surveying a large fraction of the sky, the concept of having a "Digital Sky," with multiple, TB size databases interoperating in a seamless fashion is no longer an outlandish idea. More and more catalogs will be added and linked to the existing ones, query engines will become more sophisticated, and astronomers will have to be just as familiar with mining data as with observing on telescopes.

The Sloan Digital Sky Survey, hereafter the SDSS, is a project to digitally map about 1/2 of the Northern sky in five filter bands from UV to the near IR, and is expected to detect over 200 million objects in this area. Simultaneously, redshifts will be measured for the brightest 1 million galaxies. The SDSS will revolutionize the field of astronomy, increasing the amount of information available to researchers by several orders of magnitude. The resultant archive that will be used for scientific research will be large (exceeding several Terabytes) and complex: textual information, derived parameters, multi-band images, and spectra. The catalog will allow astronomers to study the evolution of the universe in greater detail and is intended to serve as the standard reference for the next several decades. As a result, we felt the need to provide an archival system that would simplify the process of "data mining" and shield researchers from any underlying complex architecture. In our efforts, we have invested a considerable amount of time and energy in understanding how large, complex data sets can be explored.

2. Accessing Terabytes of Data

2.1. GENERAL CONSIDERATIONS

Today's approaches to accessing astronomical data do not scale into the Terabyte regime—brute force does not work! Assume a hypothetical 500 GB data set. The most popular data access technique today is the World Wide Web. Most universities can receive data at the bandwidth of about 15 kbytes/sec. The transfer time for this data set would be 1 year! If the data is residing locally within the building (access via Ethernet at 1 Mbytes/sec), the transfer time drops to 1 week. If the astronomer is logged on to the machine which contains the data, all of it on hard disk, then with SCSI bandwidth it still takes 1 day to scan through the data. Even faster hardware cannot support hundreds of "brute force" queries per day. With Terabyte catalogs, even small custom datasets are in the 10 GB range, thus a high level data management is needed. In order to identify possible solutions we need to look at how the archives will be used.

2.1.1. *Who Will Be Using the Archives?*

One can identify three major classes of users. *Power users* are the most sophisticated, with a lot of resources, whose research is centered around the archive. They spend a large fraction of their time querying the archive or performing a lot of small scale exploratory tests, before launching a moderate number of very intensive queries. These are mostly statistical in nature, with a large output volume. *The general astronomy public* will most likely do frequent but casual lookup of certain objects or regions. The archives help their research but the research is not statistical. This class of users will perform a very large number of small queries, which include a lot of cross-identification requests. *The wide public* will be an important component as they can browse a virtual telescope. This usage of the large archives can have an enormous public appeal, but it also needs special packaging. This could amount to an extremely large number of simple requests.

2.1.2. *How Will the Data Be Analysed?*

The data is inherently multidimensional, each object is represented by several fluxes, position on the sky, size, redshift, *etc.* Searching for special categories of objects, like quasars, involves defining complex domains in this N-dimensional space. Spatial relations will be investigated, like finding nearest neighbours, or other objects satisfying a given criterion within an angular distance. The output size of the objects satisfying a given query can be so large that intermediate files simply cannot be created. The only

EXPLORING TERABYTE ARCHIVES IN ASTRONOMY

way to analyze such data sets is to send them directly into analysis tools, thus these will have to be linked to the archive itself.

2.2. TYPICAL QUERIES

We expect that most of the queries will be of exploratory nature, no two queries will be exactly alike, at least for a while. Generally, scientists will try to explore the multi-color properties of the objects in the SDSS catalog, starting with small queries of limited scope, then gradually making their queries more complex on a hit-and-miss basis. Several typical types of activities need to be supported: manual browsing, where one would look at objects in the same general area of the sky, and manually/interactively explore their individual properties, the creation of sweeping searches with complex constraints, which extend to a major part of the sky, searches based upon angular separations between objects on the sky, cross-identifications with external catalogs, creating personal subsets, and creating new "official" data products.

2.3. THE MAIN PROBLEM: SWEEPING SEARCHES

2.3.1. *Geometric Indexing*

At this point we can identify the main problem in searching such archives: fast, indexed, but complex searches of Terabytes in k-dimensional space, with the added complexity that constraints are not necessarily parallel to the axes. This means that the traditional indexing techniques well established with relational databases will not work, since we cannot build an index on all concievable linear combinations of five or more attributes. On the other hand, one can use the fact that the data are very geometric in nature, every object is a point in this k-dimensional space. One can quantize the data into containers. Each container has objects of similar colors, from the same region of the sky. These containers represent a coarse grained density map of the data, and enable us to build a multidimensional index tree which can tell us which containers are fully inside, or outside our query, and which ones are partially contained. Only these latter containers have to be searched through, while the other two categories can either be accepted or rejected in full. If the containers are stored physically together, the cache efficiency in the retrieval will be very high—if an object satisfies our query, it is likely that most of its "friends" will as well.

2.3.2. *The Organization of Searches*

Most of the queries, at least the part which is on sky positions and colors, is inherently geometric. The simplest, primitive constraint is a half-space, one side of a multi-dimensional hyperplane. The Boolean combinations of these half-spaces are allowed, in such a way the queries are represented by k-dimensional polyhedra. First these queries are evaluated against the coarse-grained map, represented in the form of a tree structure, the so called k-d tree, and intersections are determined. From these intersections the time necessary to perform the query and the total output volume can be predicted. At the same time a list of containers is created, which needs to be searched for the actual data. At this point all the containers not intersecting with the query can be discarded, without ever touching the data itself. This can yield very substantial perfomance gains. Next, this list of containers is sent to the database for the final search, which is done in a quantized, container-by-container fashion. The actual searches can thus be also evaluated in parallel, if there is adequate hardware available.

3. A Case Study: The Sloan Digital Sky Survey Archive

3.1. THE SDSS

The Sloan Digital Sky Survey (SDSS) is a collaboration between the University of Chicago, Princeton University, the Johns Hopkins University, the University of Washington, Fermi National Accelerator Laboratory, the Japanese Promotion Group, the United States Naval Observatory, and the Institute for Advanced Study, Princeton, with additional funding provided by the Sloan Foundation and the National Science Foundation. In order to perform the observations, a dedicated 2.5 meter Ritchey-Chretien telescope was constructed at Apache Point, New Mexico, USA. This telescope is designed to have a large, flat focal plane which provides a 3° field of view. This design results from an attempt to balance the areal coverage of the instrument against the detector's pixel resolution.

The survey has two main components: a photometric survey, and a spectroscopic survey. The photometric survey is produced by drift scan imaging of 10,000 square degrees centered on the North Galactic Cap using five broad-band filters that range from the ultraviolet to the infrared. The photometric imaging will use an array that consists of 30 2K $\times$ 2K imaging CCDs, 22 2K $\times$ 400 astrometric CCDs, and 2 2K $\times$ 400 Focus CCDs. The data rate from this camera will exceed 8 Megabytes per second, and the total amount of raw data will exceed 40TB. The spectroscopic survey will target over a million objects chosen from the photometric survey in an attempt to produce a statistically uniform sample. This survey will utilize two multi-fiber medium resolution spectrographs, with a total of 640 optical fibers, 3" in diameter each, that provide spectral coverage from 3900–9200 Å. The telescope will gather about 5000 galaxy spectra in one night. The total number of spectra known to astronomers today is about 50,000— only 10 days of SDSS data! Whenever the Northern Galactic cap is not accessible from the telescope site, a complementary survey will repeatedly image several areas in the Southern Galactic cap to study fainter objects and identify any variable sources.

3.2. THE DATA PRODUCTS

The SDSS will create four main data sets: a photometric catalog, a spectroscopic catalog, images, and spectra. The photometric catalog is expected to contain one hundred million galaxies, one hundred million stars, and one million quasars, with magnitudes, profiles, and observational information recorded in the archive. The anticipated size of this product is about 250GB. Each detected object will also have an associated image cutout ("atlas image") for each of the five filters, adding up to about 700GB. The spectroscopic catalog will contain identified emission and absorption lines, and one dimensional spectra for one million galaxies, one hundred thousand stars, one hundred thousand quasars, and about ten thousand clusters, totaling about 50GB. In addition, derived custom catalogs may be included, such as a photometric cluster catalog, or QSO absorption line catalog. Thus the amount of tracked information in these products is about 1TB.

The collaboration will release the data to the public after an initial verification period. The actual distribution method is still under discussion. This public archive is expected to remain the standard reference catalog for the next several decades, presenting additional design and legacy problems. Furthermore, the design of the SDSS science archive must allow for the archive to grow beyond the actual completion of the survey. As the reference astronomical data set, each subsequent astronomical survey will want to cross-identify its objects with the SDSS catalog, requiring that the archive, or at least a part of it be dynamic.

3.3. THE SDSS ARCHIVES

The survey archive is split into two orthogonal functionalities and the corresponding distinct components: an *operational archive*, where the raw data is reduced and mission critical information is stored; and the *science archive*, where calibrated data is available to the collaboration for analysis and is optimized for such queries. In the operational archive data is reduced, but uncalibrated, since the calibration data is not necessarily taken at the same time as the observations. Calibrations will be provided on the fly, via method functions, and several versions will be accessible. The Science Archive will contain only calibrated data, reorganized for efficient science use, using as much data clustering as possible. If a major revision of the calibrations is necessary, the Science Archive and its replications will

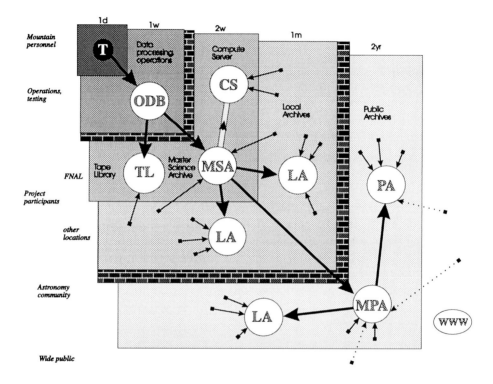

Figure 1. A conceptual data-flow diagram of the SDSS data. The data is taken at the telescope (T), and is shipped on tapes to FNAL, where it is processed within one week, and ingested into the operational archive (ODB), protected by a firewall, accessible only the personnel working on the data processing. Within two weeks, data will be transferred into the Master Science Archive (MSA). From there data will be replicated to local archives (LA) within another two weeks. The data gets into the public domain (MPA, LA) after two years of science verification, and recalibration, if necessary. These servers will provide data for the astronomy community. We will also provide a WWW based access for the wide public, to be defined in the near future.

have to be regenerated from the ODB. At any time the Master Science Archive (see Figure 1) will be the standard to be verified against.

High level requirements for the archive included (a) easy transfer of the binary database image from one architecture to another, (b) easy multi-platform availability and interoperability, (c) easy maintainance for future operating systems and platforms. In order to satisfy these, the Science Archive employs a three-tiered architecture: the user interface, the query support component, and the data warehouse. This distributed approach provides maximum flexibility, while maintaining portability, by isolating hardware specific features. The data warehouse, where most of the low-level I/O access happens, is based upon an OODBMS (Objectivity/DB), where the porting issues depend mainly on the database vendor. Objectivity's database image is binary compatible, can be copied between different platforms, since the architecture is encoded on every page in the archive.

4. Summary

We are in the middle of designing and constructing an extremely ambitious archival system, aiming to provide a useful tool for almost all astronomers in the world. We hope that our efforts will be successful, and the resulting system will substantially change the way scientists do astronomy today. The day when we have a "Digital Sky" at our desktop may be nearer than most astronomers think. Given the enormous public interest in astronomy, we hope that the resulting archive will also provide a challenge and inspiration to thousands of interested high-school students, and a lot of fun for the web-surfing public. To serve a TB archive even to the scientific community is quite a challenge today. To offer it to the wide public will be a task that we have not even started to appreciate. We hope that the pace of current hardware and software technologies in the area of large object databases will accelerate even further, and integration of applets into the standard set of scientific data analysis tools will soon begin. We are very excited to be at the front line when all this is happening, and we are quite convinced that this effort would have been orders of magnitude harder, if not impossible without our total reliance on object oriented databases, and object technology in general. We feel that the technology has matured to the point when it provides real solutions to real problems.

Acknowledgements

We would like to thank the rest of the SDSS Science Archive team at the Johns Hopkins University: Kumar Ramaiyer, Andrew Connolly, Istvan Csabai, Gyula Szokoly and Doug Reynolds. We also wish to thank Robert Lupton, Don Petravick, Steve Kent, Jeff Munn, Brian Yanny, Tom Nash and Ruth Pordes of the SDSS project for stimulating discussions.

THE WIDE-FIELD PLATE DATABASE:
A NEW TOOL IN OBSERVATIONAL ASTRONOMY

M.K. TSVETKOV[1], K.Y. STAVREV[1], K.P. TSVETKOVA[1],
E.H. SEMKOV[1], A.S. MUTAFOV[2] AND M.-E. MICHAILOV[2]
[1] *Institute of Astronomy, Bulgarian Academy of Sciences*
[2] *Computer Center of Physics, Bulgarian Academy of Sciences*

1. Introduction

Since the first applications of wide-field photography in astronomy nearly 2 million plates and films have been obtained and stored in archives all over the world. The Wide-Field Plate Database (WFPDB) provides astronomers with detailed information about the wide-field photographic observations. Its preparation started in 1991 as one of the main projects initiated by the Working Group on Wide Field Imaging at the IAU Commission 9.

2. Main Features of the Wide-Field Plate Database

An important part of the WFPDB is the List of Wide-Field Plate Archives (LWFPA), which summarizes the data for the archives and the observational instruments. Its latest version is accessible in the WFPDB WWW home page at http://www.wfpa.acad.bg. The number of archives, instruments and plates in the LWFPA is given in Table 1.

The data in the WFPDB originates from quite different plate catalogues. The original data undergoes a complex reduction procedure for standardization of the observation parameters: coordinate and time transformations to J2000 and UT, object name, emulsion and filter designations, decoding of coded data, correction of errors, supplementing of missing data, and structuring of non-structured data. The number of archives, instruments and plates incorporated in the WFPDB before August 1996 are included in Table 1.

TABLE 1. Number of archives, instruments and plates in the List of Wide-Field Plate Archives and in the Wide-Field Plate Database (August 1996)

		Archives	Instruments	Observatories	Plates
Direct	LWFPA	277	215	75	1 830 435
observations	WFPDB	87	83	26	397 324
Spectral	LWFPA	26	24	20	51 816
observations	WFPDB	13	13	11	9 321
Total	LWFPA	281	219	93	1 882 251
	WFPDB	88	83	29	406 645

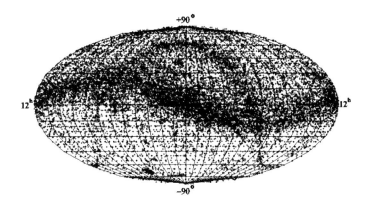

Figure 1. All-sky distribution of the plates in the WFPDB

A sky map of the observations in the WFPDB (equatorial coordinates, equal area projection) is shown in Figure 1. Figure 2 shows the distribution of observations according to the spectral band. WFPDB is currently accessible only in batch mode by user requests sent to wfpdb@wfpa.acad.bg. We plan to enlarge considerably of the WFPDB in the near future, with data from the Harvard College Observatory Plate Collection, the Bamberg and GRO plate archives, and others. Another important development of the WFPDB will be its installation on a HP 9000/712/80 workstation under the management of ORACLE DBMS V7.2 and the providing its on-line access via the INTERNET.

3. Some Applications of the WFPDB

Salvaging of Astrometric Treasures (Broshe et al. 1994). Searches in the WFPDB have been done for this project whose aim is to save the astrometric and photometric information from the old *Carte de Ciel* plates.

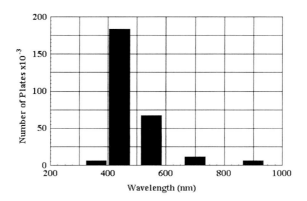

Figure 2. Number of direct plates in the WFPDB versus spectral band. Observation are roughly grouped in 5 bands corresponding to U, B, V, R, I

Existence of flare cycles in late dwarf stars. This project, suggested by R. Gershberg (Crimean Astrophysical Observatory), uses patrol observations of stellar aggregates for studies of red dwarfs. By his request 87 plates have been found in the WFPDB from the archives of the Asiago, Heidelberg, Kiso, Rozhen, Siding Spring and Tautenburg observatories, for the stars PZ Mon and V577 Mon.

Search for optical analogues of gamma ray bursts. Mutafov et al. (1995) has used the WFPDB to search for archival plates with possible optical analogues of GRB.

Light Curve of T Tauri and variable stars. For determination of the light curve of V 350 Cep (Semkov 1996) the WFPDB was searched for deep photographic plates in the field of NGC 7129. 44 plates from the Asiago observatory archive obtained in the period 1971–1977 have been found.

Acknowledgements

This work was supported by grants F-311/93 and I-529/95 of the Bulgarian National Science Fund, Alexander von Humboldt-Stiftung (Germany). M.K.T. and K.Y.S. are very thankful to the IAU and SOC of the 179 Symposium for the support to attend it.

References

Broshe, P., et al. 1994 Contract No. ERBCHRXCT940533, Salvaging an Astrometric Treasure.
Mutafov, A., et al. Proceedings of the 3rd Huntsville Symposium on Gamma Ray Bursts, 25–27 October 1995, Huntsville, USA, in press.
Semkov, E. H. (1996) IBVS, No 4339.

SKYVIEW: THE MULTI-WAVELENGTH SKY ON THE INTERNE

T. MCGLYNN[1,2], K. SCOLLICK[1,3] AND N. WHITE[1]
[1] *NASA/Goddard Space Flight Center*
[2] *Universities Space Research Association*
[3] *Computer Sciences Corporation*

1. Introduction

The *SkyView* virtual observatory provides a single, simple interface where users can retrieve images of the sky at all wavelengths from the radio through gamma rays. Below we discuss how *SkyView* works and how it represents a new paradigm in astronomical data archives. Users get to *SkyView* over the WorldWideWeb at http://skyview.gsfc.nasa.gov. Within a few moments they can have high-quality images from ground- or satellite-based surveys.

Surveys currently in *SkyView* include the optical Digitized Sky Survey, the IRAS Sky Survey Atlas, COBE DIRBE data, five radio surveys, the EUVE and ROSAT/WFC all-sky surveys, the Compton EGRET and CompTel surveys, ROSAT PSPC pointed observations and others.

SkyView addresses two problems which inhibit astronomers from using survey data. First, *SkyView* provides a single contact point and a single interface to all of these different surveys regardless of origin. Second, the system allows the user to specify how the image is to be created without regard for its native format. The user does not need to know the underlying coordinate system, equinox, scale, or orientation of the image, nor how the data for a survey may be subdivided. *SkyView* transforms, precesses, rescales, rotates and mosaics the image as appropriate. Users are relieved of understanding the geometric details of the data and may immediately begin to do astronomy.

2. A New Archive Paradigm

SkyView presages a needed shift in how we view archives: as the community is inundated with data from surveys like 2MASS and Sloan, our archive systems need to do more than point out where data is for our users, they need to assist in getting users the data in an immediately useful form. Most astronomical archives that have been built up till now have used what is essentially a library paradigm. The archive comprises a set of indivisible atoms (or books or observations). These atoms are indexed in some catalog. The users first search the catalog for the atoms of interest and then check out (or copy) those observations. *SkyView* goes beyond this. Rather than just lending the books in the archive's shelves, the *SkyView* server reads the books and writes a report based on them which specifically addresses the user's needs. To do this *SkyView* has built into it a substantial understanding of the geometry of astronomical images.

3. Building Archives in the Web

The advent of the Web has accelerated the explosion in the amount of digital astronomical information available. While individual astronomers have found the Web extraordinarily helpful in locating and distributing information, data providers are only now beginning to recognize the possibilities implicit in this new technology. While a Web form may be designed for interactive use, each form is essentially a view into the resources of the organization providing the data. It can be used not only by a user interactively, but by services at other data providers. *SkyView* is using a vast array of Web services to provide its own specialized interface for users. Already it uses the NED and SIMBAD name resolvers to translate names into positions, talks to the HEASARC catalog services to get more than 150 astronomical catalogs, and has begun to retrieve catalog data and images from the University of Minnesota. The time of monolithic data services is ending.

4. Status and Future Plans

In two years *SkyView* use has grown to between 15,000 and 20,000 images generated each month. While most of these are to support astronomical research we estimate that perhaps one-third of the images are retrieved by amateur astronomers and other interested members of the public. Our simple Web interfaces are very useful for scientists but also give the public a window into the world of astronomical research.

New surveys are constantly being incorporated into *SkyView*. Both two-dimensional images, and three-dimensional data, *e.g.*, line surveys, are included. We welcome suggestions for new surveys to be added. In the near future we shall be opening new *SkyView* archives for limited regions which have been especially well-studied in multiple wavelengths such as the Galactic Plane and M31.

A WWW DATABASE OF APS POSS IMAGES

C.S. CORNUELLE, G. ALDERING, A. SOUROV,
R.M. HUMPHREYS, J.A. LARSEN AND J. CABANELA
University of Minnesota

1. The Image Database

We have used the Automated Plate Scanner (APS) at the University of Minnesota to digitize glass copies of the blue and red plates of the original Palomar Observatory Sky Survey (POSS I) with $|b| > 20°$. The APS Image Database is a database of all digitized images larger than the photographic noise threshold. It includes all of the matched images in the object catalog, as well as those unmatched images above the noise threshold. The matched image data of the catalog has the advantage of confirming the reality of the image. This is especially important for small images near the plate limit. But these are not all of the detected real images; very blue or very red faint objects may be excluded by this matching requirement. The image database allows information on them to be retrieved, and is therefore a valuable complement to the object catalog. The operation of the APS and the scanning procedures are described in detail in Pennington *et al.* (1993). We are now processing plate data into the image database. A set of query forms, a tutorial and documentation can be found at http://isis.spa.umn.edu/IDB/homepage.idb.html.

2. The On-Line Catalog

The APS Catalog of the POSS I contains coordinates, magnitudes, colors, and other computed image parameters for all of the matched images on the blue and red plates. The catalog provides individual information for about one hundred million stars in our galaxy and tens of millions of galaxies down to 20–21st magnitude (in the blue). The stellar and non-stellar images are separated using a neural network image classifier (Odewahn *et al.* 1992, 1993), with a success rate better than 90% to within one magnitude of the plate limit. The catalog of objects is available at http://isis.spa.umn.edu/aps_catalog.html. The image database and object catalog are compared in Table 1.

TABLE 1. Comparison of APS Catalog of POSS I and Image Database

Property	Object Catalog[1]	Image Database[2]
Data Source	APS scans of POSS I O and E plates	Same
Data Filtering	Uses O-to-E plate matching to remove noise	images $> 1.7''$ in two colors
Data Returned	Image parameters in flat text table or Postscript finder chart	FITS image files with $0.33''$ pixels
Searchable Fields	Any data field	RA and Declination
Photometry	O and E magnitudes	Under Construction
Astrometry	Tied to Lick NPM	Same
Classification	'Star' or 'Galaxy'	None

[1] *http://isis.spa.umn.edu/IDB/homepage.idb.html*
[2] *http://isis.spa.umn.edu/aps_catalog.html*

3. Future Directions

Scanning in pairs of the Luyten and POSS I E-emulsion plates has begun, with the goal of generating a Proper Motion Database. When completed, this will form the third leg of an internally-consistent set of star and galaxy data produced with the APS and available on-line to the community.

Acknowledgements

The operation and maintenance of the Automated Plate Scanner and the development, production, and distribution of the APS on-line catalogs are supported by the NSF and the University of Minnesota. The APS Image Database has been supported by NASA and the University of Minnesota.

References

Odewahn, S.C., Humphreys, R.M, Aldering, G., and Thurmes, P.M. 1993. Publ. Astron. Soc. Pacific, 105, 1354.
Odewahn, S.C., Stockwell, E.B., Pennington, R.L., Humphreys, R.M., and Zumach, W. 1992. Astron. J., 103, 318.
Pennington R.L., Humphreys, R.M., Odewahn, S.C., Zumach, W., and Thurmes, P.M. 1993. Publ. Astron. Soc. Pacific, 105, 521.

THE ALADIN INTERACTIVE SKY ATLAS

F. BONNAREL[1], H. ZIAEEPOUR[1,4], J.G. BARTLETT[1],
O. BIENAYMÉ[1], M. CRÉZÉ[1], D. EGRET[1], J. FLORSCH[1],
F. GENOVA[1], F. OCHSENBEIN[1], V. RACLOT[1],
M. LOUYS[2] AND P. PAILLOU[3]
[1], Observatoire Astronomique de Strasbourg
[2] Ecole Nationale Supérieure de Physique de Strasbourg
[3] Institut Géodynamique Bordeaux
[4] Presently: European Southern Observatory

1. Introduction

The subject of this symposium, Multi-Wavelength Sky Surveys naturally invokes a discussion of methods of astronomical object identification and classification: Given a set of objects detected at a certain waveband, how does one integrate the new sources with previous data? The **ALADIN** system (Paillou *et al.* 1994) of the CDS is a software package designed to tackle this problem: It provides simultaneous access to digitized sky photographs, catalogs and databases to facilitate direct, visual comparison of user data with previously classified data, as well as automatic source extraction and calibration tools.

2. System Description

In the basic client-server architecture of **ALADIN**, the client communicates with the CDS **ALADIN** server and manages local data, such as user catalogs. The CDS **ALADIN** server in turn directs a host of servers dedicated to each of the individual CDS information services.

One of the central aspects of the system is the archive of densely sampled images the quality of which is required for optical cross-identifications in crowded regions of the sky or in areas with deep observations. This archive consists of an optical disk juke-box with a capacity of 500 Gb that includes the ESO-R and SERC images of the Galactic Plane, the Ecliptic Poles and the Magellanic Clouds all of which scanned by the MAMA or SuperCOS-

MOS facilities. In addition, the Digital Sky Survey (Lasker 1994) provides a full sky coverage at a lower density.

ALADIN displays SIMBAD objects on the current image along with their associated names and error boxes. Except for differences arising from the greater variety of formats, the acces to CDS and user catalogs works on the same principles as the access to SIMBAD.

The images displayed by **ALADIN** will not simply represent a reference map, but also quantitative data: it will be possible to perform astrometric and photometric measurements, search and extract objects (for example near external catalog positions), measure object parameters, and classify objects as either stars or non-stars (*e.g.*, galaxies).

ALADIN directly provides both the astrometric calibration, as given by the digitizing machine, and a photometric calibration, performed in Strasbourg, as decribed in Bartlett *et al.* (1995). Eventually **ALADIN** will also allow an interactive photometric and astrometric *recalibration* of an image using new catalogs of standards (Tycho, for example).

3. Status of the Project

An extended prototype capable of managing a set of about 190 densely sampled Schmidt plate images, stored in the optical disk jukebox, and the DSS-1 CDROMS has been installed at several astronomical institutions in France. We are currently testing the first public version of the **ALADIN** client, incorporating most of the functions of the proposed system, which we plan to distribute in France and to a couple of other sites by the end of 1996. A second version, including recalibration tools and contour overlays, as well as a WEB version ('ALADIN-lite') will be distributed in late 1997. The archive of densely sampled data is 70% full, presently containing 400 Gbytes of data, and will be completed in the first semester of 1997. The DSS-II data (Lasker *et al.* 1996) will be progressivly integrated starting in early 1997.

References

Bartlett, J.G., *et al.* 1995, ADASS V, ASP Serie 101,489.
Lasker, B.M. 1994, in Astronomy from Wide-Field Imaging, IAU symposium 161, H.T. MacGillivray and E.B. Thomson eds., Kluwer Academic Publ., p.167.
Lasker, B.M. *et al.* 1996, poster presented at this meeting.
Paillou, Ph., Bonnarel, F., Ochsenbein, F., Crézé, M. 1994, in Astronomy from Wide-Field Imaging, IAU symposium 161, H.T. MacGillivray and E.B. Thomson eds., Kluwer Academic Publ., p. 347.

THE SDSS SCIENCE ARCHIVE

R.J. Brunner
Dept. of Physics & Astronomy, The Johns Hopkins University

1. Introduction

Designing and implementing the Science Archive for the Sloan Digital Sky Survey (Gunn *et. al* 1992) has presented several unique architectural obstacles. First, the final archive will be large; the cumulative data products will exceed tens of Terabytes. Second, the data will be complex; extracted parameter catalogs will contain links to images, spectra, and objects in other wavelength catalogs. Third, the archive will be widely distributed allowing transparent world-wide access to the available data. Finally, the archive is expected to remain viable well into the next century.

The SDSS Science Archive will consist of four main components: a photometric catalog, a spectroscopic catalog, atlas images, and spectra. The photometric catalog is expected to contain at least one hundred million galaxies, one hundred million stars, and one million quasars. Each detected object will have measured parameters (*e.g.*, magnitudes and profiles) recorded as well as an associated image cutout for each of the five bandpasses. The spectroscopic catalog will contain identified emission and absorption lines and one dimensional spectra for one million galaxies, one hundred thousand stars, one hundred thousand quasars, and about ten thousand clusters.

During the design of this archive, several novel approaches were utilized: the object-oriented design and implementation of the persistent storage and transfer of the actual data, a geometric strategy in the organization of the inherently multidimensional spatial and flux information, and the introduction of custom analysis applets that filter the extracted data into a more manageable stream. The knowledge gained in the development of this archive will help not only anyone who is interested in the internal architecture of the SDSS Science Archive, but also those who are developing similarly large astronomical archives.

2. Object Oriented Design & Implementation

In an effort to simplify the portability and maintenance over the expected lifetime for this archival system, we have attempted to adhere to proven Object Oriented Design and Implementation techniques throughout this project. Except for the provided Graphical User Interface, all of the software is written in C++. The various subsystems have been designed and modeled using the Rumbaugh/OMT (Rumbaugh *et. al* 1991) Object Modeling Technique. This design strategy provides both a clear picture of the dynamic and static interrelationships between the objects within a subsystem and also a legacy snapshot of the actual architectural framework.

During the project's early work on archival implementation, Fermi National Accelerator Laboratory evaluated several relational, object/relational hybrids, and object oriented database systems (OODBS), primarily in regards to system performance. The general consensus was that OODBS provided the best performance as well as the optimal implementation model for scaling to the Terabyte regime. After several years of working with this emerging technology, the project settled on Objectivity/DB to satisfy all persistent storage and querying requirements. An additional benefit provided by Objectivity/DB is the ability to control the clustering of data on the storage media; a fact which is crucial to our geometrical indexing strategy.

3. Geometric Strategy

As with other types of data, Astronomical data often contain a numerical subset (*i.e.*, spatial coordinates) that are indexed in order to expedite a certain class of queries. Unfortunately, traditional indexing techniques have several shortcomings. First, they usually must be restricted to a few parameters; otherwise, they begin to match the actual data in physical size and complexity. Second, the actual index is unable to provide additional information about the underlying data, such as providing a coarse grained density map. Finally, current archive queries are limited to simple ranges of parameter values, while the desired query may be more complicated.

Using ideas from the field of Computational Geometry, we have developed an indexing strategy that while still providing the benefits of a traditional indexing scheme, also provides accurate predictions of query volumes and times, a snapshot of the spatial relationships that exist within the dataset, and aids in the quantization of the data on the storage media. Our strategy utilizes a spatial data structure (Samet 1990) to provide a coarse grained density map of the actual subset of the data that will be indexed. This density map is then encapsulated in a tree-like structure where each node on the tree conceptually represents a sub-volume within the entire volume occupied by the data. Thus, the root node represents the entire dataset, and the leaf nodes represent the terminal cells in the density map. All objects which lie within the leaf node's boundaries are then quantized and stored contiguously on the storage media in an attempt to ensure efficient cache hits by the object request broker within the data warehouse.

The geometrical indexing strategy can naturally incorporate the actual query, resulting in a more powerful search mechanism. Rather than limit a user to parameter cuts, linear combinations of attributes form the query primitive within our system. These linear combinations can then be combined using Boolean Algebra to form complex polyhedra that can carve out complicated volumes within the available parameter space. In order to simplify spatial queries, we work with a Cartesian projection of the spherical astrometric coordinates. This simplifies coordinate conversions, and reduces spherical proximities to a linear combination of the Cartesian coordinates.

4. Analysis Filters

An often over-looked problem inherent in large archives is the management of the extracted data, which can often swamp the resources of many users. This problem is compounded with the inclusion of the network latency. When all that is required is a simple plot or calculation, a user does not need the entire dataset produced during the extraction phase of a query. As a result, we have developed a toolkit of analysis applets that can filter the extracted data.

Acknowledgements

First I would like to thank the rest of the SDSS Science Archive team at JHU, especially Alex Szalay and Kumar Ramaiyer. I also wish to thank Robert Lupton, Don Petravick, Steve Kent, Jeff Munn, and Brian Yanny for stimulating discussions. In addition, I would like to acknowledge the SDSS for funding this project.

References

Gunn, J.E. and Knapp, G.R. 1992 Publ.Astron.Soc.Pacific , 43, 267
Rumbaugh, J., Blaha, M., Premerlani, W., Eddy, F., and Lorenson, W., 1991, "Object Oriented Modeling and Design," Prentice Hall.
Samet, H. 1990 "The Design and Analysis of Spatial Data Structures," Addison-Wesley

THE GSC-I AND GSC-II DATABASES: AN OBJECT-ORIENTED APPROACH

G. GREENE[1], B. McLEAN[1], B. LASKER[1], D. WOLFE[1],
R. MORBIDELLI[2] AND A. VOLPICELLI[2]
[1] *Space Telescope Science Institute*
[2] *Osservatorio Astronomico di Torino*

1. Introduction

The original GSC-I (Jenkner *et al.* 1990) which contains 25 million entries and requires approximately 1GB of storage was at the edge of technological capability at the time catalogue construction began in 1984. At that time, a custom coded database was built since the relational databases of the era were unsuited to the HST-specific access requirements. A second generation GSC is now being constructed (Lasker *et al.* 1995), with an estimated 10 billion entries and a size of 2 Terabytes. The current generation of object-oriented database (OODB) systems are more suited to the needs of large astronomical catalogues and are being adopted by many large-scale projects. In a joint effort between the Space Telescope Science Institute and Osservatorio Astronomico di Torino, we are currently designing such an OODB for the Guide Star Catalogues and are implementing a prototype using the GSC-I data.

2. GSC-I Object-Oriented Database Project

The unique experience of developing and utilizing the GSC-I database combined with modern computer architectures has lead to an evolution in the requirements for maintaining an all-sky astrometric catalogue. The GSC-I custom database is no longer very large in terms of present-day data volume. Nevertheless, it has inherent complexities not only due to the record and file system dependencies, but also fundamental relationships linking the internal catalogue objects, the underlying calibrations, and their parameters. As improved calibration methods for computing more accurate astrometry become available (Röser *et al.*, this volume, p. 420), inflexible data structures with a large number of files associated become tedious, time-consuming, and ultimately forbidding to perform quality assessment. Also, since the GSC-I continues to be an effective reference catalogue for a wide range of astronomical projects, practical solutions for correcting catalogue errors must be considered.

The object-oriented approach provides a robust method for structuring the GSC catalogue objects and parameters such that large or small scale maintainance and recalibration can be performed efficiently. This concept along with Objectivity/DB,'s commercial OODB, was first introduced to us by the SDSS Science Archive project members (Brunner *et al.* 1994). The prototype GSC-I OODB is currently being implemented using a standard C++ language interface to the commercial software based on the Class Object Model shown in Figure 1. Other features of this software which are attractive to the GSC projects are platform-independent data access and scalability. This means the OODB software infrastructure does not place limitation on increasing or rescaling the size of data. These offer a significant advantage in development of access methods from a variety of user levels.

3. GSC-II Database Development

The GSC-I object oriented design can easily be extended for the construction of the second generation GSC-II database by using class inheritance to exploit commonalities between the survey data sets while also providing an intrinsic framework for sharing class structures with enhanced functionality. The GSC-II database is estimated to be 2 Terabytes in size. This also includes in addition to the basic GSC astrometric and photometric data, colors and proper motions. The construction of the GSC-II catalogue is tightly coupled to the development of the database. Single-plate pipeline processed data will be input to the GSCPlateObject containers of the database. Then, using an "object index" scheme which links overlapping survey data, multi-plate proper motion and color index data can be computed efficiently. The final GSC database will also include basic reference catalogues such as PPM, CMC, and Tycho for direct calibrations. Efforts are currently in progress to develop both the SDSS and GSC databases with a common sky-partitioning scheme for faster cross-matching and retrieval rates between two very large scale astronomical survey data sets.

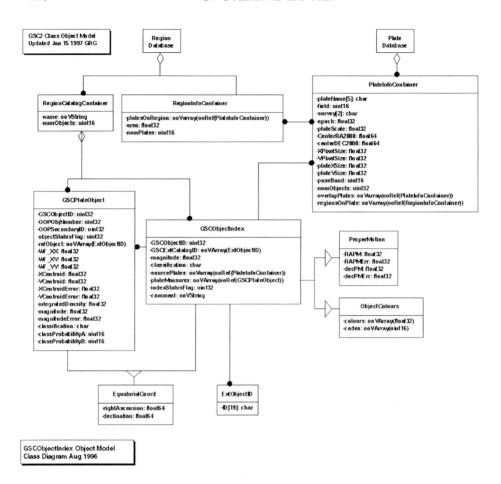
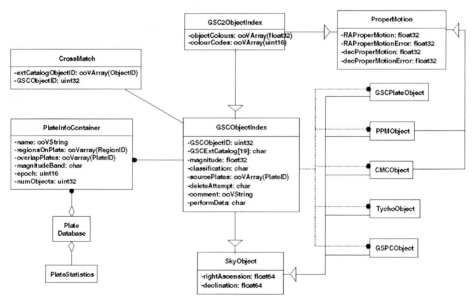

Figure 1. GSC-I Database Object Model

4. Summary

In summary, we have developed a working GSC-I OODB prototype. It has been successfully loaded and tested with a minimum number of lines of code for GSC 1.1. Quality Assurance applications, loading of GSC1.2, and the expansion for GSC-II data import are underway.

The OODB offers an exciting new capability to rapidly access and maintain 2 Terabytes of sky survey data and the potential for linking more than one large-scale sky survey project, for example, the GSC and SDSS surveys. With this new advance in the database industry, astronomical research with multiple parameter sets for large scale structure will become increasingly easier to perform.

References

Brunner, R. J., Ramaiyer, K., Szalay, A., Connolly, A.J. and Lupton, R.H., 1994 Astron. Soc. Pacific Conference Series 77.

Jenkner, H., Lasker, B. M., Sturch, C.R., McLean, B.J., Shara, M. M. and Russell, J. L., 1990 Astron. J. 99, 2082

Lasker, B.M., McLean, B.J., Jenkner, H., Lattanzi, M.G., 1995 in "Future Possibilites for Astrometry in Space" ESA SP-379 pp.137

SURVEYS IN THE ADC ARCHIVE

N.G. ROMAN
HughesSTX/ADC, Goddard Space Flight Center

1. Introduction

The ADC archive contains more than 800 catalogs and extensive tables from more than 750 journal articles. The archive is increasing steadily at the rate of about 20 data sets per month. This archive can be accessed either by anonymous FTP or by a World Wide Web (WWW) browser at
http://adc.gsfc.nasa.gov/

2. Searching the ADC Archive

In addition to the archive, which is divided into separate subdirectories for catalogs and the journal files, both methods also access several other directories including address directories for both individuals and institutions and a small collection of software for catalog searching. At each level, there is a "key" which lists the content of the directory and its subdirectories (except for the individual catalog or article level). With FTP access, this key may be downloaded and searched for subjects and authors of interest.

The WWW access provides a full text search capability in the document which describes each file. On the WWW, the document can be reviewed to determine if the files are of interest. All files except the documents are stored in compressed ASCII form. They can be retrieved either as compressed or as uncompressed files or converted to FITS.

3. Sample Search

A search of the archives with the word "survey" returned a list of 338 items. The following are some examples:

STARS (OPTICAL)	Durchmusterungen
	Michigan Spectral Classification Catalogs
	Case Low Dispersion Survey
	General Catalog of Variable Stars
	Lowell, Lick, and Luyten Proper Motion Catalogs
	Catalog of Radial Velocities
	Luminous Stars in the Milky Way
	Northern M Dwarfs
	Thick Disk Chemical Abundance Distribution
GALAXIES (OPTICAL)	Rich Clusters of Galaxies (Abell)
	CfA Redshift Survey
	Multicolor Survey of High Redshift Quasars
	APM Bright Galaxy Catalog
	Catalog of High-Redshift Quasars
	Spectroscopic Survey of Faint QSO's
	Pisces-Perseus Supercluster
	Third Reference Catalog of Bright Galaxies
X RAY	HEAO A-1 X-ray Source Catalog
	ROSAT Wide Field Camera All-sky Survey
	Wisconsin Soft X-ray Diffuse Background All-sky Survey
	X-ray Emission at the Low-Mass End
	Einstein Data Bass of Quasars
	X-ray AGN Content of the Molonglo 408 MHz Survey
	ROSAT All-Sky Bright Source Catalogue
EUVE	First EUVE Source Catalog
	Second Extreme Ultra-Violet Explorer Catalog
	EUV Explorer Bright Source List
	Far UV Point Source
INFRARED	IRAS Catalogs of Point Sources and Faint Sources
	Revised AFGL Infrared Sky Survey and Supplement
	Two-micron Sky Survey
	IRAS Faint Source Catalog
	IRAS Redshift Survey
	Equatorial Infrared Catalog
	K Survey of Ori-A Molecular Cloud
	New Infrared Survey of Northern Planetaries
RADIO	Catalogs from Deep 327 MHz Westerbork Survey
	Bell Laboratories HI Survey
	Parkes-MIT-NRAO (PMN) Surveys
	Shapley-Ames Catalog at 2.8 cm
	Fourth MIT-Green Bank 5 GHz Survey
	Molecular Outflow Sources
	53,522 Sources at 4.58 GHz
	7C Survey of Radio Sources

GLOBAL NETWORK ACCESS AND PUBLICATION OF SURVEY DATA

N.E. WHITE
High Energy Astrophysics Science Archive Research Center,
Laboratory for High Energy Astrophysics, GSFC

1. Introduction

The results of astronomical surveys include 1) catalogs containing anywhere from a few to many million objects, 2) data products used to generate the catalogs (*e.g.*, images or spectra), 3) publications and 4) object based compilations of information from many sources. The ubiquitous growth in the Internet and the dramatic reduction in the cost of mass storage systems now allows instant global access to this information. Astrophysics on-line services have grown up with the Internet, and represent an invaluable resource whose access is a routine part of any research project. Unfortunately users are also faced with searching and accessing multiple sites with different content, access and response methods. There can also be inconsistencies amongst the various systems, which can involve extra effort to resolve. A network-integrated astrophysics system has long been sought because it would remove multiple user interfaces and allow much simpler integration of services. In an era of shrinking budgets, the issue is how to achieve this in a cost effective manner. This review gives an overview of the current situation and discusses the likely evolution towards a network-integrated astrophysics system.

2. Current Capabilities

There are currently three levels of access to the results of scientific investigations 1) the literature search, 2) on-line catalog and data services, and 3) value added services. The literature search provides the top level published knowledge base, which typically denotes the start of any research project. The catalog and data services provide access to the raw catalogs and data products that are the result of an investigation. A value-added service takes elements of the first two and combines them to produce a new product, *e.g.*, combining catalogs into a single object based compilation. In this section the current on-line astronomical services are discussed in terms of these classifications. This is not a comprehensive list of services available, but rather examples used to illustrate the three levels of service currently provided.

2.1. THE LITERATURE SEARCH

2.1.1. *Electronic Journals*

Over the past year the publishers of astronomical journals have started experimenting with providing WWW access to their publications. The Astrophysical Journal at http://www.journals.uchicago.edu/ApJ/ provides one of the first examples of direct online access to a journal. This has proved a very effective and popular method for rapid dissemination of papers. We expect that all the major journals will follow this lead. One major advantage of these electronic publications is that they can include the original tables from the papers in a machine readable form, so they can be more readily integrated into databases. Ultimately, one might expect that links to the original data may also be included.

The Los Alamos National Laboratory (LANL) and other servers distribute preprints ahead of publication in the journals. They too have become very popular. However, in the near future one might hope that the turn around-time between acceptance in a journal and electronic publication will be reduced to a few days. This would make the preprint a relic of the past.

2.1.2. *The Astrophysics Data System Abstract Service*
(http://adsabs.harvard.edu/ads_abstracts.html)

The NASA Astrophysics Data System (ADS) abstract service based at the Smithsonian Astrophysical Observatory (SAO) provides WWW access to astronomical journal abstracts going back to 1975. This system is becoming the glue that ties together the on-line journals. It is also taking the lead in scanning back issues of journals, so they are network accessible. The ADS currently includes 240,000 astronomical articles and 410,000 space instrumentation and engineering articles. A search engine allows a search by author, object name, words in the title, and/or abstract text. The results are ranked by how closely the paper matches the query and in many cases contain links to the scanned images of the original journal articles, SIMBAD object pages, and even in some cases to the original data in the archive. The ADS has given a major boost to the productivity of astronomers. It is rapidly making obsolete the need to visit the library to make a literature search.

2.2. CATALOGUE AND DATA SERVICES

2.2.1. *CDS*
(http://cdsweb.u-strasbg.fr/CDS.html)

The Centre de Donnees astronomiques de Strasbourg (CDS) collects and distributes astronomical catalogs related to observations of stars and galaxies, as well as other galactic and extragalatic objects. At the time of this writing it includes 1622 catalogs of which 1182 are available via ftp in ASCII or FITS format. Very recently a new catalog browser called VizieR has made it possible to search 995 catalogs via the a WWW interface. The search options include by coordinates, object name, mission name, wavelength and other catalog parameters.

2.2.2. *Astronomical Data Center (ADC)*
(http://adc.gsfc.nasa.gov)

The Astronomical Data Center (ADC) at the NASA GSFC acquires, verifies, formats, documents, and distributes files containing astronomical data in computer readable format. The current holding is 800 astronomical catalogs plus 750 data tables from various astronomical journals. It is possible to make a search via a Web interface for catalogs and then download them directly from the ftp site. The ADC has produced two CD ROMs with the most often requested catalogs. These CD ROMS have proved very popular.

2.2.3. *HEASARC*
(http://heasarc.gsfc.nasa.gov)

Access to data and catalogs from 22 different X-ray and Gamma-ray astrophyiscs satellite observatories is provided by the High Energy Astrophysics Science Archive Research Center (HEASARC) at NASAs GSFC. There is ftp access to over 500 Gb of data in FITS format from both active missions (ASCA, CGRO, ROSAT, RXTE) and newly restored data (*e.g.*, HEAO-1, Einstein, EXOSAT, Ariel V, Vela 5B, Cos B and SAS-2). The archive can be searched via a user interface called Browse (command line) and W3Browse (WWW interface). The HEASARC database also contains many of the most popular astronomical catalogs. The command line and WWW versions both allow a remote user to search by coordinates, name and all database parameters. A key feature of the HEASARC browse system is to link the catalogs to the data. This allows the user to view data, make quicklook analysis and retrieve selected items that were used to generate the catalog. Also possible in the command line version are cross-correlations between catalogs, plots and filtering of parameters (these will appear in the Web version in early 1997).

2.2.4. *Astrophysics Multi-spectral Archive Search Engine (AMASE)* (http://amase.gsfc.nasa.gov)

The Astrophysics Multi-spectral Archive Search Engine (AMASE) is a WWW accessible multi-mission and multi-spectral catalog being jointly developed by the GSFC Astrophysics Data Facility and the University of Maryland. It allows the search of space mission data in the NASA public archives. It is built using an object database methodology. This is currently a prototype sytem and so only has IRAS, ROSAT and a few other catalogs. It is a hybrid system where some value is added by trying to use an object orientated system to link catalogs from different wavelength regimes to the data archives. This system can be searched using a WWW interface by coordinates and object name.

2.3. VALUE ADDED SERVICES

2.3.1. *The SIMBAD* (http://cdsweb.u-strasbg.fr/SIMBAD.html) *and NED* (http://www.ipac.nasa.gov/ned/ned.html) *Object Databases*

SIMBAD provided by CDS was the first and is the most comprehensive of the online astronomical services. It is an object based compilation of basic data, cross-identifications, observational measurements and bibliographic information. It contains all non-solar system objects (stars, galaxies and nonstellar objects). SIMBAD provides information on an object by object basis, and in doing so has integrated the information in lower level catalogs together with an ongoing object based literature search. SIMBAD can be queried via telnet, an X-windows client-server interface or by e-mail. Very recently a Web interface was also made available. Using these interfaces it is possible to search by object name or position. The results contain detailed information about the object, as well as a bibliographic listing.

The NASA Extragalactic Database (NED) contains an object based compilation of extragalactic objects based on combining catalogs and published results in the literature. Also available are the notes from catalogs and abstracts from 1988 onwards. The NED database has rapidly become an essential tool for extragalactic astronomers. It can be searched by name, near name, position, literature reference and various object parameters. Recently a skyplot and spectral energy distribution plot functions were added. NED is accessible via telnet, X-window and e-mail. A WWW interface is planned for the near future.

While NED and SIMBAD are invaluable tools, they are not well optimized for use with large surveys, *e.g.*, the ROSAT X-ray sky survey where the identification of a large number of sources is required. A typical starting point is to cross-correlate a new source list against SIMBAD and NED. But the output from such a search presents a number of problems, that can be very time consuming to resolve.

- NED and SIMBAD are dynamic systems, that are constantly being updated and corrected. This makes it difficult to obtain a reproducible result at different times. A static catalog available via ftp that represents a snapshot of their current object catalogs at a particular time would make it much simpler to track objects that have been deleted, changed or added.
- The two databases can give inconsistent results for the same object. It is essential that the two teams establish a procedure to make regular consistency checks and correct errors.
- The output from the two systems are poorly structured. A standardized tabular form output would make it much easier to utilize these results within a large survey.

2.3.2. *Skyview*
(http://skyview.gsfc.nasa.gov/skyview.html.)

Skyview is a virtual telescope that can provide, on demand, images from any portion of the sky at wavelength regimes from radio to gamma-ray. It is the first example of an astronomical service that dynamically generates data on the fly in the form that is required by the astronomer. It makes trivial the once complex operation of obtaining views and overlays of the sky in different wavebands. Skyview currently contains 21 surveys accessible. The user selects which part of the the sky, the scale, the coordinate system, map projections, and equinoxes. In a more advanced interface they can also make contour, image and catalog overlays. The catalogs are provided via a client-server interface to the HEASARC database. The latest release allows image manipulation using a Java based interface.

3. Source Classifications

A typical goal of a survey is object classification. This can be done in a number of ways including 1) directly determining the class via source properties, *e.g.*, taking a spectrum, 2) making a cross-correlation with existing catalogs, and 3) by requesting telescope time to optically identify a source detected in another band. The major problem with source classification is that there is no standard classification scheme. Also most existing schemes are character based, which is not efficient for a database search engine. This makes class based searches of databases very difficult and gives inhomogeneous results. To circumvent these problems different numerical schemes have been introduced by both the IUE and the EXOSAT projects. The EXOSAT class scheme (originally developed by P. Giommi) is currently in use at the HEASARC. It assigns each catalog entry a four digit numeric code to represent its classification. The first digit describes the global classification (*e.g.*, AGN or star). The following digits assign further classifications or properties such as spectral type, or AGN sub-class. Each sub-class is chosen to contain a unique set of properties. This class system, which is more fully described at http://heasarc.gsfc.nasa.gov/docs/xray/class.html provides the ability to both search using numerical methods and to allow combinations of properties. It is however limited to a few sub-classes and does not allow complex class combinations. To make further progress in this important area requires an international agreement on a numerical based classification system. This system could then be adopted by the IAU in a similar fashion to the standards adopted for source naming conventions, for example.

4. A Network Integrated Astronomy System

It has been long recognized that having a single access point to astronomical catalogs and data potentially could increase the productivity of astronomers. In the late 1980s NASA created the Astrophysics Data System (ADS) and ESA the European Space Information System (ESIS) with the objective to create a layer that connected together the diverse astronomy services available on the network at that time. But both projects failed to meet their objectives and were not considered a success. In the case of ESIS the project was discontinued. The ADS program was descoped, leaving only the abstract service, the only feature of the system that became widely used. A shortcoming of these global services was the fact that they attempted to compete against the existing, and well established, user interfaces. There was insufficient incentive to use them over the original more familiar interface. Another issue was the lack of standards to connect to data files with the database tables that are used to access them, which made it difficult to access data through any generic interface. ADS was eventually overtaken by the WWW, which effectively made redundant the machine dependent and vendor specific user interface that had been developed. After spending substantial sums of money, both NASA and ESA retreated from trying to provide a unified service as an expensive moving target that was not widely supported by the community they were designed to serve.

Despite the failure to integrate the existing astrophysics services together by NASA and ESA we (and the funding agencies) should not be discouraged. The Web browser addresses the problem that ADS and ESIS were unable to solve, *i.e.*, a simple multi-platform interface that the astrophysics service providers are willing to adopt. The current WWW services are not yet as capable as the originals, because of the stateless nature of the WWW client-server interaction. But this is being solved with the development of Java and plug-ins, which will allow fully interactive interfaces. The development of the URL, which underpins the success of the WWW, provides the mechanism required to achieve a network integrated astrophysics system. The URL is a networked extension of the filename concept. The HEASARC has already assigned each dataset in its archive a unique URL, which specifies the path to the data which can be directly accessed by any user interface. In addition to this, the URL can be used to access programs and scripts over the network giving transparent access to specialized services. By making available URLs that access given services, both data centers *and* users can develop user interfaces that are best suited to their applications. The most simple example is the name-coordinate resolver services already provided by NED and SIMBAD. This is a simple network service that given an object name returns the coordinates, and is commonly found integrated within other on-line services. In the future one can expect an increasing number of services that are cross linked to provide access to a resource on the network, *e.g.*, calling up images from skyview, within NED. This approach will naturally lead to a user-driven network-integrated astrophysics system that will both increase data access and boost astronomer productivity.

Acknowledgements

I thank Tom McGlynn, Sherri Calvo, and Gunther Hasinger for useful comments and discussions.

Part 11. Conference Summary and Resolutions

THE NATURE AND SIGNIFICANCE OF SURVEYS

Summary of Panel Discussion

V. TRIMBLE
Astronomy Department, University of Maryland,
and Physics Department, University of California

Abstract. We present here some of the ideas and questions mentioned by the panelists and other participants during the discussion that immediately preceded Ofer Lahav's concluding remarks. Official panelists were George Djorgovski, Michael Disney, Ofer Lahav, and Virginia Trimble (chair). The topics of the posters are very briefly summarized as well.

1. Do We Really NEED Surveys

What is a survey? A large, systematic assemblage of data for which someone else will get more glory than the people who did the assembling and systematizing. It is with this in mind that the recommendation was made to dedicate the proceedings to George Ogden Abell and Albert G. Wilson, who took virtually all the plates for POSS I.

How many surveys are there? According to N.G. Roman's poster, there are already well over 300 in the ADC data base. S. Okamura's summary for IAU Commision 28 (triennial report) records 499 extragalactic atlases and surveys published in the three years ending 30 June 1996 (though some of these are quite specialized, like isolated pairs of galaxies in the southern hemisphere and intracluster gas temperature).

How many surveys should there be, or, as R.E. Williams asked it, what fraction of telescope time should be devoted to surveys? This is exactly the sort of question that Working Groups exist to answer (or, at least, discuss), as are the next several. Most participants seeme to agree that the answer is "more"—but not at the expense of our own more specialized projects.

How many surveys do we really need? M. Disney suggested 19, based on dividing up the electromagnetic spectrum into suitable slices. But, realistically, this is very much a lower limit, since one cannot as a rule take care of all the necessary ranges of temporal, spectral, and angular resolution, lines vs. continuum, and point vs. extended sources with a single survey at a given wavelength. This increases the necessary number to 42 or 63 or whatever your favorite might be.

Which "windows" are currently more opaque than they need to be and so most in need of surveys? This was raised by M. Harwit, and suggested answers including HI 21 cm (from the southern hemisphere), the lowest reachable radio frequencies (from space for less than 1 MHz, mentioned by Lahav), the submillimeter (300 μ) region, and the vaccum ultraviolet near 1000Å. In fairness, however, one should remember that there are periodic opportunities in both the USA and Europe to put proposals to do these into the potential pool along with other, perhaps more popular wavelength bands.

Was the universe created at optical wavelengths? This was Jasper Wall's phrasing of the feeling we all have that an astronomical object doesn't really exist until there is an optical identification. The relevant passage in Genesis indeed says: *Vayomer Elohim, y'hi* or *(and God said, let there be light)*. That this should surely be interpreted to include all forms of electromagnetic radiation (and probably static magnetic fields as well) does not vitiate the point that we continue to find more information (spectral lines for measurement of composition, redshift, and all the rest) at the energies where most atoms have some excited, but bound, electrons.

How much redundancy is appropriate? POSS I has been scanned and digitized at least three times. Is this too many, too few, or just right? And, looking ahead, for instance, how much support should be provided to balloon groups that want to map the 3K background not quite so well as will be done by MAP and COBRAS/SAMBA, but earlier?

2. Science from Survey Archives

How can we promote access to archived data? Should data bases be centralized or distributed among sites? And how can we make good use of the multiplicity of surveys at many wavelengths that are or soon will be available? No one provided any very profound answers to these questions (though we all admire the problem, and some NASA experience indicates that 10% of a project cost needs to be set aside for long term storage and accessing). Notoriously, archiving is cheap, retrieval is expensive.

Are there lessons to be learned from other massive data bases? Disney mentioned the European cancer registry (apparently assembled at considerable cost and not yet much used for anything, though national analogs are heavily exploited to look for all kinds of correlations with demographic variables). Images from particle colliders are not a good model. They are normally looking for needles in haystacks, while our surveys are practically all needles. And then there is the Human Genome Project, about which we all agree that it is a Good Thing, though not perhaps about why.

What are the implications of massive, electronically archived surveys for the way astronomy is done and the kinds of people who will be successful astronomers in the future? A number of participants expressed thoughts and worries that are widespread in the community. Will we give PhDs to people who know only how to handle a given image processing system very skillfully? Will there still be astronomers who know how to build things and make them work? There had better be, but how do we reward these people, given the publication-oriented structure of academic science? Very possibly there has already been a shift from a pre-dominance of solitary observers to younger astronomers who prefer to work in groups? Will they still generate the kinds of new ideas that we historically associate with mavericks? How can the inventor of an idea be identified and rewarded if all papers are published as "Aardvark *et al.*, on down to Zyzygy" And, finally, given that an astronomer anywhere in the world now has access to much the same data as staff members at NRAO, Keck, or GSFC, there is surely an opportunity for people who are skilled in handling "large, systematic assemblages of data" to do their own thing, wherever they may be, somewhat leveling the traditional playing field.

3. Posters

At any given conference, the poster contributions provide a glimpse of the near future, since many represent work in progress, quite often work by graduate students and postdoctoral fellows. Of the 105 posters I read (all but a couple that were either never put up or were taken down in the first two days), the distribution of subject matter was roughly the following:

- Eight concerned reprocessing or other reconsidering of old surveys (including the use of the Carte du Ciel catalogue for proper motions).
- Applications of completed surveys (Einstein Medium Deep, IRAS, *etc.*) to finding new objects or classes of objects appeared in 25.
- Surveys under way were the topic of 31 (DENIS, SDSS, and many others).
- Seven posters dealt with techniques for processing, archiving, or retrieving survey data (SkyView is a particularly interesting case).
- Multiwavelength applications (beyond merely finding optical identifications) appeared in 15 posters, including some with the most spectacular graphics.
- Surveys that could conceivably be carried out from ground or space (if only the money/equipment/satellite/*etc.*, existed) were the topics

of 12 posters. (My mother used to say about such things that, if the sky falls, we'll all catch larks.)
– Seven dealt with other, non-survey, topics.

Perhaps the most striking aspect of the poster presentations was the very high technical quality. Almost no-one simply tacked up his preprint. And many of the color images were impressive as art as well as science. Future conferences on topics like this one should perhaps consider publication of a CD/ROM as well as a book of proceedings to accomodate these presentations.

SUMMARY TALK : MULTI-WAVELENGTH SKY SURVEYS

O. LAHAV
Institute of Astronomy, Cambridge

1. Introduction

An astronomer's career can be viewed in a 3-dimensional space where the (nearly orthogonal) axes are :

– the objects of interest (from planets to the Universe),
– techniques (from instrument design to analytic calculations),
– the wavelength (from the radio to gamma rays).

This interdisciplinary conference brought together experts from different bands of the 'wavelength axis.' It has been an interesting meeting, with a lot of cross-talk, excellent review talks and high-quality posters.

I shall begin by summarizing the highlights (as described in the oral presentations) in different wavelengths, and then discuss the need for statistical techniques, some key astrophysical questions which should be addressed by multi-wavelength (MW) approaches, and the changing sociology of the field.

2. Across the Spectrum

For most of the history of mankind our picture of the universe was restricted to the visual band. However, in particular since the Second World War there has been rapid progress in instrumentation and space technology which allows us to have a nearly complete MW picture of astronomical objects and background radiations. Below is a brief summary of what we have heard at the conference, ordered from long to short wavelengths.

2.1. RADIO

Being a 'modern' wavelength, many new astronomical phenomena were discovered in the radio. In fact, all Nobel prizes in observational Astronomy were given to radio-astronomers: for the discoveries of the Cosmic Microwave Background (CMB) Radiation, the pulsars, and the binary pulsar. At this meeting we have mainly heard about radio extragalactic surveys (Becker, Condon, Disney, Fürst, Sokolov, Wall). Present radio surveys (*e.g.*, 87GB and PMN) have median redshift $\bar{z} \sim 1$, hence providing an unusually deep picture of the universe at early epochs. Recent studies have indicated that radio sources are strongly clustered, suggesting that they reside in high density regions. Several on-going and future surveys in the continuum (FIRST, NVSS, WENSS, UTR) are most promising, although a crucial problem is how to get optical redshifts to these distant radio objects, in order to map their 3-dimensional distribution. The Parkes multibeam survey in 21 cm promises to give an unbiased view of the neutral hydrogen content in the local universe, with implications for identification of new galaxy populations (*e.g.*, dwarfs and low surface brightness galaxies). Moving to much higher redshifts, radio measurements of temperature fluctuations on the last scattering surface of the CMB by COBE and future experiments (COBRAS/SAMBA, MAP, VSA) reveal the seeds of cosmic structure.

2.2. INFRARED

The relatively recent development of detectors in the infrared provides a new window to Galactic and extra-galactic objects (as discussed by Beichman, Oliver, Ruphy). The IRAS data base (in 12, 25, 60 and 100μ) has become a major tool for studies of the Galactic structure and the local universe. In particular, follow-up redshift surveys (IRAS1.2Jy, QDOT, PSCZ) have yielded nearly whole-sky 3-dimensional catalogues for studies of the density and peculiar velocity fields. The new surveys at 2μ (2MASS, DENIS) are in particular useful for tracing the old 'stable' stellar population in galaxies, and for overcoming the problem of Galactic extinction. Together with ISO, they will give a new picture of stellar populations within the Galaxy and a deeper view of large scale structure.

2.3. OPTICAL

This traditional band remains most useful. Photographic surveys (Djorgovski, Morrison, Reid) such as POSS I+II and UK Schmidt still provide the most important wide-angle data bases, *e.g.*, as target lists for redshift surveys. In the old days plates were examined by eye (*e.g.*, Zwicky, Abell, Nilson, Lauberts), but at present the scanning is done by machines (APM, COSMOS, APS, DDS, DPOSS, PMM), with clever software and heroic efforts to calibrate and match plates. It is worth stressing that there is still no whole uniform sky optical catalogue (the nearest to that is a compilation of UGC, ESO, ESGC, followed up by redshift surveys, *e.g.*, SSRS and ORS).

Existing redshift surveys (*e.g.*, CfA, SSRS, ORS, IRAS, APM, LCRS, CFRS) contain each no more than 30,000 galaxies. A major step forward using multifibre technology will allow us in the near future to produce redshift surveys of millions of galaxies. In particular, two major surveys will probe a median redshift of $\bar{z} \sim 0.1$. The American-Japanese Sloan Digital Sky Survey (SDSS) will yield images in 5 colours for 50 million galaxies, and redshifts for about 1 million galaxies over a quarter of the sky (see reviews by Bahcall, McKay and Szalay). It will be carried out using a dedicated 2.5m telescope in New Mexico. A complementary Anglo-Australian survey (discussed by Parker/Taylor), called the 2 degree Field (2dF), will produce redshifts for 250,000 galaxies selected from the APM catalogue. The survey will utilize a new 400-fibre system on the 4m AAT. These surveys will probe scales larger than $\sim 30h^{-1}$ Mpc. It will also allow better determination of Ω and bias parameter from redshift distortion. Surveys like 2dF and SDSS will produce unusually large numbers of galaxy spectra, providing an important probe of the intrinsic galaxy properties, for studying, *e.g.*, the density-morphology relation. Several groups recently devised techniques for automated spectral classification of galaxies utilising, *e.g.*, Principal Component Analysis and Artificial Neural Networks. The SDSS and 2dF projects will also carry out important surveys of quasars (see report by Smith).

We have heard about progress in Astrometry (Egret, Jones, Mignard, Seidelmann). Present and future missions (Hipparcos, Tycho, SIM, GAIA, VSOP) have various applications to a diverse range of topics: *e.g.*, planets, the distance scale and tests of General Relativity. Gravitational lensing (Cook, Schneider) is another innovative area where a major part of the activity is in the optical band. Finally, the impressive pictures of the Hubble Deep Field (reviewed by Dickinson) reveal morphologically disturbed galaxies, probably in the process of formation.

2.4. UV, EUV

Th UltraViolet and extreme UV bands (reviewed by Brosch) have been explored by the missions TD1, UIT, GLAZAR and FOCA and future projects include, *e.g.*, TAUVAX, MSX and ARGOS. Here we only point out two important implications of the UV band. The first is that in order to compare local galaxies with the images at high redshift (*e.g.*, from the Hubble Deep Field) one has to image the local galaxies in the UV. The second issue is the relevance of detection of the UV background, which plays an important role, *e.g.*, in the evolution of Lyman-α clouds. It is also worth emphasizing the lack of a deep full-sky survey in the UV.

2.5. X-RAY

The study of the X-ray band (discussed by Hasinger, Stewart, Trümper) is less than 40 years old, but it opened a new window to high energy phenomena. Satellites like Uhuru, HEAO1, ROSAT, Ginga and ASCA and future missions (ABRIXAS, AXAF, XMM, SRG) probe a wide range of astronomical objects in the X-ray, including comets, binary stars, AGNs, clusters of galaxies and the yet unexplained X-ray Background. We have learned that 50% of the ROSAT resolved objects are AGNs, 10% are galaxies and clusters, 35% stars and 5% other objects. X-ray clusters are important cosmological probes: the X-ray temperature tells us the depth of the cluster potential well, and hence the total mass, while the X-ray emission indicates a large fraction of baryonic mass, with implications for the density parameter Ω (see below). Optical follow up observations of the X-ray selected clusters are important in order to get redshifts (so far obtained for 600 clusters).

2.6. γ-RAYS

This is the most 'energetic' band (reviewed by Gehrels and Miller) which in particular became active following the Compton GRO observations. GRO detected Galactic objects, Blazars (which are most probably due to inverse Compton effect in beamed jets), and about 1500 γ-ray bursts. The origin of γ-ray bursts is still a mystery and a MW identification programme is essential in order to associate them with other astronomical objects at known distances. One current popular theoretical idea is that they are due to merging of 2 neutron stars at cosmological distances. It seems less likely that the γ-ray bursts reside within the Milky Way. A new innovative project (Milgarno) utilises air showers/Cerenkov radiation to detect very high energy γ-rays.

3. Statistical Techniques

We have heard many talks about cross identification of objects observed at different wavelengths (Bartlett, Becker, Brunner, Hasinger, Helou, Prandoni, Szalay, Wagner, Wall, White). Typical problems are the accuracy of coordinates and overlap of objects. In principle, one should take into account not only the angular proximity, but also the radial distribution of the objects in the different catalogues. For example, the median redshift of the radio survey 87GB is $\bar{z} \sim 1$, much deeper than local optical and IRAS surveys ($\bar{z} \sim 0.02$), hence it is not surprising that Condon finds only 1% of sources in common. One can formulate the cross-identification by taking into account this prior probability for the radial distribution, but it depends to what extent one is prepared to be 'a Bayesian.' It is also possible to calculate the non-zero lag cross-correlation between catalogues. As most objects are known to be clustered, this could yield valuable extra information. An example is cross-correlation of the diffuse X-ray Background with known galaxies.

Other statistical issues discussed at the meeting are selection effects in catalogues (Disney), and the need for innovative methods of classification and pattern recognition (Feigelson, Harwit). Clearly we should keep an open eye on developments of statistical tools in other fields, as many similar problems have already been solved. Special attention was given at this meeting to data bases of astronomical objects and user-friendly software packages (*e.g.*, ADS, CDS, SIMBAD, NED, ALADIN, SkyView) and on how to handle Terabytes of data (Szalay, White). The progress is impressive, and we are all thankful to those who compile the data sets and develop public domain software.

4. Astrophysical Problems

Here we outline several key problems in Astrophysics which are best tackles by MW approaches (with a bias of the reviewer towards cosmological problems).

4.1. GALACTIC STRUCTURE

The MW measurements and new techniques such as microlensing have renewed interest in the structure and evolution on the Milky Way. Many speakers (Bienaymé, Boulanger, Fürst, Fukui, Gehrels, Larsen, Majewski, Méndez, Parker, Price, Ruphy) discussed unsolved problems related to Galactic structure and dynamics. Here we only list some of them: what is the stellar initial mass function? what is the extent of the thin and thick discs? is the halo lumpy? does the Milky-Way have a bar? how strong is the interaction between the Milky-Way and nearby dwarfs?

4.2. GALAXIES BEHIND THE ZONE OF AVOIDANCE

Recent discoveries of galaxies (*e.g.*, the Sagittarius dwarf and Dwingeloo 1) and clusters (*e.g.*, A3627 at the centre of the Great Attractor) illustrated how a combination of eye-balling of plates, 21 cm, infrared and X-ray measurements can be combined to unveil galaxies hidden behind the 'Galactic fog.' Although this topic was not directly discussed at this meeting, it is clear that collaboration of Galactic and extragalactic astronomers on this topic can benefit both groups. There is also a great need for a better extinction map of the Galaxy (*e.g.*, by combining optical, UV, HI, IRAS and COBE maps). We were reminded that there is substructure in the ISM even at the north and south Galactic poles.

4.3. GALAXY FORMATION

A MW approach is required to produce the 'H-R diagram' of galaxies. Rather than talking just about a luminosity function in one colour, one should consider a multivariate function which includes luminosities in various spectral bands and dynamical properties of galaxies. Deeper galaxy images such as the HDF hold the key to issues of galaxy formation and evolution (*e.g.*, Dickinson's review), but to make sense of the high redshift measurements it is crucial to improve our knowledge of galaxy properties in the local universe. Another constraint on galaxy formation can be obtained from the 'maximum' redshift of QSOs (Osmer, Padovani).

4.4. LARGE SCALE STRUCTURE & COSMOLOGY

Recent work on understanding clusters of galaxies is a good example of MW measurements. Optical, X-ray, CMB (for the Sunyaev-Zeldovich effect) and gravitational lensing are combined to quantify the extent and matter content of clusters (discussed by Bahcall, Trümper, Schneider). For example, the high fraction of baryons in clusters suggests that $\Omega \approx 0.2$, in conflict with the popular $\Omega = 1$ value. The derivation of Ω from comparison of density and velocity fields and from redshift distortion is affected by 'biasing,' *i.e.*, the way galaxies trace the underlying mass distribution. Galaxies observed at different wavelengths (*e.g.*, optical and IRAS) have different clustering properties, hence a different 'bias parameter.' This conceptual issue will remain crucial in the interpretation of the 2dF and SDSS surveys, and in connecting the power-spectrum of density fluctuations from galaxy surveys and the CMB.

MW studies are also important for understanding the background radiations. Most of the background energy is in the CMB ($0.25\ eV\ cm^{-3}$), compared with, *e.g.*, the XRB ($5 \times 10^{-5} eV\ cm^{-3}$). Only upper limits (from direct measurements) are available for the optical and UV backgrounds. An interesting example of MW approach is how the hot intergalactic medium model for the XRB was ruled out by the lack of distortion of the blackbody CMB spectrum. It is also important to detect the dipole in various background radiations, to confirm that the CMB dipole is due to motion. Sub-degree measurements of fluctuations in the CMB by new experiments (COBRAS/SAMBA, MAP, VSA) will provide a new insight into the nature of the dark matter, and will yield estimation (in a model-dependent way) of the cosmological parameters (*e.g.*, Ω and H_0) to within a few percent.

5. Changing Sociology

We have heard at this meeting about future big telescopes, big collaborations and big data bases. Several speakers made the remarks "it takes at least 2–3 times the most pessimistic estimate to begin/complete/analyse a survey" and "to make a big impact, a new survey must be 10–1000 times better in sensitivity/resolution/number of objects." This indicates that it is becoming more and more challenging to make an impact by conducting big surveys. It also raises some questions about the changing sociology of doing research in astronomy: what will be the individual's contribution in a big collaboration (cf. particle physics experiments)? what skills should be acquired by the next generation of astronomers? Will the increase in projects and data sets be followed by more jobs for young astronomers? How would the community deal with public domain data (*e.g.*, HDF)? and how to communicate the knowledge resulting from the surveys to the tax payer? While focusing on the technological aspects of the MW surveys, the human aspects should not be forgotten.

Acknowledgements

Finally, many thanks to the scientific and local organizing committees, in particular to Barry Lasker and Marc Postman, for organizing this interesting and stimulating meeting.

IAU SYMPOSIUM 179
"New Horizons from Muliwavelength Sky Surveys"
26–30 August 1996
The Johns Hopkins University, Baltimore, MD

RESOLUTION
Astronomical Photographic Emulsions

The participants of the IAU Symposium 179 on New Horizons from Multi-Wavelength Sky Surveys:

Recognizing the continued need for specialized astronomical photographic emulsions, which are essential for wide field and high resolution imaging, for astrometry, and for other purposes,

Being aware of new developments in high resolution, fine grain Technical Pan Blue (TPB) or similar emulsions by Kodak, which are suitable for astronomy,

Commend Kodak on their long-standing support for astronomy and on their sensitivity to the concerns of the community, as expressed at IAU Symposium 161, and,

Urge Kodak to bring these developments to fruition as soon as possible.

IAU SYMPOSIUM 179
"New Horizons from Muliwavelength Sky Surveys"
26–30 August 1996
The Johns Hopkins University, Baltimore, MD

RESOLUTION
Access to Distributed Astronomical Data Bases

The participants of the IAU Symposium 179 on New Horizons from Multi-Wavelength Sky Surveys:

Recognizing the advantage of distributed terabyte data bases for astronomical research,

Identifying the availability of data links and their data transfer speed as the major obstacle to the wide-spread usage of such data bases,

Urge Commission 5 of the IAU:

1. to work towards ensuring the availability of at least **ONE** dedicated wide-band, high-speed link in every country (country node) into the Global Data Network, to serve for high-volume data retrievals for scientific purpose and,
2. to prioritize the availability of such links preferrably to developing nations and,
3. to work towards supplying data retrieval hardware and software for each country node.

Author Index

Aldering, G., **273**, 467
Amram, P., 186
Annis, J., **268**
Appenzeller, I., 449
Argyle, R.W., 389
Artamonov, B.P., **121**
Aschenbach, B., 433

Bässgen, G., 395
Babu, G.J., 363
Bade, N., **444**
Bahcall, N.A., **317**
Baker, A., 234
Baldwin, J., 329
Balkowski, C., 346
Bardelli, S., **342**
Bartlett, J.G., **437**, 469
Bastian, U., 395, 409
Batuski, D., **344**
Beckwith, S., 293
Beichman, C.A., **27**
Bienaymé, O., **209**, 221, 469
Birkinshaw, M., 266
Blanchard, A., 346
Bloch, J., 144
Boller, Th., 433
Bonnarel, F., **469**
Borgman, T., 379
Borne, K.D., **275**
Boroson, T., 302
Botashev, A.M., 139
Boulanger, F., **153**, 194
Boyle, B.J., 348
Bräuninger, H., 433
Briel, U., 433
Briggs, D.S., 89
Brinkmann, W., **447**
Brosch, N., **57**
Brosche, P., 415
Brunner, R.J., 455, **471**
Bucciarelli, B., 379, 420

Burkert, W., 433
Bushouse, H., 275

Cabanela, J., 467
Cabrera-Guerra, F., 270
Cannon, R.D., 135
Canzian, B., **422**
Cappi, A., 346
Carrasco, B.E., 278
Casalegno, R., 379
Castets, A., 186
Cayatte, V., 346
Cesarsky, C., 112
Chen, J.-S., **123**
Chester, T., 450
Chincarini, G., 346
Chu, Y., **131**
Chubey, M.S., **125**
Clemens, D.P., 109
Cohen, M., 115
Colin, J., 415
Colina, L., 275
Collins, C.A., 339, 346
Condon, J.J., **19**
Connolly, A.J., **376**
Conrow, T., 450
Copet, E., **172**
Corbin, T.E., 418
Cordis, L., 444
Cornuelle, C.S., **467**
Costa, E., 379
Couch, W.J., 234
Crézé, M., 469
Croom, S.M., 348

Danese, L., 112
Davis, R.J., 389
de Carvalho, R.R., 424
de Marchi, G., 234
de Ruiter, H.R., 353
Dennerl, K., 433

Dennison, B., **182**
Digel, S.W., **175**, 194, 237
Disney, M., **11**
Djorgovski, S.G., **424**
Doggett, J., 379
Doi, M., 287
Douglass, G.G., 418
Drimmel, R., 217

Ebeling, H., 308
Egret, D., 395, 437, 450, 469
Ekers, R.D., 353
Engels, D., 299, 444
Englhauser, J., 433
Epchtein, N., **106**
Evans, T., 450

Fürst, E., **97**
Fabricius, C., 395
Fazio, G.G., **109**
Feigelson, E.D., **363**
Florsch, J., 469
Fockenbrock, R., 293, 296
Foster, R.S., 89
Franceschini, A., 112
Frattare, L., 302
Frenk, C.S., 278
Fresneau, A., 415
Fried, J., 293
Fukui, Y., **165**
Fullmer, L., 450

Gal, R.R., 424
Gardner, J.P., **278**
Garrington, S.T., **389**
Gautier, T.N., 118
Geffert, M., 415
Gehrels, N., **69**
Geitz, S., 237
Genova, F., 469
Genzel, R., 112
Georgelin, Y.M., 186
Georgelin, Y.P., 186

Giommi, P., 310
Girard, T.M., 223, 384
Gorshanov, D.L., 125
Graf, P., 118
Greene, G., **474**
Gregorini, L., 353
Großmann, V., 395
Gruber, R., 433
Gulyaev, A., 409
Guzzo, L., 346

Haase, S., 344
Haberl, F., 433
Hacking, P.B., 118
Halbwachs, J.L., 395
Hartner, G., 433
Harwit, M., **3**
Hasegawa, T., 189
Hasinger, G., 312, 433
Hawkins, G., 431
Hazard, C., 329
Hecht, S., 415
Henry, R.C., 115
Herrero, J., 302
Herter, T., 118
Hiesgen, M., **415**
Hill, J.M., 344
Hippelein, H., **293**, 296
Hopp, U., 293, 299, **335**
Horner, D., **308**
Houck, J.R., 118
Huang, J., **281**
Humphreys, R.M., 467
Hunter, S.D., 175
Høg, E., **395**

Ichikawa, T., **285**
Il'In, A.E., 125
Inoue, H., 312
Ishisaki, Y., 312
Itoh, N., 285
Izotov, Y., 299, 302

Author Index

Jenkner, H., 431
Jonas, J.L., **95**
Jones, D.H.P., **406**
Jones, L.R., 308, 310

Kümmel, M.W., **337**
Kürster, M., 433
Kallenbach, E., 415
Kanayev, I.I., 125
Kashikawa, N., 287
Kassim, N.E., **89**
Kawasaki, W., 287
Kelleher, C., 182
Kii, T., 312
Kirian, T.R., 125
Kneer, R., 449
Kniazev, A., 299, **302**
Komiyama, Y., 287
Kopylov, I.M., 125
Kozhurina-Platais, V., 384
Krautter, J., **449**
Kroll, P., 127
Kuimov, K., 409
Kuzmin, A., 409

López, C.E., 384
la Dous, C., **127**
Lahav, O., **493**
Larsen, J.A., **225**, 467
Lasker, B.M., 228, 420, 431, 474
Lattanzi, M.G., 217, 228, 431
Lawrence, A., 112
le Coarer, E., 186
Leinert, C., 293
Leisawitz, D., **237**
Lemke, D., 112
Lipovetsky, V., **299**, 302
Longo, G., 424
Lonsdale, C.J., 118, **450**
Louys, M., 469
Lu, C.-L., **384**
Lucas, R.A., 275

Méndez, R.A., **223**, **234**
Maccagni, D., 346
MacGillivray, H.T., 346, 415, 450
Majewski, S.R., **199**, 223
Makarov, V.V., 395
Makishima, K., 312
Malkan, M., 308
Mann, R.G., **339**
Marcelin, M., 186
Masheder, M.R.W., 177
Massone, G., 228
Maurogordato, S., 346
McCullough, P.R., **184**
McGlynn, T., **465**
McKay, T.A., **49**
McLean, B.J., 228, 420, **431**, 474
McMahon, R., 112, 329
Meisenheimer, K., 293, 296
Merighi, R., 346
Michailov, M.-E., 462
Michaud, K., 344
Mignard, F., **399**
Mignoli, M., 346
Miley, G., 112
Miller, C., 344
Miller, L., 348
Mingaliev, M.G., **139**
Minniti, D., 234
Miyaji, T., 312
Miyazaki, A., 189
Monet, D.G., 384
Moody, J., 302
Moran, E.C., **358**
Morbidelli, R., 474
Morgan, D.H., **374**
Morrison, J.E., **381**, 420
Morrison, L.V., 389
Moseley, S.H., 118
Moshir, M., 115, 450
Mujica, R., 449
Munari, U., **142**
Mutafov, A.S., 462

Nan, R., 91, 93
Nass, P., **305**
Nesterov, V., **409**
Newberg, H.J., **291**

Ochsenbein, F., 469
Odenkirchen, M., 415
Ogasaka, Y., **312**
Ohta, K., 312
Ojha, D.K, **221**
Oka, T., **189**
Okamura, S., **287**
Oliver, R.J., **177**
Oliver, S., **112**
Ortiz Gil, A., 415
Osmer, P.S., **249**

Pérez-Fournon, I., 270
Padovani, P., **257**, 310
Pahre, M.A., 424
Paillou, P., 469
Parker, Q.A., 135, **179**, 374
Parma, P., 353
Passuello, R., 142
Paxton, L.J., 115
Peng, B., **91**, **93**
Perlman, E.S., 308, **310**
Pfafman, T., 144
Pfeffermann, E., 433
Phillipps, S., 179, 266
Pietsch, W., 433
Platais, I., 384
Popescu, C., 299
Postman, M., **379**
Prandoni, I., **353**
Predehl, P., 433
Price, S.D., **115**
Proust, D., 346
Puget, J.-L., 112

Röser, H.-J., 296
Röser, S., 381, 409, **420**
Raclot, V., 469

Ramella, M., 346
Ramirez-Castro, L., **270**
Read, M., 348
Reich, P., 97
Reich, W., 97
Reid, I.N., **41**
Reimers, D., 444
Reynolds, J., 310
Rich, R.M., 223
Richards, E.A., **356**
Richter, G., 299
Robin, A.C., 221
Rocca-Volmerange, B., 112
Roman, N.G., **478**
Rosso, C., 433
Roussel-Dupré, D., **144**
Rowan-Robinson, M., 112
Ruphy, S., **231**
Russeil, D., **186**
Röser, H.-J., 293

Salzer, J., 302
Sambruna, R., 310
Sato, F., 189
Scaramella, R., 342, 346, 424
Schäffel, C., 415
Scharf, C.A., 308
Schmitt, J.H.M.M., 433
Schneider, P., **241**
Schwekendiek, P., 395
Scollick, K., 465
Seidelmann, K.P., **79**
Sekiguchi, M., 287
Sementsov, V., 409
Semkov, E.H., 462
Serrano, A., 449
Shanks, T., 348
Sharples, R.M., 278
Shimasaku, K., 287
Shupe, D.L., **118**
Siebert, J., 447
Simonetti, J.H., 182

Author Index

Slinglend, K., 344
Smart, R.L., **217**
Smette, A., 329
Smith, R.J., **348**
Snowden, S.L., 175
Soifer, B.T., 118
Sokolov, K.P., **100**
Sourov, A., 467
Spagna, A., **228**, 431
Stacey, G.J., 118
Stavrev, K.Y., **332**, 462
Stecher, T.P., 147
Stirpe, G., 346
Stolyarov, V.A., 139
Sturch, C., 379
Szalay, A.S., 376, **455**

Takahashi, T., 312
Taylor, K., **135**
Tedesco, E.F., 115
Thadddeus, P., 177
Theiler, J., 144
Thiering, I., 449
Thommes, E., 293, **296**
Thuan, T., 302
Topasna, G.A., 182
Trümper, J., 433
Trimble, V., **489**
Trushkin, S.A., **103**
Tsuboi, M., 189
Tsvetkov, M.K., **462**
Tsvetkova, K.P., 462
Tucholke, H.-J, 415
Tzioumis, A., 310

Ueda, Y., 312
Ugryumov, A., 299
Uyaniker, B., 97

van Altena, W.F., 223, 384
van den Bergh, S., 223
Varosi, F., 194
Vettolani, G., 342, **346**, 353

Viale, A., 186
Visvanathan, N. , **351**
Voges, W., **433**, 444, 449
Volpicelli, A., 474
Vydrevich, M.G., 125

Wagner, K., 395
Wagner, S.J., 337
Walker, A.R., **129**
Walker, R.G., 115
Wall, W.F., **191**
Waller, W.H., **147**, 191, **194**
Wegner, G., 308
Werner, M.W., 118
White, N.E., 465, **480**
White, R., 431
Wicenec, A., 395
Wielebinski, R., 97
Wieringa, M.H., 353
Williger, G., **329**
Windridge, D., **266**
Witteborn, F.C, 115
Wolf, C., 293
Wolfe, D., 474
Wolstencroft, R., 450
Wycoff, G.L., 418

Yagi, M., 287
Yamada, T., 312, 351
Yamasaki, H., 189
Yan, Y., 91
Yanagisawa, K., 285
Yanny, B., 291
Yasuda, N., 287
Yentis, D., 450
Yershov, V.N., 125
Yonekura, Y., 165
Yuan, W., 91, 447

Zacharias, M.I., **418**
Zacharias, N., **386**, 418
Zamorani, G., 342, 346
Zhao, Y.-H., 131

Ziaeepour, H., 469
Zickgraf, F.-J., 449
Zimmermann, U., 433
Zucca, E., 342, 346